Solidification and Casting

Series in Materials Science and Engineering

Series Editors: **B Cantor**, Department of Materials, University of Oxford, UK
M J Goringe, School of Mechanical and Materials Engineering, University of Surrey, UK

Other titles in the series

Microelectronic Materials
C R M Grovenor
Department of Materials, University of Oxford, UK

Physical Methods for Materials Characterisation
P E J Flewitt
Magnox Electric, Berkeley, UK
and R K Wild
University of Bristol, UK

Aerospace Materials
B Cantor, H Assender and P Grant
Department of Materials, University of Oxford, UK

Forthcoming titles in the series

Fundamental of Ceramics
M Barsoum
Department of Materials Engineering, Drexel University, USA

Topics in the Theory of Solid Materials
J M Vail
University of Manitoba, Canada

Computer Modelling of Heat, Fluid Flow and Mass Transfer in Materials Processing
C-P Hong
Yonsei University, Korea

Fundamentals of Fibre Reinforced Composite Materials
A R Bunsell and J Renard
Centre des Matériaux, Pierre-Marie Fourt, France

Metal and Ceramic Composites
B Cantor, F P E Dunne and I C Stone
Department of Materials, University of Oxford, UK

High Pressure Surface Science
Y Gogotsi and V Domnich
Department of Materials Engineering, Drexel University, USA

Series in Materials Science and Engineering

Solidification and Casting

An Oxford–Kobe Materials Text

Edited by

Brian Cantor and Keyna O'Reilly
Department of Materials, University of Oxford, UK

Institute of Physics Publishing
Bristol and Philadelphia

© IOP Publishing Ltd 2003

All rights reserved. No part of this publication may be reproduced, stored in a retrieval system or transmitted in any form or by any means, electronic, mechanical, photocopying, recording or otherwise, without the prior permission of the publisher. Multiple copying is permitted in accordance with the terms of licences issued by the Copyright Licensing Agency under the terms of its agreement with Universities UK (UUK).

British Library Cataloguing-in-Publication Data

A catalogue record for this book is available from the British Library.

ISBN 0 7503 0843 5

Library of Congress Cataloging-in-Publication Data are available

Series Editors: **B Cantor and M J Goringe**

Commissioning Editor: Tom Spicer
Production Editor: Simon Laurenson
Production Control: Sarah Plenty
Cover Design: Victoria Le Billon
Marketing: Nicola Newey and Verity Cooke

Published by Institute of Physics Publishing, wholly owned by The Institute of Physics, London

Institute of Physics Publishing, Dirac House, Temple Back, Bristol BS1 6BE, UK

US Office: Institute of Physics Publishing, The Public Ledger Building, Suite 929, 150 South Independence Mall West, Philadelphia, PA 19106, USA

Typeset by Academic + Technical Typesetting, Bristol
Printed in the UK by Antony Rowe Ltd, Chippenham

Contents

Preface ix

Acknowledgments xii

SECTION 1: INDUSTRIAL PERSPECTIVE 1
Introduction

Chapter 1 3
Direct chill casting of aluminium alloys
Martin Jarrett, Bill Neilson and Estelle Manson-Whitton
Alcoa

Chapter 2 26
Continuous casting of aluminium alloys
Philip Thomas
Holton Machinery

Chapter 3 48
Continuous casting of steels
Hisao Esaka
National Defense Academy

Chapter 4 59
Castings in the automotive industry
Kuniaki Mizuno
Aisin Takaoka

Chapter 5 69
Cast aluminium–silicon piston alloys
John Woodthorpe, R Thomson, C Daykin and P Reed
Federal-Mogul

Contents

SECTION 2: MODELLING AND SIMULATION   77
Introduction

Chapter 6   79
Mold filling simulation of die casting
Koichi Anzai, Eisuke Niyama, Shinji Sannakanishi and
Isamu Takahashi
Tohoku University

Chapter 7   87
The ten casting rules
John Campbell
Birmingham University

Chapter 8   106
Grain selection in single crystal superalloy castings
Philip Carter, David Cox, Charles-Andre Gandin and Roger Reed
Cambridge University

Chapter 9   121
Defects in aluminium shape casting
Peter Lee
Imperial College

Chapter 10   142
Pattern formation during solidification
John Hunt
Oxford University

Chapter 11   160
Peritectic solidification
Hideyuki Yasuda, Itsuo Ohnaka, Kentaro Tokieda and
Naohiro Notake
Osaka University

SECTION 3: STRUCTURE AND DEFECTS   175
Introduction

Chapter 12   177
Heterogeneous nucleation in aluminium alloys
Peter Schumacher
Oxford University

Contents vii

Chapter 13 199
Control of grain size in solidification
Lindsay Greer
Cambridge University

Chapter 14 248
Step casting
Masayuki Kudoh, Tatsuya Ohmi and Kiyotaka Matsuura
Hokkaido University

Chapter 15 257
Intermetallic selection during solidification of aluminium alloys
Keyna O'Reilly
Oxford University

Chapter 16 271
Cooling rate and the structure of commercial aluminium alloys
Jim Kellie
London & Scandinavian Metallurgical

Chapter 17 286
Solidification structure control by magnetic fields
Itsuo Ohnaka and Hideyuki Yasuda
Osaka University

Chapter 18 298
Direct observation of solidification and solid phase transformations
Toshihiki Emi and Hiroyuki Shibata
*Formerly CISR (TE) and Institute of Multidisciplinary
Research for Advanced Materials (HS)*

Chapter 19 326
Interfacial energy and solidification structures in aluminium–silicon alloys
Hideo Nakae and Hiroaki Kanamori
Waseda University

SECTION 4: NEW MATERIALS AND PROCESSES 339
Introduction

Chapter 20 341
Rheocasting of TiAl alloys and composites
Kiyoschi Ichikawa and Yoshiji Kinoshita
MITI

Chapter 21 359
Solidification of metallic glasses
Nobuyuki Nishiyama and Akihisa Inoue
Tohoku University

Chapter 22 373
New eutectic ceramics
Itsuo Ohnaka, Yoshiharu Waku, Hideyuki Yasuda and
Yoshiki Mizutani
Osaka University

Chapter 23 390
Rapid solidification of peritectics
Kuzuhiko Kuribayashi, Yuzuru Takamura and Kosuke Nagashio
Institute of Space and Aeronautical Science

Chapter 24 402
Peritectic solidification of superconducting oxides
Yuh Shiohara, Makato Kambara, Teruo Izumi and
Yuichi Nakamura
International Superconductivity Technology Center

Index 415

Preface

This book is a text on Solidification and Casting, arising out of presentations given at the second Oxford–Kobe Materials Seminar, held at the Kobe Institute, on 21–24 September 1999.

The Kobe Institute is an independent non-profit-making organization. It was established by donations from Kobe City, Hyogo Prefecture and more than 100 companies all over Japan. It is based in Kobe City, Japan, and is operated in collaboration with St Catherine's College, Oxford University, UK. The Chairman of the Kobe Institute Committee in the UK is Sir Peter Williams, Master of St Catherine's College; the Director of the Kobe Institute Board is Dr Yasutomi Nishizuka; the Academic Director is Dr Helen Mardon, Oxford University; and the Bursar is Dr Kaizaburo Saito. The Kobe Institute was established with the objectives of promoting the pursuit of education and research that furthers mutual understanding between Japan and other nations, and to contribute to collaboration and exchange between academics and industrial partners.

The Oxford–Kobe Seminars are research workshops which aim to promote international academic exchanges between the UK/Europe and Japan. A key feature of the seminars is to provide a world class forum focused on strengthening connections between academics and industry in both Japan and the UK/Europe, and fostering collaborative research on timely problems of mutual interest.

The second Oxford–Kobe Materials Seminar was on Solidification and Casting, concentrating on developments in science and technology over the next ten years. The co-chairs of the Seminar were Professor Itsuo Ohnaka of Osaka University, Professor Brian Cantor and Dr Keyna O'Reilly of Oxford University, and Dr Kaizaburo Saito of the Kobe Institute. The Seminar coordinator was Ms Pippa Gordon of Oxford University. The Seminar was sponsored by the Kobe Institute, St Catherine's College, the Oxford Centre for Advanced Materials and Composites, London and Scandinavian Metallurgical Co Ltd, and The Luxfer Group. Following the Seminar itself, all of the speakers prepared extended manuscripts in order to compile a text suitable for graduates and for researchers entering the field. The contributions are compiled into four sections: industrial perspective, modelling and simulation, structure and defects, and new materials and processes.

Acknowledgments

Brian Cantor and Keyna O'Reilly

The editors would like to thank the following: the Oxford–Kobe Institute Committee and St Catherine's College, Oxford University, for agreeing to support the Oxford–Kobe Materials Seminar on Solidification and Casting; Lord Plant, Sir Peter Williams, Professor Itsuo Ohnaka and Drs Lincon Wallen, Helen Mardon and Kaizaburo Saito for help in organizing the Seminar; and Ms Pippa Gordon and Ms Sophie Briant for help with preparing the manuscripts.

Individual authors would like to make additional acknowledgements as follows.

Toshihiki Emi and Hiroyuki Shibata

The authors would like to thank Mr Y Arai, Sumitomo Metal Industries Ltd, Mr H Chikama, West Japan Railways Co, and Mr S Yoshinaga, Sumitomo Electric Co, who contributed greatly to carrying out the experiments while they were with the Institute of Advanced Materials Processing, Tohoku University, Sendai, Japan.

Lindsay Greer

The author's research in this area is supported mainly by the EPSRC (UK), but has also been supported by NEDO (Japan) by the EU Brite-Euram Programme (contract BRE2-CT92-0159), and by the EU Network MEBSP (Microstructural Engineering by Solidification Processing). Further support, and access to experimental facilities, has been provided by Alcan International Ltd (Banbury, UK), London and Scandinavian Metallurgical Co Ltd (Rotherham, UK), Pechiney CRV (France) and Trikon Technologies Ltd (Newport, UK). Collaborations with these companies, and with the University of Oxford (Department of Materials), are gratefully acknowledged. The author thanks the following for useful discussions and for contributing to the work reviewed in this chapter: D J Bristow, A M Bunn, C D Dobson, P V Evans, F Gärtner, T F Gloriant, D M Herlach, R Jakkaraju, X-Y Jiang, M W Meredith, A F Norman, P Schumacher, A Tronche, M Vandyoussefi and Z-C Zhong.

Kiyoschi Ichikawa and Yoshiji Kinoshita

The authors would like to thank the sponsors of the Industrial Science and Technology Frontier Program, the Agency of Industrial Science and Technology, MITI.

Jim Kellie

The author would like to thank staff at Nottingham University, Sheffield Hallam University and Imperial College and colleagues at LSM and Alpoco for all the help and suggestions that they have provided.

Kuzuhiko Kuribayashi, Yuzuru Takamura and Kosuke Nagashio

This work has been supported mainly by a Grant-in-Aid for Scientific Research from The Ministry of Education, Science, Sports and Culture, and partly by the New Energy and Industrial Technology Development Organization for the R & D of Industrial Science and Technology Frontier Program. This work is also part of the project of Core Research for Evolutional Science and Technology (CREST).

Peter Lee

The author would like to thank his colleagues at both Imperial College (R C Atwood and D See, to name but a couple) and Aisin Takaoka (especially S Nishido) for their contributions to the work described in this paper. The author gratefully acknowledges the financial support of Aisin Takaoka, together with the EPSRC for providing partial funding for the computer facilities used under grant GR/L86821. The author would also like to thank EKK Inc for the use of their macromodelling codes and their assistance in incorporating the mesomodels.

Itsuo Ohnaka and Hideyuki Yasuda

The authors would like to acknowledge support from a Grant-in-Aid for Scientific Research of Priority Areas given by the Ministry of Education, Science, Sports and Culture, Japan. A part of the experiments was performed under the Visiting Researcher's Program of the High Field Laboratory for

Superconducting Materials, Institute for Materials Research, Tohoku University.

Itsuo Ohnaka, Yoshiharu Waku, Hideyuki Yasuda and Yoshiki Mizutani

This work is supported in part by NEDO and a Grant-in-Aid for Scientific Research given by the Ministry of Education, Science, Sports and Culture, Japan. The authors would like to thank Mr N Nakagawa, Ube Industry, Japan, for valuable assistance in the experiments.

Keeyna O'Reilly

The author thanks the EPSRC (UK), Alcan International Limited, London Scandinavian Metallurgical Company Ltd, The Luxfer Group and The British Council, Madras (UK/India) for supporting research in this area.

Yuh Shiohara, Makato Kambara, Teruo Izumi and Y Nakamura

This work was supported by the New Energy and Industrial Technology Development Organization (NEDO) as Collaborative Research and Development of Fundamental Technologies for Superconductivity Applications.

Peter Schumacher

The author would like to thank Alcan International Ltd, Banbury, London and Scandinavian Metallurgical Co Ltd, Rotherham, and the EPSRC for support through an Advanced Fellowship.

Hideyuki Yasuda, Itsuo Ohnaka, Kentaro Tokieda and Naohiro Notake

The authors would like to acknowledge support from a Grant-in-Aid for Scientific Research on Priority Areas and for Encouragement of Young Scientists given by the Ministry of Education, Science, Sports and Culture, Japan.

SECTION 1

INDUSTRIAL PERSPECTIVE

Casting technology is developing rapidly, driven by a combination of improvements in understanding of underlying solidification theory, increased computer capacity for design simulation and on-line control, and enhanced industrial competition in increasingly globalized world markets. Wrought products are often dominated by the quality and yield of primary solidification of ingot and billet feedstock. Moreover, many industrial sectors are developing rapid prototyping and agile manufacturing processes to reduce time to market and speed up responsiveness to customer demands. Casting has much to offer as a near net shape technology, but important issues of reproducibility and quality need to be improved. The overall industrial perspective on casting technology is discussed in this section, concentrating on recent innovations and challenges for the future.

Chapters 1 and 2 discuss respectively semi-continuous and continuous methods of casting aluminium alloys, and chapter 3 discusses continuous casting of steel. Chapters 4 and 5 discuss different aspects of the importance of casting in the important automotive sector.

Chapter 1

Direct chill billet casting of aluminium alloys

Martin Jarrett, Bill Neilson and Estelle Manson-Whitton

Introduction

This chapter details the operational and technological developments of the direct chill (DC) casting process for a high volume commercial extrusion business. Other continuous casting processes are discussed in chapters 2 and 3. Modelling of DC casting is discussed in chapter 6. Improved extrusion manufacturing efficiency is driving the need for better and more consistent billet quality. This has necessitated significant technological and process development of DC billet production, in order to produce extrusion ingots of predictable performance. A thorough understanding is required of the interaction of the equipment and process technologies that impact on the metallurgical macro- and microstructure of the DC cast ingot, and its subsequent performance in the extrusion process.

The first commercial semi-continuous casting machine for aluminium alloys was opened in Germany in 1936, following development of vertical continuous casting techniques for other metals, such as lead, since the mid-nineteenth century [1]. Over the past 50 years, the DC casting process has been developed to become the predominant process for extrusion billet production [2, 3], with 6xxx alloy extrusion billet accounting for a large percentage of the throughput of worldwide aluminium DC casting production.

Together with market driven advancement, environmental considerations are driving significant change in DC casting technology. The provision of consistent high quality DC cast billets of predictable performance is of fundamental importance in operating extrusion presses at maximum efficiency, while meeting the stringent quality requirements of the market place. Several key factors affecting billet quality, that impact on extrusion performance, have been previously described by Weaver [4] and

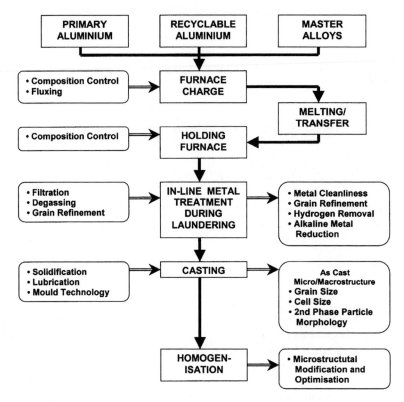

Figure 1.1. Schematic representation of a typical direct chill casting process.

Langerweger [5], and more recently by Bryant and Fielding [6]. These emphasize the importance of characterization and control of the total process. A schematic representation of a typical DC casting process is shown in figure 1.1. The key process stages form the basis of the subsequent sections of this chapter, which describe the critical aspects of both equipment and process technology.

Direct chill billet production

The DC casting process from melting, through melt in-line treatment, casting and homogenization is discussed in terms of the present best practice and promising novel techniques. The impact of best practice on billet quality and ultimately the impact on extrusions are also then discussed, utilizing the extrusion limit diagram concept, the critical aspects of which have been summarized by Parson *et al* [7]. All discussion is in the context of a high throughput DC casting operation, concentrating on 6xxx series alloys.

Melting

Current standard DC casting facilities use melting furnaces which are charged with a mixture of primary and secondary scrap aluminium depending on target billet specification. The molten metal is transferred to a holding furnace before casting. The use of a holding furnace maximizes efficiency by fully utilizing melting time. The ideal configuration balances the melting capacity of the furnaces with the casting capacity of the DC casting machine. Alloying additions are made in the melting furnace, in the holding furnace, or during laundering to the holding furnace. Better mixing is achieved if alloying additions are made earlier, although there is a danger of significant loss if additions are made to the melting furnace, and, unless additions can be made with the furnace door closed, production time may be lost.

Melting and holding furnaces can be either induction or gas heated. Gas is preferred for efficiency, and for melt cleanliness as the churning resulting from induction heating can drag oxide particles from the dross back into the melt. More recently oxyfuel furnaces, burning gas with approximately 10% oxygen, have been introduced, although a well-controlled gas furnace, incorporating either regenerative or recuperative air heating systems to maximize fuel efficiency, remains the industry preference.

Furnaces can be fixed hearth or tilting. Fixed hearth furnaces have lower capital cost but tilting furnaces are preferred for metal cleanliness, process control, and safety, as at any point the flow can be stopped by resetting the furnace, whereas a fixed hearth furnace requires manual plugging. Older fixed hearth furnaces are generally being replaced with tilting furnaces. A further advantage of tilting furnaces is that they can be fully drained, allowing greater flexibility for alloy changes.

Together with other areas of the DC casting process, the environmental impact of furnaces is being minimized by the reduction of particulate emissions through more efficient furnace design (for example, the use of regenerative burners) and, where necessary, the use of equipment for the capture of both particulate and noxious gaseous emissions.

Temperature control is of paramount importance in casting, having a direct bearing on product quality and production efficiency. The optimum melt delivery temperature for 6xxx alloys is in the range 690–750°C, and is product and plant specific. Temperature measurement is predominantly through the use of thermocouples. Other methods of temperature measurement such as optical pyrometry are used, but encounter problems with aluminium due to oxide skin formation. Immersed thermocouples, however, remain the most accurate, robust and reliable technology. Heating equipment is universally thermostatically controlled, to maintain thermal efficiency and process control.

Molten metal pre-treatment

Molten metal pre-treatments, carried out in the melting or holding furnace, can be distinguished from in-line treatments given during laundering of the melt to the casting machine. The main purpose of fluxing [8] is to clean the melt by degassing and to remove oxides and other inclusions. Other advantages of fluxing include the production of a dry dross, which minimizes metal losses during skimming. Fluxing, together with more recent developments in dross reprocessing (pressing and recycling), has led to improved recoveries in the industry, whilst maintaining cleaner and more efficient melting units.

Currently, two methods of fluxing are available, using chlorine gas or chlorine-based salt. Injection of gas (sometimes using a spinning nozzle) below the surface of the melt has until recently been the preferred method. However, environmental legislation is now necessitating the phasing out of chlorine gas use at the melting stage. The use of fluxing salts, although a more mature technology than gas fluxing, is now being re-evaluated as a replacement for chlorine gas. However, there are still environmental concerns over the chlorine and fluorine reaction products produced which remain in the dross, and the uncertain determination of potentially hazardous products of reaction emitted to the environment from the use of these fluxes. The need to reduce and eventually eliminate the use of chlorine in fluxes has led to a gap in the market for an alternative environmentally friendly fluxing method. Currently there are no processes yet capable of commercial operation.

In-line metal treatment

Certain melt treatments, namely grain refinement, degassing and filtration, must be given in-line during laundering to the casting machine to accrue maximum benefit. Grain refining inoculants have in the past been added to the holding or even melting furnace. The disadvantages of this method include fade (where the effectiveness of the grain refiner decreases with time), and the formation of a boron-particle-rich sludge in the bottom of the furnace which contaminates the metal, and leads to through-length variation in cleanliness and composition of the DC cast log. Degassing and filtration are performed in the launder such that there is minimum turbulent flow, which can cause the reintroduction of oxides before the casting machine. Recent technological developments of these techniques have been reviewed by Fielding and Kavanaugh [9], who emphasize the criticality of degassing and filtration in the DC casting process.

Grain refinement

The purpose of grain refinement is to produce a refined, equiaxed grain structure with modified second phase particle morphology through the thickness of the DC cast log. Nucleation and grain refinement are discussed in detail in chapters 12 and 13. Current practice is to use approximately 10 ppm of titanium to give the most desirable grain size of approximately 80–150 μm. There are a number of alternative grain refiners available, all based on Al, the most common being 6%Ti, 3%Ti, 5%Ti–0.1%B, 5%Ti–1%B and 5%Ti–0.2%B. All these compositions are in commercial use for a variety of applications and products. Best metal cleanliness is achieved if grain refiner is added before de-gassing or filtration. Figure 1.2 shows a system for injection of grain refiner rod into the launder, which can be controlled automatically to give the desired rate of addition.

Universally the 5%Ti–1%B grain refiner is the preferred inoculant [10]. As a result of this it has been extensively researched, leading to a number of publications discussing in detail its capability as a grain refiner under various conditions [11,12]. Its difficulty, however, is the high boron content. Boron is an insoluble element and boron particles are prone to flocculate, forming large clusters which can be deleterious to products in the form of pick-up, and as an abrasive to dies. To counter the problem of boron inclusions, it is advised to inoculate at a reduced rate that accounts for the boron content of the recycled aluminium.

Figure 1.3 shows the effect of titanium content on grain size for a typical 5:1 grain refiner achieving the target grain size. In addition to grain size modification [13], experiments have shown that the morphology of the insoluble iron-rich phase can be influenced, not only by the grain refiner composition, but also by the manufacturing route used by different suppliers. (The presence of the iron-rich phase in the form of α-script has been shown to affect adversely extrusion surface quality by increasing pick-up [14]). This would indicate that the nucleation mechanism of the iron-rich particles is strongly influenced by the grain refinement process. This effect is currently being investigated to gain an understanding of the process. Figure 1.4 is a plot of α-script occurrence for different grain refiners, and shows that the choice of grain refiner can have a significant effect on the amount of α-script in the microstructure.

The most promising new grain refiner currently available is titanium carbide [15–17], which, in addition to eliminating boron contamination, is reported to overcome problems of poisoning by zirconium- and

Figure 1.2. System for automatic injection of grain refiner rod into the melt during laundering to the casting machine.

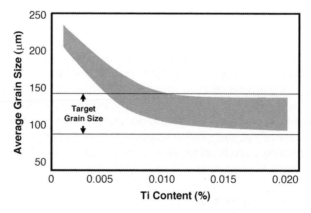

Figure 1.3. Average grain size (determined at mid-radius) as a function of total melt titanium content for a 180 mm diameter 6063 billet following in-line inoculation with 5% Ti–1% B rod.

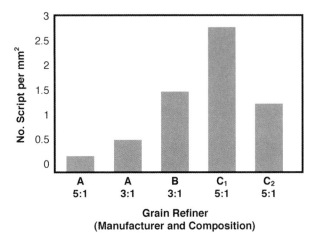

Figure 1.4. Number of α-script particles per mm^2 for different grain refiners produced by three different suppliers.

chromium-containing alloys, yet is as effective in reducing grain size as titanium boride [15]. Closer temperature control is, however, required in using titanium carbide and thus commercial implementation may require adjustment to processing control. Individual plants are currently evaluating the use of titanium carbide. Other novel methods of grain refinement include Nb additions [18], and physical methods (which retard formation of and break up dendrites) such as ultrasonic vibration [19], sump displacement and electromagnetic stirring [20].

Degassing

Most high-quality cast houses now use in-line spinning nozzle degassing systems for the removal of hydrogen. The most common systems are the Alpur (Pechinney) and SNIF (Foseco). However, some patents are now ending, and enterprising companies are designing their own systems based on existing and new technology [9]. An example of an Alpur degasser is shown in figure 1.5, and a schematic of a SNIF degasser is shown in figure 1.6. Efficiencies of spinning nozzle type degassers are very high, and they can deliver hydrogen levels of less than approximately 0.1 cc/100 g, compared to previous levels of approximately 0.3–0.4 cc/100 g. Figure 1.7, for example, shows the hydrogen content of a melt measured before and after passing through the Alpur degasser shown in figure 1.5.

Both Alpur and SNIF systems, when using chlorine gas, have also been shown to reduce oxide inclusions by approximately 50%. They will reduce the overall volume fraction of inclusions, and through the stirring action

Figure 1.5. The Alpur degasser in use at British Aluminium Extrusions, Banbury, UK.

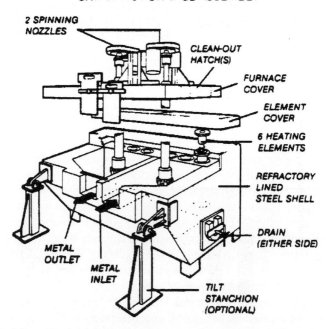

Figure 1.6. Schematic of the SNIF R-140 degasser showing two chambers.

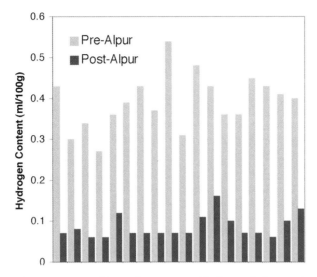

Figure 1.7. Hydrogen content of 6xxx alloys measured before and after passing through an Alpur degasser.

will break up coarse borides. However, as a result of this, the number density of particles can increase. Improved process control using multi-chamber SNIF and Alpur systems, incorporating two-way metal flow and controlled argon and chlorine gas mixtures have, however, demonstrated significantly improved levels of particle removal. Figure 1.8 shows inclusion levels of

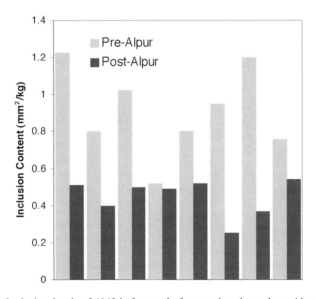

Figure 1.8. Inclusion levels of 6063 before and after passing through an Alpur degasser.

6063 casts before and after passing through the Alpur degasser shown in figure 1.5.

Filtration

There are a number of filtration methods in use throughout the industry [9]. The most common methods are those that rely upon oxide entrapment by the tortuous route the aluminium must take as it travels though the filtration medium. These methods can be classified as ceramic foam filters (CFF), Alcan bed filter (ABF) and porous tube filter (PTF). Glass cloths are also used but remove only the coarsest particles, and their primary purpose is to give a uniform distribution of metal during the start of casts using the dip tube and float configuration.

The Alcan bed filter is viewed as being a particularly efficient system, and comprises a layered bed of graduated ceramic spheres through which the metal passes. Particles are trapped at the interstices between the spheres. Ceramic foam filters work in much the same way as the Alcan bed filter, but are less effective because of the shorter distance through which the metal flows. They allow flexibility as they are discarded after each cast, and are widely used as secondary filters. The newest filtration development, shown in figure 1.9, is the porous tube filtration system [21], which filters more effectively than the Alcan bed filter or the ceramic foam filter systems. Porous tube filtration is currently used for quality-critical finished products such as photocopier tubes. Figure 1.10 shows a comparison of inclusion content following filtration with (a) a ceramic foam filter and (b) a porous tube filter, showing an order of magnitude difference. The Alcan bed filter and the porous tube filter will eventually block, and therefore a continuous maintenance programme is required in order to sustain maximum performance, and these are best used in conjunction with an in-line degasser.

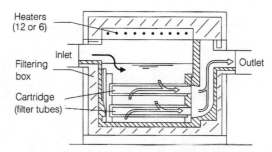

Figure 1.9. Schematic of a porous tube filter. Metal flows into the box and is filtered through the walls of a cartridge of porous tubes.

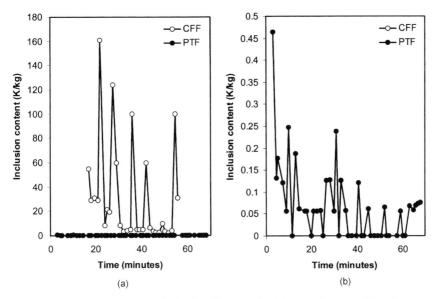

Figure 1.10. Inclusion content following filtration by (a) ceramic foam filter (average inclusions ~40 K/kg), and (b) porous tube filter (average inclusions ~0.1 K/kg).

Casting configuration

The three main casting configurations in commercial use are dip tube and float, hot top, and airslip [3]. Within each of these general systems, many variations have been individually developed. Figure 1.11 shows an illustration of the casting configuration for a typical dip tube and float system [22]. Although this system has been in use for much longer than hot top and airslip and is gradually being replaced, thousands of tonnes of good quality aluminium billet for a wide diversity of markets are still cast through this system annually. The dip tube and float system remains popular because of its low capital cost, ease of maintenance and fast turnaround in casting preparation. Billet can consistently be produced with desirable macro- and microstructure. Good surface quality can be achieved using either grease or continuous lubrication. Mould type is critical, and aluminium moulds must have a sustained maintenance programme to ensure consistent surface quality, whereas graphite-lined moulds require minimal maintenance to achieve good surface quality, and improved shell. The drawbacks of the dip tube and float system are partly a result of the turbulent metal flow, leading to oxide entrainment, and occasional blocking of the dip tube. The other major disadvantage of the system is the unavoidable presence of a shell zone, which must be minimized to reduce the impact on extrusion recovery.

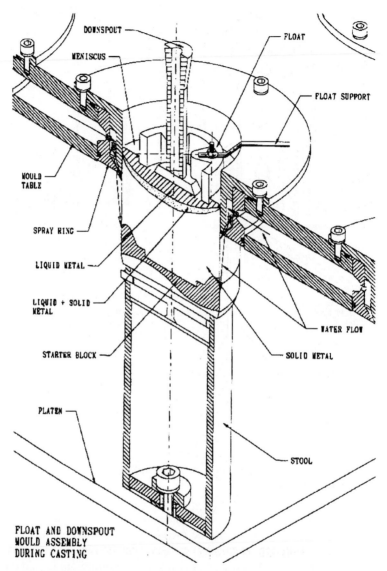

Figure 1.11. Schematic of the dip tube and float configuration for DC casting [22].

The hot top system, shown in figure 1.12 [22], reduces turbulent flow but does not eliminate the shell zone and its associated problems. Moulds tend to be either graphite or ceramic for ease of maintenance, and are again lubricated by either grease or continuous lubrication systems depending on cast length. In both dip tube and float and hot top configurations, the deleterious shell zone can be minimized by casting with a low metal height in the mould

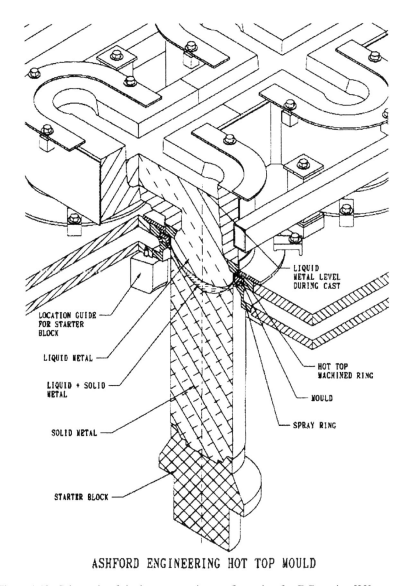

Figure 1.12. Schematic of the hot top casting configuration for DC casting [22].

and using an appropriately high casting speed. Under optimum conditions the shell zone can be controlled to a thickness of less than approximately 10 mm, but cannot be eliminated.

The shell zone, shown in figure 1.13 for conventional dip tube and float casting, results from the region of slow cooling within the mould, between the meniscus and point of water impingement, where the air gap acts as effective

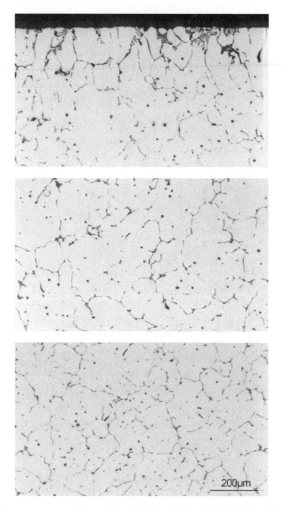

Figure 1.13. Microstructure of an as direct chill cast (dip tube and float) 6082 billet (a) at the edge showing inverse segregation, (b) ~5 mm from the edge showing large grains in the shell zone, and (c) in the centre of the billet.

insulation. Generally, the deeper the mould the longer the air gap and the deeper the shell zone. The airslip system, shown in figure 1.14, was developed in response to this issue, and is so named because the annulus around which the billet is formed is a pressurized system, and the metal flowing into the mould does not touch the mould but is suspended on an air/lubricant film over which it slips. Solidification in an airslip mould then is not influenced by heat transfer through the mould but through the air/lubrication gap and, together with a short mould, the system effectively eliminates the shell zone, and minimizes surface inverse segregation. Figure 1.15 shows the

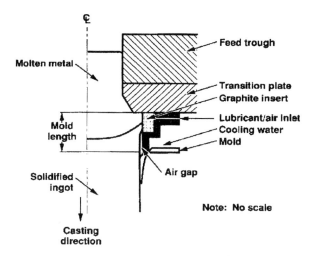

Figure 1.14. Schematic of the airslip DC casting configuration.

shell zone in conventional dip tube and float DC cast billet, and elimination of the shell zone using airslip technology. Figure 1.16 shows the shell zone thickness as a function of effective mould depth [23].

Increased billet quality, with resultant improved extrusion recoveries from reduced discards, through the use of airslip is achieved at the expense of a higher capital cost and maintenance. Balancing of the pressurized system can, however, be problematic, and gas, whilst always shown as escaping downwards, can under some conditions escape upwards, leading to

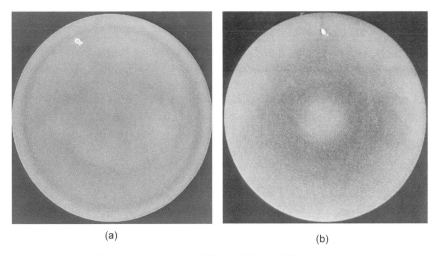

Figure 1.15. Anodized sections through 180 mm DC cast billet: (a) conventional dip tube and float showing shell zone, and (b) airslip with no shell zone.

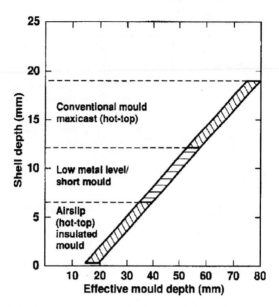

Figure 1.16. Shell depth as a function of effective mould depth for conventional configurations, and airslip which can eliminate shell entirely [23].

oxidation of the molten metal stream feeding the mould. The system is undergoing continuing development by suppliers and will certainly play an increasingly major role in DC casting in future.

Casting speed

Casting speed is optimized for both productivity and quality. Cast speeds vary with ingot size and alloy type from 50 to 150 mm min^{-1} in the general ingot size range from 400 to 150 mm diameter. Start speeds are nominally 10% lower than the final cast speed and are controlled to lessen the effect of high-stress-induced centreline cracking at the start of the cast. This is also controlled by careful design of stool shape. Ramping to the final speed is PLC controlled.

Cooling

Water is the universal coolant for all DC casting systems. It must be supplied at a rate which continuously removes the film of boiling water adjacent to the billet skin, promoting nucleate boiling to give a continuous, uniform cooling of the ingot [24]. Systems have been developed to slow cool at the start of the cast and reduce billet internal stresses [1]. Alcoa, Alcan and Reynolds have devised their own methods (CO_2, pulsed water, and air mixture respectively) [1, 3]. These systems rarely extend beyond the influence of their plants. Universally, the water flow rate for billet production is in the range 0.8 to 1 gallon per minute per inch (1.42 to 1.77 litres per minute per cm) of mould circumference. Temperature ranges of delivered water can vary depending on supply from ambient (0–25°C depending on season) to constant temperatures of approximately 25°C where control or recycling is used. Water temperature is an important variable in achieving microstructural control. Water composition is rarely viewed as a controlled factor, but the influence of composition is considered to be a potentially important factor. Contamination by lubricants is controlled for environmental reasons only, although it may have an influence on billet cooling rate, and therefore on microstructural development.

Analytical systems and process control

All casting systems are now controlled by PLC systems. The list of factors which can be monitored and controlled is extensive. As computing power increases, so does the amount of information that can be collected and stored. This information can be used, not only for on-line monitoring, but as a valuable management tool for continuous improvement.

An example of what can be measured as metal flows though a casting plant is as follows: charge weight and make up; charging time; melting time and time at temperature; transfer temperature, time and weight; casting temperature; metal level; stool height in the mould (ram position); degassing parameters such as gas pressure, speed, and temperature; furnace tilt; casting speed including start speed and ramp time; cast time; launder temperature; water temperature and flow rate; inoculation rate; log length, weight, and number; homogenizer time in, temperature, time out; and cooling rate following homogenization. Modern systems also provide condition monitoring of key equipment so as to provide early warning of potential equipment failure and to aid maintenance and total productive maintenance (TPM) activities.

Melt composition and temperature are probably the two most crucial parameters in DC casting which must be measured and controlled. Compositional control starts with charge composition, which varies with the amount and quality of scrap and primary metal used. For cost efficiency, the greatest quantity of scrap of a known quality is used which still maintains final billet quality. Primary aluminium (to purity specification based on a combination of iron and silicon content, e.g. 99.7% Fe + Si = 0.3% max.) and sometimes commercial scrap, make up the rest of the charge. Following melting, a sample is taken from the furnace and analysed; compositional adjustment is then made by introduction of primary aluminium, to dilute elements in concentrations greater than alloy compositional limits impose, and/or master alloys which contain high concentrations of alloying elements, to bring the composition up to specification.

Master alloys are available as element-rich aluminium ingot, element-rich tablets, or 'pure' alloys (e.g. magnesium, silicon). The choice is dependent on furnace size, and method and position of addition. To increase melting capacity, additions can be made in the launder or holding furnace, although this requires an initially dilute charge, as alloying element concentrations can only be increased using this method. Control of elements within most alloys is nominally ±0.03%. However, extrusion customers are increasingly specifying both tighter tolerances and capping of deleterious impurity elements to attain their own process consistency.

Homogenization

Homogenization of DC cast billet is necessary to achieve an optimized microstructure for ease of extrusion. An ideal homogenization treatment eliminates microsegregation, modifies the insoluble particle morphology, and on cooling precipitates a fine dispersion of Mg_2Si which, by removing magnesium and silicon from solution, lowers the flow stress of the billet but allows full redissolution during extrusion. Homogenization is typically performed at temperatures between 500 and 595°C for times of 1–4 h, and depends upon the degree of microstructural refinement and homogeneity

Figure 1.17. Typical batch homogenizer for aluminium DC cast billets [27].

Figure 1.18. An example of a continuous homogenizer, allowing greater control and flexibility than a batch homogenizer [27].

required. Following the high temperature soak, air cooling or water quenching is used, and a controlled cooling rate is critical in achieving the ideal Mg_2Si size, morphology and distribution [25, 26].

Traditionally, batch homogenization is used, shown in figure 1.17, with the disadvantage of temperature variations from log to log and along the same log with position in the furnace, leading to variations in microstructure and mechanical properties from billet to billet. Recently, commercial fully automated continuous homogenization furnaces have been introduced, shown in figure 1.18 [27], which virtually eliminate many of the problems associated with batch homogenization. In the future, continuous homogenization will allow delivery to the extruder of billet with microstructure and flow stress controlled to within very tight limits.

Summary

Extruders demand not only high quality DC cast billet, but also consistency of quality and mechanical properties (most notably flow stress) from billet to billet and batch to batch. All technological development is aimed at achieving this, while keeping costs low and ensuring environmental targets are met. Consistently high billet quality achieved through controlled production allows maximum recovery through elimination of quality based rejections and minimization of top and tail discard and scalping (if required).

Developments in equipment and process technology have provided the potential for significant improvements in billet quality. The most significant of these developments are:

- airslip mould technology
- in-line multi-chamber degassing systems
- grain refinement, e.g. TiC, ultrasonics
- porous tube filtration
- continuous homogenization.

The optimization of DC cast billet requires a holistic approach to the continual development of the technology through

- the mechanistic quantification of the critical process stages through laboratory scale optimization and in plant trials
- an effective technology transfer process translating innovative technology to robust industrial practice
- a proactive in-plant continuous improvement culture
- precise characterization of extrusion equipment and process capabilities with effective evaluation and performance measures.

Microstructural optimization

The ability to manipulate the as cast microstructure through controlled solidification and grain refinement techniques, coupled with an ability to apply accurate heating and cooling cycles to individual logs during continuous homogenization, gives the capability to optimize billet microstructure for specific products or processes. Ultimately the need for a homogenization treatment, separate from billet pre-heat, may be negated if sufficient microstructural refinement is achieved during casting. This will only be achieved, however, through a fundamental understanding of the nucleation and growth mechanisms taking place during solidification under the influence of (i) temperature and time, (ii) alloy chemistry, (iii) inoculant chemistry and (iv) physical turbulence, e.g. deliberate dendrite breakage. Moreover, the effective removal of the shell zone using airslip technology provides the extruder with an increased capability of achieving maximum recovery by drastically reducing extrusion billet discard size.

In-line metal cleanliness

Removal of hydrogen, residual alkaline metals, and non-metallic inclusions using the next generation in-line degasser systems may remove the need for final filtration other than for the most critical applications. Control of molten metal turbulence throughout the casting cycle provides further opportunity for cleanliness improvement, by reducing the formation and entrainment of oxide inclusions. Considerable benefits, for example in structural integrity, have been achieved for foundry casting by minimizing oxide formation through carefully designed runners and gating systems. Full characterization of the DC casting process is required to understand fully the potential improvement to be derived from better metal flow. In order to produce the cleanest metal, however, best available technology would suggest the incorporation of the porous tube filter in conjunction with an in-line degassing system.

Impact on extrusion

With a supply of reliable high quality billet with tailored microstructure and consistent flow stress, the extruder can operate presses at maximum efficiency and minimum cost. Figure 1.19 shows a typical speed and surface quality limit diagram for extrusion of 6063. Such extrusion limit diagrams are a valuable aid in evaluating extrusion and maximizing productivity within the bounds of avoidance of surface defects, specific pressure requirements, and attainment of mechanical properties. Figure 1.19 shows that maximum productivity is achieved by extruding under conditions at the apex of the triangle bounded by insufficient pressure, and inadequate surface. Thus, high quality and consistent billet can increase extrusion productivity in two ways. First, the position of the apex of the triangle in figure 1.19 can be raised by improved billet microstructure leading to lower susceptibility to tearing and pickup. Second, more consistent billet properties allow press

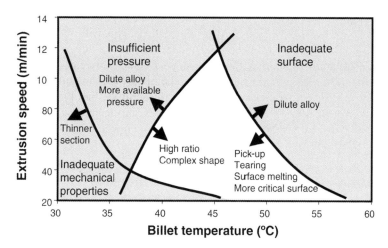

Figure 1.19. A typical extrusion limit diagram for 6063.

operation much closer to the theoretical maximum speed without the risk of defects leading to expensive quality rejections.

In addition to maximizing productivity, consistent high quality billet allows the extruder to supply a more consistent product to their customers, while maximizing recovery due to improvements in planned and unplanned scrap. Assurance of quality also allows the extruder to design to tighter limits, for example reducing wall thickness while maintaining mechanical property requirements and improving through-batch consistency for the automotive market. Mechanical property requirements of ± 5 MPa proof stress variations within a one tonne batch of extrusions can be required for certain critical applications. These stringent requirements are becoming increasingly common, putting increased demands on the extruder and billet suppliers' ability to improve process control and to achieve predictable performance.

Extensive programmes involving continuous improvement, total productive maintenance, total quality and manufacturing excellence schemes underpin many of today's successful businesses. These must also be coupled with a strong research and development capability in order to provide the fundamental science necessary for full exploitation of the technology. Indeed, without these processes, optimization of the technological advances will be stifled. It is therefore essential, in order to maximize the impact of technology within the business, that these processes be undertaken simultaneously with effective communication and technology transfer processes in place.

References

[1] Furrer P 1989 *Continuous Casting of Non-Ferrous Metals and Alloys* eds H D Merchant, D E Tyler and E H Chia (TMS) p 87
[2] Grandfield J F 1993 *Aluminium Melt Treatment and Casting* ed M Nilmani (TMS) p 351
[3] Grandfield J F 1997 *5th Australasian Asian Pacific Conference on Aluminium Cast House Technology* eds M Nilmani, P Whiteley and J Grandfield (TMS) p 231
[4] Weaver C H 1988 *4th International Aluminium Extrusion Technology Seminar* (The Aluminium Association) p 385
[5] Langerweger J 1984 *3rd International Aluminium Extrusion Technology Seminar* (The Aluminium Association) p 41
[6] Bryant A J and Fielding A P 1999 *Light Metal Age* April p 78
[7] Parson N C, Hankin J D and Bryant A J 1992 *5th International Aluminium Extrusion Technology Seminar* (The Aluminium Association) p 13
[8] Utigard T A, Friesen K, Roy R R, Lim J, Silny A and Dupuis C 1998 *JOM* November p 38
[9] Fielding A P and Kavanaugh C F 1996 *Light Metal Age* October p 46
[10] Bryant M and Fisher P 1993 *Aluminium Melt Treatment and Casting* ed M Nilmani (TMS) p 281

References

[11] Backerud L 1983 *Light Metal Age* October p 6
[12] Vatne H E 1999 *Aluminium* **75** 84
[13] DeBall D J and Kidwell B L 1992 *5th International Aluminium Extrusion Technology Seminar* (The Aluminium Association) p 129
[14] British Aluminium Extrusions, Banbury, UK, unpublished work
[15] Schneider W, Kearns M A, McGarry M J and Whitehead A J 1998 *Light Metals 1998* ed B Welch (TMS) p 953
[16] Granger D A 1998 *Light Metals 1988* ed B Welch (TMS) p 941
[17] Hoefs P, Reif W, Green A H, van Wiggen P C, Schneider W and Brandner D 1997 *Light Metals 1997* ed R Huglen (TMS) p 777
[18] Pontes P S, deGaliza J A, Robert M H and Cupini N L 1983 *Solidification Technology in Foundry and Casthouse* (Metal Society, London) p 70
[19] Mason T J, Lorimer J P, Paniwnyk L and Pollet B 1999 *Materials World* **7**(2) 71
[20] Froes F H 1998 *Light Metal Age* December p 32
[21] Mitsui Kinzoku, TKR Division, Sanyo Building, 2-26-6 Higashi-Nihonbashi, Chuo-ku, Tokyo 103, UK Patent No 1428437
[22] Ashford Engineering Services, Dolfor, Llanfwrog, Holyhead, Anglesey, Gwynedd LL65 4YF, UK
[23] Anderson A N 1992 *5th International Aluminium Extrusion Technology Seminar* (The Aluminium Association) p 43
[24] Grandfield J F, Hoadley A and Instone S 1997 *Light Metals 1997* ed R Huglen (TMS) p 691
[25] Reiso O, Hafsas J E, Sjothun O and Tundal U 1996 *6th International Extrusion Technology Seminar* (The Aluminium Association) vol 1, p 1
[26] Zajac S, Bengtsson B, Johansson A and Gullman L O 1996 *Materials Science Forum* **217–222** 397
[27] IUT Industriell Ugnsteknik AB, Industrivagen 2, SE-43892 Harryda, Sweden

Chapter 2

Continuous casting of aluminium alloys

Philip Thomas

Introduction

It is now over twenty years since Edward Emley published his review 'Continuous casting of aluminium' [1] and, although considerable progress has been made during this time, many of the doubts raised in his review still exist and many of the technical issues remain unresolved. One major change that has taken place is the attitude of the major aluminium companies, not only to continuous casting, but also to the whole philosophy of organizing and operating their multinational businesses. In the past few years we have seen Alcoa and Reynolds, for example, forming partnerships with companies operating in mainland China. Another major change that has taken place is the entry of the major aluminium companies into the mini-mill business with the intention of producing specific end products in purpose-built facilities which will be located close to the end user.

For the purpose of this chapter continuous casting is defined as any casting process where the end product is produced more or less continuously in coil form. By definition this excludes direct chill (DC) casting, which is discussed in detail in chapter 1, as well as its variants such as electromagnetic casting. DC casting is still the method used for producing the bulk of the feedstock for all flat rolled and extruded aluminium products, as explained in chapter 1, and this position is unlikely to change in the foreseeable future. On this basis, one might imagine a proliferation of DC caster-based plants, but it is difficult to justify the kind of investment required. The death knell of the big hot mill has been sounded before but there is still a tremendous kudos attached to owning and operating an integrated large scale facility. Traditionally this kind of facility has been the preserve of the international major companies but in the recent past, while they are shutting down hot mills, the developing world is keen to fill the vacuum.

Casting scale

The technical and economic comparisons between DC casting and the truly continuous casting processes are well documented [2], but suffice it to say that if the starting material for the rolling process is a slab more than 0.5 m thick it is obvious that a considerable amount of downstream equipment is required to produce the finished product. Also, to be economical, huge volumes of material have to be produced. These large scale facilities usually produce a minimum of 200 000 tonnes per annum and a number of facilities produce in excess of a million tonnes per annum.

Stepping down in gauge from DC cast slabs, the next group of casters produces material typically 20 mm thick and the product is normally referred to as thick strip. This type of facility is normally owned and operated by independent producers and has an annual capacity of 50 000–100 000 tonnes per annum. These plants are often referred to as mini-mills in an attempt to mirror the changes that have taken place in the steel industry over the past 20 years or so. The casting technology employed in the production of thick strip is invariably the twin-belt process developed by Hazelett or the block caster invented by Alusuisse and commercialized by Golden. Twin-belt and articulated block casters are shown in figure 2.1. The criteria for this kind of facility are again well documented [3] but there is a fairly heavy dependence on the availability of secondhand equipment—particularly the tandem warm mill that is required to roll the continuously cast feedstock. This type of facility is usually dedicated to a limited range of end products, which are often consumed internally. A case in point is the Golden Aluminum operation that supplied can stock materials into its parent company, Coors. It is difficult to know whether the end products from these thick strip technologies are inferior to DC cast/hot rolled material—the only thing that is certain is that the end product is different.

The next stopping point in cast gauge is the range 6–12 mm produced from roll casters, and a twin-roll caster is shown in figure 2.2. Although not usually referred to as mini-mills there are a large number of roll caster based facilities, producing typically 30 000–40 000 tonnes per annum. After the introduction of twin-roll casting by Hunter in the mid-1950s a large number of plants were built, particularly in the USA. With a cast gauge of near 6 mm the cast product could be cold rolled without the need for a hot mill, and although the range of alloys that could be cast was limited, this type of facility remains successful. With the introduction of the bigger diameter roll casters, the cast gauge increased to typically 10–12 mm but the principle of eliminating the hot mill remained the same. So successful was this technology for the production of foilstock, the vast majority of foil produced in North America emanates from roll cast material. In spite of its undoubted success in foilstock production, traditional roll casting has been limited by its inability to produce a full range of alloys at acceptable

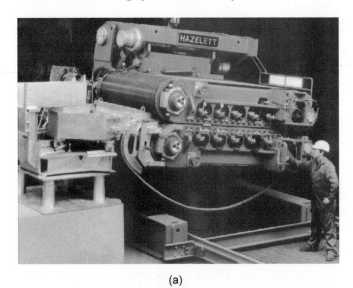

Figure 2.1. Thick strip casters. (a) Hazelett twin-belt caster. (b) Alusuisse articulated block caster.

production rates. In twin-roll casting the production rate is a function of roll diameter and the alloy being processed, and is usually between 1 and 1.5 tonnes per hour per metre (t/h/m) of cast width.

Before moving on to the next generation of roll casters it is worth discussing an alternative method being developed by Kaiser. The Kaiser

Figure 2.2. Conventional twin-roll caster.

Micro-Mill process [4] uses a modified Hazelett twin-belt caster to produce material typically 3 mm thick that is sequentially hot rolled, flash annealed, quenched and cold rolled in line to produce coils of finished product. Apart from the superior mechanical properties claimed by Kaiser, the major benefit of this approach is the dramatic reduction of 'work in progress' times. The concept is interesting in that, whereas most other producers are developing technologies for producing material at widths 2 m or above, the Kaiser approach is to produce narrow material—typically 500 mm or less. The technology is aimed at the third world/developing countries but it remains to be seen whether the narrow approach will be acceptable. Again the economics of the process appear to rely heavily on the availability of used equipment.

The current status of continuous casting, particularly in respect of the alloys and end products that are currently achievable, has been illustrated by CRU International as part of a study proposal [5] and this is shown in figure 2.3. CRU estimate that there are over 200 continuous casting facilities in operation world-wide and that, in the USA alone, over a million tonnes per annum are being produced by these techniques. The bulk of this production is concentrated in foilstock and building products, but there are a number of technical issues that have to be resolved for other products. The key issue of their study was an evaluation of the new generation of high speed, thin strip casters to establish if this new technology can compete

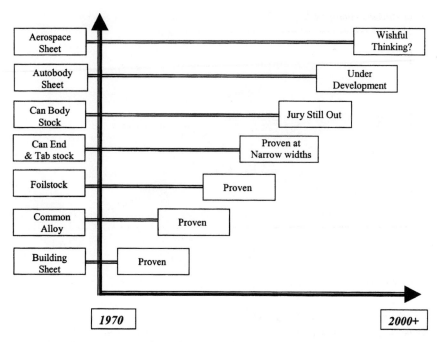

Figure 2.3. Current status of continuous casting in terms of alloys and end products (source: CRU International).

with conventional mills, in the same materials, and achieve the same flexibility.

New generation roll casters

Numerical modelling work carried out in the late 1980s as part of the collaboration between Davy International and Oxford University predicted that significant gains in productivity were possible by reducing the cast gauge [6]. Figure 2.4 is typical of this type of prediction and demonstrates that if the cast gauge is reduced from 6.5 to 2 mm, productivity is approximately doubled. As the cast gauge is reduced even further, the increases in productivity become really interesting. Although this kind of tabulation is widely used, there are a number of popular misconceptions about the approach. Most importantly, the predictions are based on a thermal analysis and tell you nothing about the as-cast microstructure or the suitability of the cast product for downstream processing. Also, the predictions are only valid for one set of conditions; for example, if the tip set and casting speed are altered, different productivity figures will result.

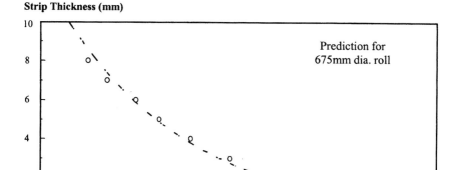

Figure 2.4. Theoretical prediction of productivity as a function of cast gauge.

Apart from the obvious and quantifiable benefit of increased productivity, there is at least the possibility of other benefits. These include

- enhanced properties due to increased solidification rates
- new alloys (and end products)
- reduced capital and operating costs.

If the cast gauge is reduced from say 6.35 to 2.5 mm, at least one cold mill pass can be eliminated and, although the cost savings from this operation may not be dramatic, when one considers that the output of the caster has approximately doubled as a result of this down-gauging, the benefits can be considerable. Even more dramatic savings can be predicted if one considers the setting up of a new facility where the feedstock to the mills is from high speed, thin strip caster(s).

Experimental development

The details of the programme of work and the results of the experimental programme at Oxford have been reported in detail elsewhere [7]. To date well over 1000 experimental casts have been carried out on the two experimental roll casters and, although the vast majority of casts have been with conventional wrought aluminium alloys a range of other materials have been examined. Table 2.1 shows the results for a variety of alloys, and from this it can be seen that not only is it possible to cast at reduced gauges, but significant gains in productivity are also possible. The results

Table 2.1. Summary of experimental results (Oxford caster)

Alloy	Gauge (mm)	Speed (t/h/m)	Productivity (t/h/m)	Productivity (piw)
AA1100	1.00	12	1.95	107
AA8111	1.20	15*	2.93	161
AA3004	2.30	5.7	2.17	125
AA5182	2.30	15*	5.60	306

* Current machine limit.

for AA5182 are particularly interesting as this alloy is generally regarded as uncastable.

In terms of cast product evaluation, a considerable amount of effort has been expended in characterizing the material produced on the experimental machines and in determining its suitability for downstream processing and specific end uses. Although effort has been concentrated on foil and finstock applications, many other materials have been investigated, such as can-end and automotive alloys. It is clear from the considerable database emanating from this work that the high speed, thin strip variant of roll casting is capable of producing a range of microstructures, some of which are quite different from those produced by traditional casting methods.

Alloy development remains a key area for investigation as the vast majority of wrought aluminium alloys currently used were developed for the direct chill casting/hot mill process route and, as such, are not necessarily compatible with the requirements of the high speed, thin strip casting route. A number of alloys have been developed specifically to overcome the problem of sticking, but a comprehensive alloy development programme was beyond the scope of the Oxford phase of the development. If, by casting thinner and faster, materials with significantly different microstructures and properties are produced, it is inevitable that a major alloy development programme will be necessary. To quote from Emley's review paper of 1976, 'hand in hand with improvements in cast products and processing should go alloy development, for it may be that the best compositions or strip casting are yet to be devised, let alone optimized' [1].

Industrial development

Following the success of the Oxford programme, it became apparent that for the next phase of the development the equipment would have to be scaled up and that this would require the involvement of an industrial partner. To this end, a collaborative development agreement was signed between Davy and

Industrial development 33

Figure 2.5. The semi-commercial caster installed at Gränges' Finspong facility.

Gränges AB in early 1992. The principal objective of this phase of the development was to prove the high speed, thin strip casting concept on an industrial scale so that a full size production version of the equipment could be specified and engineered. The casting line forms part of a state-of-the-art cast house and this is shown in figure 2.5.

The heart of the caster is the System 21 control system and this incorporates the majority of the features normally associated with rolling mill control. For example, the machine can run in gap control, speed control and load control modes and incorporates features such as steer, front tension and the capability for roll eccentricity compensation. In addition to the above features there are some items that are caster specific and these include a torque sharing capability that has proved beneficial in controlling sticking. This is particularly important as the cast gauge is reduced.

The major process problems encountered in the development centred on the conflict between the conditions required to start the caster successfully with those required to produce thin strip at high speeds. Figure 2.6 summarizes the results to date and from this it can be seen that as the cast gauge of the sheet is reduced there is an increase in strip productivity and the productivity increases are broadly in line with those predicted by the numerical models. As has been mentioned earlier, the caster is becoming increasingly like a rolling mill and this is understandable if the high speed, thin strip casting process is examined in detail.

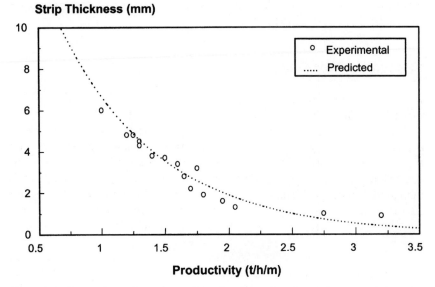

Figure 2.6. Comparison of experimental results achieved on the semi-commercial caster at Finspong and at Oxford with predictions from the Oxford modelling work.

Table 2.2 illustrates the relationship between the operating parameters for strip produced at a variety of gauges and from this it can be seen that as the cast gauge is reduced, in general, there is a significant increase in both the specific loading and, by inference, the amount of reduction taking place in the process. At very thin gauges (approximately 1 mm), however, the casting speed increases dramatically and there is a reduction in the

Table 2.2. Relationship between operating parameters and strip thickness

Cast gauge (mm)	Roll speed (m/min)	Specific load (t/mm)
5.92	0.88	0.42
5.23	1.22	0.44
4.08	1.79	0.46
3.83	1.86	0.54
3.55	2.00	0.50
3.16	2.70	0.50
2.92	2.81	0.60
2.20	3.65	0.91
1.78	4.89	0.95
1.13	15.13	0.64

separating force. This suggests that at very thin gauges there is a fundamental change in the nature of the process.

Segregation behaviour

In conventional twin-roll casting, centreline segregates occur if casting is carried out under hot conditions (i.e. small setback, high speed or high metal feed temperatures) and can normally be eliminated by reverting to colder conditions by increasing the strip separating force. As the gauge of the cast material is reduced, the geometry of the roll bite changes and centreline segregates can only be eliminated by casting with significantly increased separating forces. It has been observed, however, that during thin strip casting other forms of segregation behaviour are observed and it is possible to categorize these in terms of a segregation limit diagram. A typical segregation limit diagram is shown in figure 2.7.

As shown in figure 2.7, at thick gauges it is possible to produce microstructures free from centreline defects under all conditions except for relatively low separating forces. As the cast gauge is reduced, higher specific loads are required for the structure to remain free from centreline channels. If the gauge is reduced without increasing the specific loading the degree of segregation increases and the likelihood of having a structure that is incompatible with downstream processing requirements and end product requirements is also increased. Detailed metallography [8] reveals a

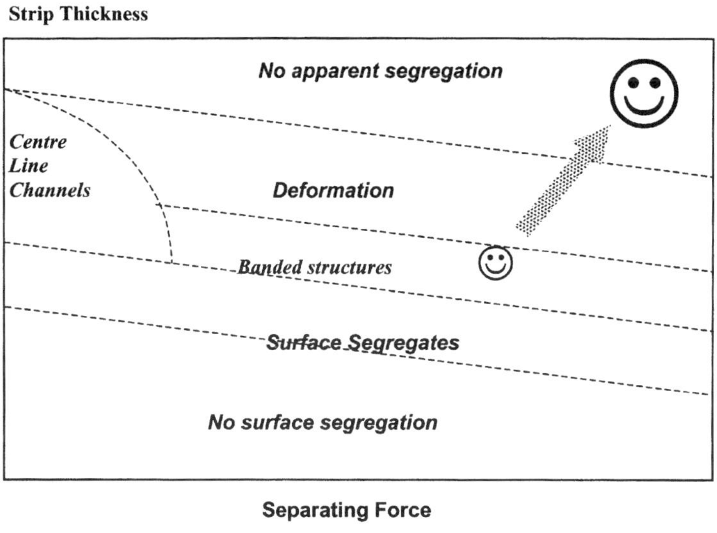

Figure 2.7. Schematic segregation limit diagram.

gradual transition from coarse centreline segregates through dispersed channels (deformation segregates) to banded structures where the second phase is uniformly distributed throughout the microstructure. At even thinner gauges and higher casting speeds, the type of segregate changes quite markedly and deformation and surface segregates predominate.

On a practical level, the segregation limit diagram can be used to make qualitative assessments regarding the suitability of as-cast material for subsequent processing into specific end products. Clearly, if well-developed centreline channels are visible in the as-cast microstructure, the chances of making critical end products such as thin gauge converter foil are remote. If, on the other hand, the as-cast microstructure is segregation free it is likely that even critical end products can be made. In other regions of the diagram, the nature of the end product will dictate whether or not a particular form of segregation pattern is acceptable. What is at issue is whether a particular segregation pattern in the as-cast microstructure rules out the possibility of producing particular ranges of finished products.

Buckling defects

At thin gauges there is a change in the type of sheet defect that occurs and the defects closely resemble those encountered during rolling. An example is shown in figure 2.8. It has been demonstrated that the onset of these buckling defects is alloy dependent and is a function of cast gauge and separating force [9]. For a given alloy there is a range of operating conditions under which the defects will develop and after considerable investigation it has been established that these buckling defects occur during the transition between different regimes of casting. It is possible to map the behaviour of a given alloy in terms of its cast thickness and specific load and this can be represented graphically using a transition or buckling limit diagram. An example of this type of diagram is shown schematically in figure 2.9.

Figure 2.8. An example of sheet buckling.

Buckling defects 37

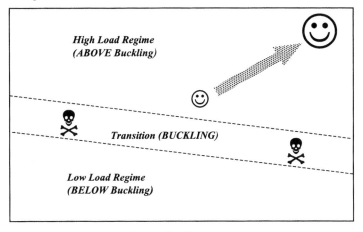

Figure 2.9. Schematic buckling limit or buckling transition diagram.

It is possible to produce visually acceptable sheet above and below the buckling transition although the characteristics of the material produced differ quite markedly. Above the buckling transition the surface of the sheet is bright and lustrous, and the surface characteristics of the casting rolls are faithfully reproduced on the cast sheet. This is entirely consistent with the as-cast microstructure, which shows a high degree of deformation and freedom from centreline channels. To all intents and purposes, material produced above the buckling transition is identical in appearance to conventional twin-roll cast material. Strip produced below the buckling transition, on the other hand, exhibits a dull grainy texture, and the microstructure is much more variable both through the strip thickness and across the surface. The microstructures indicate much lower levels of deformation and higher levels of segregation. In terms of caster productivity, the productivity gains attainable with conventional alloys in the high load regime (above the buckling transition) are fairly modest—typically 1.5–1.8 times that for conventional roll casting. In the low load regime (below the buckling transition) productivities are dramatically higher and limited in many cases only by machine parameters. The precise location of the buckling transition varies from alloy to alloy. For some alloys the transition occurs at thicknesses of 6 mm or above and in these alloys it is difficult to produce sheet that exhibits the characteristics of material produced above the buckling transition unless extremely high specific loads are employed.

For products other than foil, heat exchanger fin and a limited number of specialized applications such as lithographic sheet, strong alloys (with their associated longer freezing ranges) are conventionally used. If these alloys are to be cast successfully at thin gauges, high specific loads are required to ensure that casting takes place in the high load regime. If this is not achieved, the cast product is likely to be unsuitable for rolling to end product. As with the segregation patterns described earlier, the suitability of the cast product for downstream processing is significantly more important than the appearance of the cast sheet and from this point of view it is apparent that material produced above the buckling transition is better suited for critical applications. Hunt *et al* [10] have developed a series of productivity limit diagrams for a range of aluminium alloys that combine segregation behaviour and the development of defects in a single diagram.

Numerical modelling

Numerical modelling work considers fluid flow in the liquid, plastic deformation in the solid and a generalized energy equation that is valid in the solid, liquid and semi-solid regions in the wedge-shaped moving mould [11]. The evolution of latent heat is governed by a function for fraction solid based on the Scheil equation and the heat transfer coefficient between the strip and the roll surface depends on the calculated normal pressure at the interface. An accurate description of the wedge-shaped mould geometry is obtained by transforming the constitutive equations to general curvilinear coordinates and using a body-fitted grid.

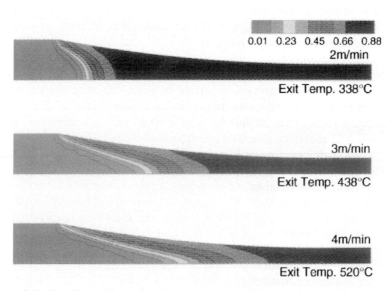

Figure 2.10. Fraction solid as a function of casting speed (4 mm strip).

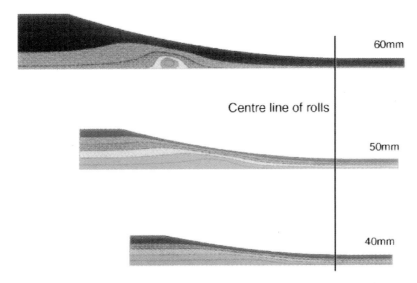

Figure 2.11. Streamlines as a function of tip setback (4 mm strip cast at 4 m/min).

All components of the model have been verified by comparison with analytic and numerical solutions and the fully coupled model is in good agreement with results from the experimental twin-roll caster at Oxford. The model has been used to investigate the effect of varying casting parameters on the thermal efficiency of the process, heat line formation, sticking, segregation and productivity. It can provide information on rolling load and torque, forward slip and field variables such as velocity, pressure and temperature. Figure 2.10 illustrates, for example, the change of sump shape with casting speed.

One significant observation is that contrary to popular opinion roll diameter does not have a major effect on productivity, and that the limit to productivity is the onset of defects. For casting conditions that match the onset of buckling the model predicts that molten and semi-solid metal in the sump recirculates as shown in figure 2.11. To avoid recirculation in the sump it is necessary to use small setbacks and high load casting conditions that correspond to casting conditions above the buckling transition. Clearly with large casting rolls it is necessary to use large setbacks that result in recirculation and below buckling casting conditions.

Development caster

From the above observations with these techniques it is clear that to produce material above the buckling transition the caster must be capable of sustaining high specific loads and it is this fact that has been responsible for the radical thinking behind the design of Kvaerner's (formerly Davy International) commercial high speed, thin strip caster [12]. Many of the features of the development machine have been incorporated into the commercial machine installed in the Luxembourg facility of Gränges Eurofoil. The caster is capable of producing material up to 1800 mm wide in the range 1–6 mm. Due to the high loads developed at thin gauges and the level of automation required to control the process, it is inevitable that the commercial high speed, thin strip caster increasingly resembles a rolling mill both in terms of its appearance and level of sophistication.

The world's first 4-Hi caster for aluminium alloys was installed in the Luxembourg facility of Eurofoil in late 1995 and this is shown in figure 2.12. The move to a 4-Hi configuration was brought about by the necessity to cast in the high load regime (with its associated high specific loads) if usable material is to be produced at commercially viable widths. Once this basic requirement had been established there was a simple choice to be made: the casting rolls could either be sized to withstand the extremely high loads and torques developed during high speed, thin strip casting; or the casting rolls could be sized to suit the process requirements, with backup rolls to accommodate the high loads and torques. From the work done at Oxford and Finspong, it has been established that to be sure of

Figure 2.12. 4-Hi caster installed at Gränges Eurofoil (Luxembourg).

operating in the high load regime the commercial caster should have a specific load capability in the range 1.25–1.35 tonnes/mm of strip width.

To accommodate such a high level of load with a conventional 2-Hi configuration would require a roll diameter in excess of 1200 mm. Using such massive rolls presents a number of operational problems—particularly in setting up and starting the caster, and the large setbacks that are unavoidable with large diameter rolls, which are likely to cause casting defects and buckling. Moreover, the weight of such a massive roll assembly would be approximately 40 tonnes, and the knock-on effects in terms of building size and cranage are dramatic. Finally, since the casting rolls are essentially a consumable item, the impact on operating costs is also significant.

With the 4-Hi construction, on the other hand, much smaller casting rolls can be employed, making tip setting much easier, and more importantly the machine can be operated with tip setbacks compatible with producing good quality (buckle free) material. The ability to control the profile of the cast product on-line is extremely important. It is well established that the profile of the cast product is directly related to the separating forces generated during casting. As the separating force increases, the amount the casting rolls deflect (both bending and flattening) increases, with the result that the sheet profile deteriorates. To overcome this, it is normal for the casting rolls to be ground with a pre-set camber so that the cast strip emerges with a profile of typically 0.5–1.0% positive. Consequently, for a given ground camber there will only be a narrow range of operating conditions that produce the correct sheet profile. If the casting conditions change (for example casting speed, tip setback or metal feed temperature) it is likely that an unacceptable sheet profile will result.

Even if casting conditions are held stable, the ground camber will only be correct for a particular alloy and strip width. Finally, as has been mentioned previously, if large cambers are required to compensate for the high separating forces encountered when casting thin and fast, the shape of the roll gap for startup can make starting difficult. With a 4-Hi machine, on the other hand, the casting rolls can be parallel which ensures a parallel roll gap at startup ensuring precise tip fitting and easy starting. Once casting has been established, adjusting the amount of roll bend can compensate for any deflection of the roll stack. This feature is shown schematically in figure 2.13.

Bend can be either positive or negative and this gives a large operating window for the machine. It means that a wide range of casting conditions can be accommodated as can a variety of strip widths and alloys, simply by adjusting the amount of roll bend. The effect of roll bend on sheet profile is shown in figure 2.14, and from this it can be seen that the required profile can be achieved by adjusting the amount of roll bend.

Another feature of the 4-Hi machine is the ability to adjust the side-to-side variations in gauge using the steer (tilt) facility. By adjusting the steer

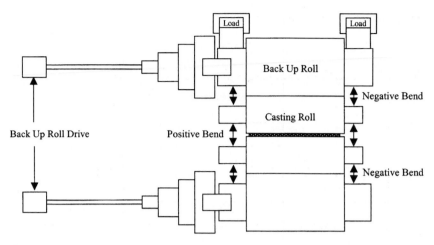

Figure 2.13. Schematic of a 4-Hi caster showing roll bend capability.

control the roll gap is opened or closed until there is no side-to-side variation in strip thickness. On conventional machines it is usual to adjust side-to-side gauge variations by pulling the tip back on one side or the other. This is an imprecise method and can introduce buckling defects due to differences in the amount of rolling from side to side. Figure 2.15 shows how the roll gap can be adjusted using the steer facility.

As a result of the development programmes carried out at Oxford, Finspong and more recently Eurofoil, it has been established that high speed, thin strip casting is a viable process at commercial widths. Thin

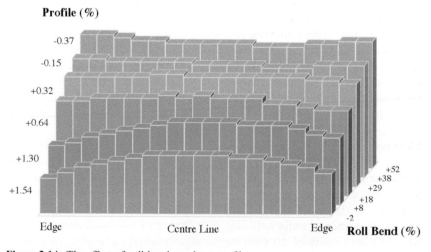

Figure 2.14. The effect of roll bend on sheet profile.

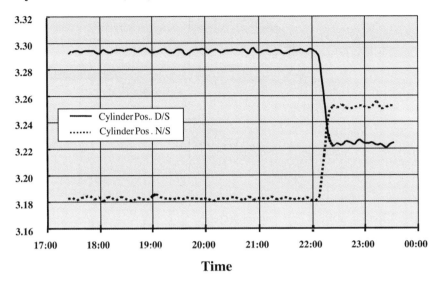

Figure 2.15. An example of dynamic steer capability.

gauge material has been produced up to 1800 mm wide and the productivity gains promised from the earlier laboratory and pilot scale work have been realized. It is apparent that high speed, thin strip casting varies significantly from conventional processes and that for the process to be commercially viable sophisticated automation and control equipment will be necessary. It has been demonstrated that in order to cast a full range of alloys capable of being rolled to finished products, high specific loads are necessary. The work to date has been limited to foilstock and finstock applications but the 4-Hi commercial casting line is capable of casting a wide range of alloys in either high or low load regimes and it is here that the real benefits are to be gained. The caster fits well into the revised concept of mini-mills and it is simply a question of time before the objective of being able to cast a full range of alloys will be fulfilled.

Round products

The same kind of logic can be applied to the casting and processing of round products where there is a similar requirement to produce a cast product that is nearer net shape and to reduce the number of process operations.

Conform^{TM} is a well established processing route for the production of a wide range of extruded profiles and coaxial end products. The process has been described in detail elsewhere [13] and the principle of the process is

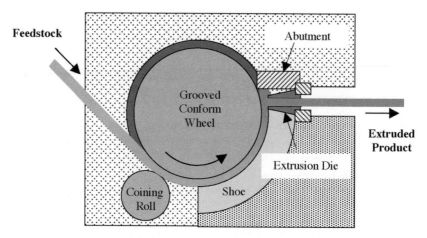

Figure 2.16. Schematic of the Conform process.

shown schematically in figure 2.16. The feedstock is fed into the profiled groove of the Conform wheel by means of a coining roll and the groove closed by a close fitting shoe. The material is prevented from continuing its passage around the wheel by means of an abutment. As a result, high temperatures and pressures are developed in the material, which becomes plastic and finally emerges from the machine through an extrusion die. The product can take a variety of forms including tubes, solids, complex profiles and coaxial products. The feedstock depends on the size of the Conform wheel but for small machines is usually 9.5 mm diameter rod, whereas for larger machines the feedstock can be cast bar with a cross section of approximately 1200 mm^2. In both cases, the material will have been cast into some intermediate shape but these operations are energy intensive and it has been demonstrated [14] that considerable cost savings are possible if the nature of the feedstock is changed.

The most energy-efficient solution would be to feed the wheel directly with molten metal and in this configuration the single extrusion shoe is replaced by a solidification shoe and an extrusion shoe. This arrangement is shown in figure 2.17. A machine of this construction has been built and the results reported [15]. The results showed that although this approach is technically feasible, the performance of the machine was compromised by a number of problems. Due to problems of feeding molten metal and the relatively small diameter of the Conform wheel, the production rates were unacceptably slow. Moreover, even at these slow production rates, the quality of the extruded product was frequently unacceptable due to the excessively high exit temperature of the extrusion which resulted in significant amounts of cracking. One solution to the problem would have been to increase the diameter of the Conform wheel but even with a large diameter

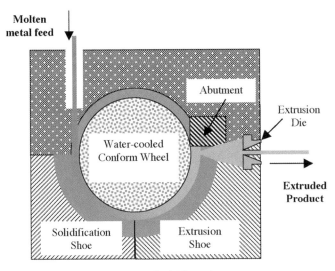

Figure 2.17. Conform with molten metal feed (Castex).

wheel it was felt that the solidification distance would have been too short to produce a high quality product at an acceptable production rate.

To overcome these difficulties it has been decided to separate the solidification and extrusion operations. In this second-generation Castex machine the feedstock machine is produced immediately upstream of the wheel and the cast bar matched to the entry requirements of the wheel in terms of both cross section and productivity. For the smallest Conform machine, which has a diameter of 300 mm, the cross section of the cast bar feedstock is nominally 250 mm^2 and the casting rate specified as 500 kg h^{-1}. A Castex machine is shown in figure 2.18. The machine employs a series of aluminium moulds connected together to form an endless belt. Each mould is 40 mm square and has a mould cavity suitable for casting material of the correct size and cross section. The length of each mould segment is 300 mm and the effective (total) mould length is approximately 4 m. Molten metal is fed into the moulds close to one end of the belt and the cast bar fed directly into a Conform machine.

By separating the casting and extrusion functions it is possible to have an in-line process but this has been limited by the presence of a number of characteristic defects. The origin of defects has been established and by modifying the machine hardware it has been possible to eliminate the problems. Based on experience gained to date, a second-generation Castex machine has been designed and built that satisfies the production requirements of the small Conform machines. This development has provided a low cost feedstock for Conform machines and an integrated production route for a variety of extruded sections and coaxial products.

Figure 2.18. Prototype Castex machine in operation.

Summary

For both flat and round products there has been a gradual move to nearer net shape production methods and it is anticipated that this trend will continue. Although the transition has taken place slowly, the pace of change is likely to accelerate now that the major aluminium companies are becoming more involved.

References

[1] Emley E F 1976 'Continuous casting of aluminium' *International Metals Reviews* June p 75
[2] 'A technical and economical comparison of conventional and continuous casting and rolling methods for the production of cold rolled aluminium strip', ISBN 3-88355-175-9, Deutsche Gesellschaft für Materialkunde e.V
[3] Diener T J 1994 'The Aluminium Mini-Mill', presented at Alumitech '94, Atlanta
[4] Evans R B, Wyatt-Mair G F and Harrington D G 1995 'Aluminum Canstock MICROMILL Process', presented at BEV-PAK AMERICAS '95, Tarpon Springs, Florida
[5] 'Continuous caster minimills in aluminium', Revolution or evolution, a study proposal, CRU International Ltd, 1995

References

[6] Browne D 1989 'The measurement of heat transfer coefficients in roll casting' MSc Thesis, University of Oxford

[7] Edmonds D V, Hunt J D, Monaghan D J, Yang X, Yun M, Browne D, Cook R and Thomas P M 'An experimental study of twin-roll casting' *Proceedings of the International Symposium on 'Extraction, Refining and Fabrication of Light Alloys'*, Ottawa, Ontario 1991

[8] Lockyer S, Yun M, Hunt J D and Edmonds D V 1996 'Microstructural defects in thin sheet, twin roll cast aluminium alloys' *Material Science Forum* **217/222**

[9] Cook R 1993 'Twin roll casting of aluminium alloys on the Davy–Gränges thin strip caster' Internal Report TD116/R240

[10] Hunt J D 1999 Private Communication

[11] Bradbury P J 1994 'A mathematical study for the twin-roll casting process', DPhil Thesis, University of Oxford

[12] Thomas P M, Grocock P G and Bouzendorffer J-M An update on the operation of the first of a new generation of roll casters, Alumitec '97, Atlanta

[13] Maddock B 1981 'Company builds around new extrusion technology' *Metallurgia* July

[14] Atkins W S 1993 Development of a liquid metal feedstock for extrusion processes, ETSU Future Practice Report 34

[15] Gibbs A M 1994 'Castex^{TM}—Continuous casting and Conform extrusion of aluminium' *6th International Machine Design and Production Conference*, METU, Ankara, Turkey

Chapter 3

Continuous casting of steels

Hisao Esaka

Introduction

The continuous casting ratio, i.e. the percentage of steel manufactured by continuous casting, is nowadays over 98% in Japan. The ratio has already effectively reached its maximum value. However, it is still necessary to develop technology for continuous casting. The reason for this is that the demand for quality and cost is always increasing, and steelmaking, especially the casting process, is the key technology to determine the quality of all steel products. Cracks, segregation (including porosity) and inclusions were the three major problems in the 1970s, when the continuous casting ratio increased steeply, as shown in figure 3.1 [1]. The three major problems that are faced now are exactly the same, but the level of technical ability is completely different. Furthermore, the formation mechanisms of these defects have been studied and understood to a considerable extent and it is now possible to produce perfect steel products under steady-state conditions.

There are two subjects to be solved in the near future. One of these is high speed casting technology. This technology presupposes maintaining quality at high production rates and needs to be developed in order to increase productivity. The other is active casting technology. When the casting conditions vary, this leads to the formation of a variety of defects in the final steel products. Thus, technology which can correspond to the variation of casting conditions needs to be developed to cope with these changes.

In this chapter the technology of continuous casting of steel is briefly reviewed and some subjects to be studied in the near future are discussed. Continuous casting of aluminium alloys is discussed in chapters 1 and 2.

Surface cracks

One of the major subjects that makes it difficult to develop high speed casting technology is longitudinal surface cracks [2]. Unevenness of the solidified

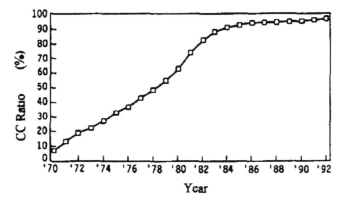

Figure 3.1. Variation of continuous casting ratio in Japan.

shell is thought to be the origin of surface cracks, and the cracks seem therefore to form in the mould, i.e. in the very early stages of solidification. Thus, the homogeneity of initial solidification is of considerable importance. From the standpoint of productivity, a high cooling rate in the mould is desired. However, the unevenness of the solidified shell usually increases with increasing cooling rate [3]. Therefore, in order to obtain a more uniform shell, slow cooling is desirable. In order to achieve slow cooling in the mould, various technologies have been developed. These approaches include mould surface treatment [4–7] and mould flux technology [8–13].

Some mechanically treated moulds have been used in casting experiments: high roughness moulds [4] (treated by shot blasting), grooved moulds and grooved and plated moulds [5–7]. One example of a mechanically treated mould is shown in figure 3.2 [5]. Using this mould, the heat flux just below the meniscus in the mould is decreased, as shown in figure 3.3 [5]. Therefore,

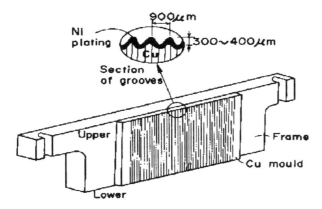

Figure 3.2. Mechanically treated mould [5].

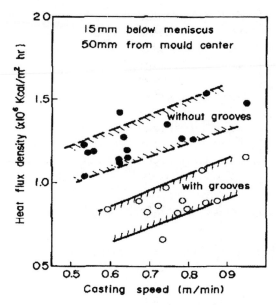

Figure 3.3. Comparison of heat flux density with and without grooves on the mould [5].

the homogeneity of the shell is improved and the number of longitudinal cracks decreased, as shown in figure 3.4 [5]. However, since the period of these experiments was quite short, it was not possible to evaluate the duration of efficiency and lifetime of these specially treated moulds. Thus, this type of mould has not yet been installed.

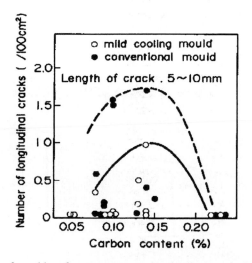

Figure 3.4. Effect of mould surface treatment on longitudinal cracks [5].

Table 3.1. Roles of mould flux.

1. Prevention of re-oxidation in the mould
2. Absorption of non-metallic inclusion
3. Heat-insulating near meniscus
4. Lubrication between mould and solidified shell
5. Control of heat flux in the mould

The roles of a mould flux are listed in table 3.1. To achieve slow cooling in the mould, the fifth item in table 3.1, i.e. the control of heat flux, has recently been paid considerable attention. Mould fluxes with a high crystallization temperature decrease the heat flux in the mould as shown in figure 3.5 [11]. To design a mould flux with a high crystallization temperature, it is necessary to increase the basicity ($=CaO/SiO_2$) as shown in figure 3.6 [11]. This type of mould flux is effective for preventing longitudinal surface cracks even in medium carbon steels. The detailed mechanism of mould cooling is, however, still unclear [8–13]. The chemical composition of the mould flux changes with time due to Al_2O_3 pick-up [14] and therefore physical properties such as viscosity and crystallization temperature vary [15–18]. In order to allow a wide margin for the variation of casting conditions, other methods of ensuring slow cooling also need to be established.

Another subject of initial solidification concerning the quality of steel products is oscillation marks. Hooks form sometimes due to the shape of the meniscus and this leads to the formation of defects. To prevent clogging

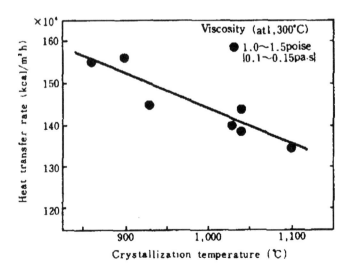

Figure 3.5. Relationship between the crystallization temperature and heat flux ($V_c = 1.2$ m/min).

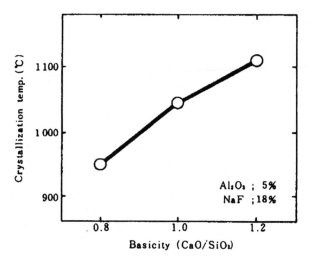

Figure 3.6. Influence of basicity on crystallization temperature.

of the submerged entry nozzle, argon gas is usually introduced. Bubbles with or without inclusions may be trapped by the solidified shell. Figure 3.7 shows a bubble trapped by a hook [19]. In order to reduce the formation of hooks at the meniscus, a mould flux with high heat insulating ability can be used. Furthermore, to keep the temperature uniform and high at the meniscus, electromagnetic stirring in the mould may also be applied. If uneven solidification is the cause of inclusions, electromagnetic stirring in the mould is also useful to obtain a uniform solidified shell. Application and control of flow in the mould by electromagnetic forces is very effective and needs to be investigated in full detail.

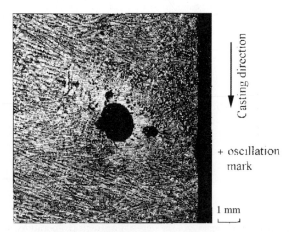

Figure 3.7. A bubble trapped by a hook [19].

Macroscopic segregation

In the case of steel, the volume change from liquid to solid is approximately 3.5%. This is the main driving force for the motion of residual liquid, which exists in interdendritic regions and is enriched with solute elements and with harmful impurities. Macrosegregation levels are also affected by post-solidification rolling deformation and by cooling effects such as the δ to γ transformation. In the case of continuous casting of slab, where the aspect ratio is large, i.e. the width of the slab is large relative to its thickness, bulging between the rolls is another mechanism which leads to significant macroscopic segregation.

To prevent macroscopic segregation, continuously cast slab is gently reduced by deformation. This is carried out when the range of fraction solid at the centre (f_s) is about 0.3–0.8 and is performed by rolls or bars. One example of this improvement is shown in figure 3.8 [20]. Usually the equipment available for reduction of this type is difficult to install in the

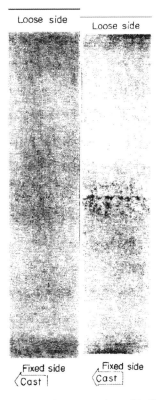

Figure 3.8. Comparison of macroscopic segregation with (left) and without (right) soft reduction [20].

centre region. The casting condition must be controlled so as to maintain the correct solid fraction range in the reduction zone, and this control is quite critical. If the fraction solid in the reduction zone is too high, the material may be impossible to reduce since the solidified shell is too hard to deform. This results in enhanced macroscopic segregation. When the rigidity of the reduction equipment is high, A-type segregation may form. On the other hand, if the fraction solid in the reduction zone is too low, macroscopic segregation is again enhanced since the reduction is performed too early.

Longitudinal irregularities in the final part of solidification lead to substantial porosity accompanying the macroscopic segregation. This phenomenon is known as 'bridging'. On the other hand, transverse irregularities lead to W-shaped craters and in this case the residual liquid moves in the lateral direction, again enhancing macroscopic segregation, which is observed on the cross section. The control of solidification is also very important in the final stages of solidification as well as in the initial stages. Fine equiaxed grains may solve this problem, and electromagnetic stirring is one of the key technologies which can be used to control the solidified structure. Further investigation of how to produce effective equiaxed grains is needed.

Defects in non-steady-state regions

The three major problems of cracks, segregation and inclusions have essentially been solved under normal conditions of steady-state casting. The residual problems to be solved are mainly under non-steady conditions, i.e. the initial and final parts of the continuous casting process and the period of ladle exchange.

In the initial part of casting, the cleanliness of the molten steel is usually poor, due to re-oxidation in the tundish and in the mould. In the period of ladle exchange, the cleanliness of molten steel is again poor, due to outflow of slag from the ladle, etc. Therefore, the problems of inclusions are apt to happen in these periods. The amount of inclusions can be measured by the electron beam remelting method [21]. Small specimens taken from various parts of the slab are partially melted using an electron beam. The inclusions contained in each specimen float on the molten surface. The surface area of inclusions which form a raft on the molten surface can then be measured to determine the amount of inclusions. The variation of raft area with position along the casting length is shown in figure 3.9 [21]. As expected, the amount of inclusions is high not only in the initial regions of the casting but also in the ladle exchange region.

Some examples of counter-measures to prevent inclusions are shown schematically in figure 3.10. These counter-measures include atmosphere control in the tundish and mould, enlarged tundish, holding start, submerged

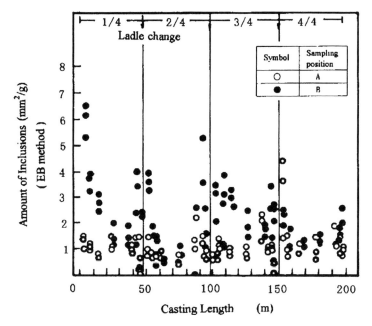

Figure 3.9. Variation of amount of inclusions with casting length [21].

ladle opening at ladle exchange, etc. Since each counter-measure is not perfect by itself, the different techniques should be combined to maximize their impact. Furthermore, active control technology depending upon the variation of cleanliness of the molten steel should be developed. Monitoring of the cleanliness of molten steel is clearly a key technology for this kind of control system.

Sometimes, longitudinal surface cracks on the cast slab are observed in the initial part of the casting. In this region, the mould surface is quite cold and the condition of the mould flux is different from that under steady state conditions. The unevenness of the solidified shell increases due to strong cooling and this leads to longitudinal surface cracks. To obtain the steady-state heat flow condition as quickly as possible, a so-called 'front powder' is often used in the initial part of the casting. This powder can increase temperature due to the oxidation of Ca–Si or Ca–Al alloys contained in the mould flux. This increases meniscus temperature and facilitates mould flux melting. Thus, the number of longitudinal surface cracks decreases, as shown in figure 3.11 [22].

The mechanisms of forming transverse surface cracks have also been investigated and the embrittling temperature range has been made clear [23]. To avoid this temperature range, the secondary cooling is controlled so as to raise or lower the surface temperature of the cast product. The surface temperature is low in the initial stage of casting, as mentioned

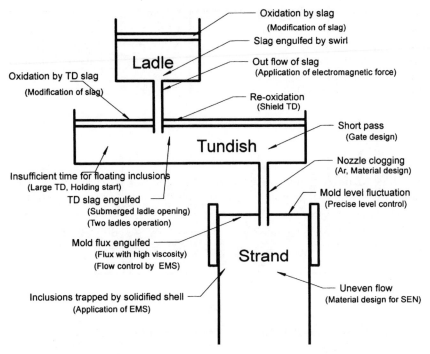

Figure 3.10. A schematic drawing of counter-measures for inclusion defects.

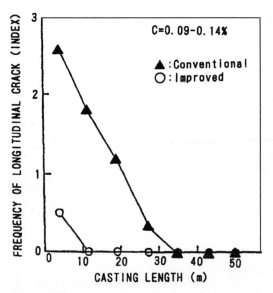

Figure 3.11. Relation between casting length and longitudinal cracks [22].

above. Thus, it is difficult to avoid the embrittling temperature range. In order to reduce surface cracks, it is also important to control the sequence of steel grades during continuous casting. Steel grades which are sensitive to crack formation should be cast later in the casting sequence. The technology should be developed so as not to decrease the casting velocity, even in periods of ladle exchange of different steel grade.

Summary

In order to develop high speed and active continuous steel casting technologies, a range of important subjects need to be studied in the future. Development of these technologies may not directly increase the continuous casting ratio, but will increase the productivity and quality of the steel products.

References

[1] Ohji M 1994 153rd–154th Nishiyama Memorial Seminar, ISIJ, Tokyo, p 1
[2] Hiraki S, Nakajima K, Murakami T and Kanazawa T 1994 *Steelmaking Conf. Proc.* **77** 397
[3] Sugitani Y, Nakamura M and Watanabe T 1981 *Tetsu-to-Hagane* **67** 1508
[4] Tanaka T, Nuri Y, Egashira T, Arima R and Ohashi T 1982 *Tetsu-to-Hagane* **68** S159
[5] Nakai K, Sakashita T, Hashino M, Kawasaki M, Nakajima K and Sugitani Y 1987 *Tetsu-to-Hagane* **73** 498
[6] Yamaguchi R, Suzuki M and Murakami K 1995 *Met. Sci. Technol.* **13** 3
[7] Shimizu H, Yamaguchi R, Suzuki M, Suzuki M, Nishimachi R and Murakami H 1996 'Improvement of surface quality and control of initial solidification' *CAMP-ISIJ* **9** 39
[8] Watanabe K, Suzuki M, Murakami K, Kondo H, Miyamoto A and Shiomi T 1997 *Tetsu-to-Hagane* **83** 115
[9] Kawamoto M, Nakajima K, Kanazawa T and Nakai K 1994 *ISIJ Int.* **34** 593
[10] Chikano T, Ichikawa K and Nomura O 1988 *Shinagawa Tech. Rep.* **31** 75
[11] Ichikawa K and Nomura O 1989 *Shinagawa Tech. Rep.* **32** 147
[12] Yamauchi A, Sorimachi K, Sakuraya T and Fujii T 1993 *Tetsu-to-Hagane* **79** 167
[13] Shibata H, Emi T, Waseda Y, Kondo K, Ohta H and Nakajima K 1996 *Tetsu-to-Hagane* **82** 504
[14] Fukuda J, Kondo T and Tsutsumi K 1992 *CAMP-ISIJ* **1** 281
[15] Mills K C 1988 *Ironmaking Steelmaking* **15** 175
[16] Grieveson P, Bagha S, Machingawuta N, Liddell K and Mills K C 1988 *Ironmaking Steelmaking* **15** 181
[17] Taylor R and Mills K C 1988 *Ironmaking Steelmaking* **15** 187
[18] Mills K C, Olusanya A, Broks R, Morrell R and Bagha S 1988 *Ironmaking Steelmaking* **15** 257

[19] Ogibayashi S, Yamaguchi K, Mukai T, Takahashi T, Mimura Y, Koyama K, Nagano Y and Nakano T 1987 *Nippon Steel Tech. Rep.* **34** 1
[20] Mizoguchi S, Kajioka H, Ogibayashi S, Mukai T and Tezuka M 1985 Japan–Canada Seminar, Tokyo J-6-1
[21] Nuri Y 1993 PhD Thesis, The University of Tokyo
[22] Emi M 1991 Abstract of Several Topics on Mold Flux, Japan Institute of Metals, Tokyo, p 10
[23] Suzuki H G, Nishimura S and Yamaguchi S 1979 *Tetsu-to-Hagane* **65** 2038

Chapter 4

Castings in the automotive industry

Kuniaki Mizuno

Introduction

In recent years the automotive industry has witnessed the formation of a number of global alliances designed to help limit the vast sums being invested in the development of environmentally friendly vehicles and the construction of product strategies. Individual development of structural components is being switched to the development and supply of modules and systems aimed at lower costs and higher performance. Moreover, the car industry is now working on a reduction in development lead time in order to respond to customer needs in a more timely fashion.

Car technologies are almost invariably influenced by the prevailing economic situation and, in this period of slow economic growth, the era of high performance is changing to an era of cost reduction. There is no doubt that an era of pursuing safer and more environmentally friendly vehicles is close at hand. To keep up with these changes, the development of compact, lightweight and highly rigid vehicles is desirable. In addition, weight optimization balanced between structural design and materials can be expected. The industry is developing innovative materials and processes for weight reduction, particularly in the area of casting which excels in flexibility of shape.

To achieve the maximum advantage of casting material and process, the product shape, material and processing method are of particular significance. Finite element analysis has been performed to reduce material usage, and for the mould filling and solidification analysis for optimum quality. Today, we are incorporating production engineering specifications from the stage of product planning by using three-dimensional data, performing finite element analysis to specify product shapes, controlling mould filling and solidification simulation to engineer the casting process and making moulds for large-scale production by direct machining. The result of these changes has been improved quality and shortened development lead time.

Car makers world wide are looking for ways to form suitable global alliances in order to reconstruct the strategies of product and global marketing, save on investments and spread the risks of development and production. Two of the major changes under consideration are as follows.

- Purchasing systems for components are poised to switch from individual parts to modules and systems.
- Alliances are being set up in which car makers will plan highly value-added products and then have the products made by global car parts manufacturers.

While taking into consideration safety and environmental issues, a car development strategy which satisfies customer demands for low costs and individuality is considered essential, and product and production technologies are therefore facing massive changes. To respond to safety and environmental issues, vehicles are developed to be both lightweight and compact. Consequently, the application of lightweight materials is essential, and casting products with flexibility of design are considered superior to those products which utilize other materials and processes. To enhance this superiority, optimization of the design, material and processing technique is vitally important. Finite element analysis by means of computer-aided engineering gives us the ability to optimize these factors, thus enabling a shortened development lead time and providing a higher quality, higher performance product for our customers.

Trends in the automotive field

World car production

Car production in the major industrialized countries of the world is falling, as shown in figure 4.1.

Manufacturers' market share

The automotive market share of many leading manufacturers has undergone major changes, as shown in figure 4.2, because of the formation of world wide alliances, mergers and acquisitions, targeting the enhancement of environment related technology, developing investments and commercialization and expanding market share. Naturally this situation will favour alliances forged between those companies which incorporate strong capital bases, emphasize research and development plus expanding sales network.

Change in automotive technologies

Automotive technologies are greatly influenced by the economic situation. The automotive industry has now entered a new era of pursuing lower costs, following the era of seeking high performance. In the future, development

Trends in the automotive field 61

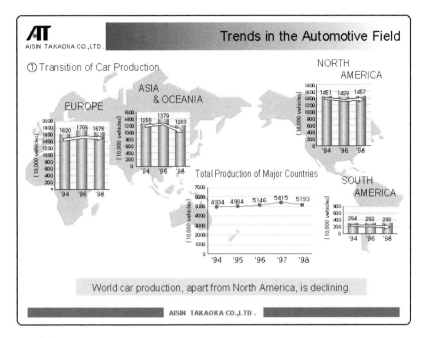

Figure 4.1.

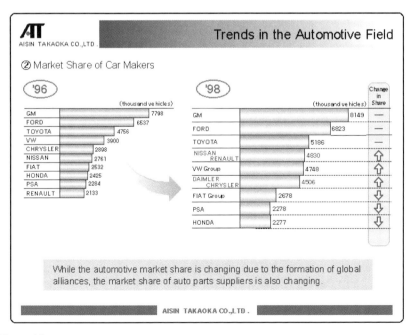

Figure 4.2.

of more advanced technologies for safety, environmental production and the introduction of enhanced communication systems will be actively implemented, and it is considered likely that intelligent transport systems will be developed and incorporated in vehicles.

Change in structural components

Engines with high energy efficiency and exhaust emission controls will be more widely used, such as fuel cell engines and other engines using alternative fuels. The application of high transmissibility continuously variable transmission for powertrains and the use of electromagnetic brakes for the chassis will expand. In car bodies, thin wall and lightweight body construction using high tensile strength materials will be developed, and vehicles as a whole will be developed for compactness and lightness of weight. These trends are summarized schematically in figure 4.3.

Change in materials

While the usage of steel materials has been declining in response to the weight reduction of vehicles, the use of lightweight materials such as aluminium, magnesium and plastic is increasing. According to the analysis and forecasts

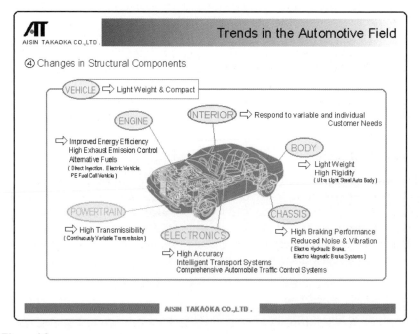

Figure 4.3.

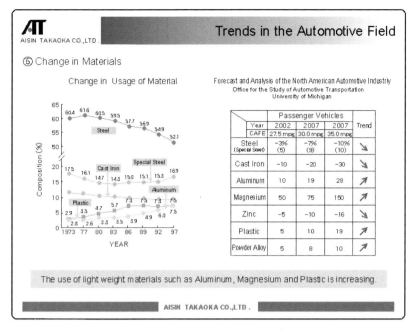

Figure 4.4.

produced by Delphi, in the case that average fuel consumption is restricted from 27.5 mpg at present to 30.0 mpg, it is likely that usage of cast iron will fall by 20%, while the usage of aluminium and magnesium will increase by 19% and 75% respectively. If restricted to 35.0 mpg, the decrease in cast iron will be 30%, while aluminium and magnesium will increase by 28% and 150%. These trends in composition are shown in more detail in figure 4.4. Detailed examples of the use of casting to manufacture an automotive piston and an automotive tyre mount hinge are described in chapters 5 and 9 respectively.

Vehicle development

The development of vehicles has been changing due to the incorporation of production engineering specifications from the planning phase, and the designing of products utilizing computer aided engineering technology. The globalization of car makers and parts makers continues, with new alliances being forged between makers at a quickening pace. The design process is shown as a block diagram in figure 4.5.

Target vehicle production

Environmental control, recycling, safety, comfort and communication systems are the key needs for the vehicles of the future, and they incorporate

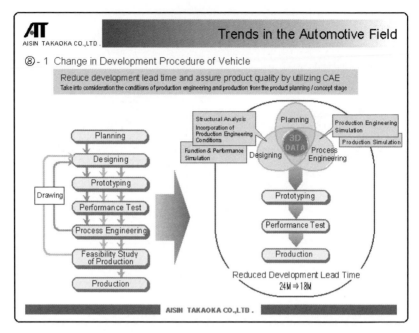

Figure 4.5.

high performance, light weight and low cost. We have to work on the conversion to lightweight materials, modularization and systemization, and the reduction of development lead time.

Cast production

Casting products with high forming flexibility must also have the advantages of light weight, compactness and design flexibility. Therefore, optimum performance casting products are to be developed by one-piece forming of complex shapes, optimization of rib and cored hole structure for light weight, and the development and design of high rigidity alloys.

For optimum performance, balancing the three factors of product shape, material and processing method is essential. Aisin Takaoka's newly developed lightweight disc rotor which reduces squeal and vibration is an excellent example of balancing these three factors. For product shape, the ventilated fin was modified to a spiral fin, cooling capacity was improved and the mass of the rotor was reduced. For material, by enhancing thermal diffusivity, occurrence of heat spots was controlled. Also, by improving machining accuracy and the total performance of the rotor, the quality and reliability of the product were improved significantly.

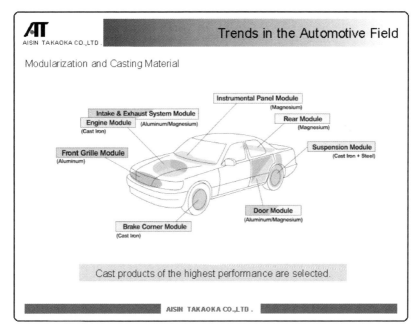

Figure 4.6.

Three-dimensional design is a solution to optimization, as shown in figure 4.6. Three-dimensional models enable engineers to design products in consideration of production conditions, to design and make moulds according to an optimized casting plan, and to build prototypes and evaluate them with actual vehicles. It is necessary to run this cycle more rapidly and with greater accuracy. The translation of three-dimensional engineering data between each step is the priority requirement.

Cast product design and development

Traditionally product developments have been launched after the receipt of orders from customers, but this method is inconsistent with meeting customer needs for reduction in development lead time. For designing product shape, therefore, it is necessary to take into consideration casting performance from the concept stage. Also by using three-dimensional computer aided design and rapid prototyping as required, products are developed in cooperation with customers and related divisions.

For product design, a product shape is designed for high functional performance, durability and casting performance by finite element structural analysis and casting simulation with three-dimensional models created at the concept stage. After that, the detail shape is designed with three-dimensional computer aided design. The shape of the raw material is designed within machining allowances and the parting line and draft angle are added to the original shape approved by the customer. The casting condition and runner system are then examined using casting simulation and the most appropriate design selected.

For mould making, the necessary machining tool locus is calculated on the basis of the mould shape designed by three-dimensional computer aided design. The data are then translated to numerical control data and a mould is directly machined. Before machining, the calculated machining tool locus is checked by numerical control simulation. The completed resinous mould is installed on the permanent mould of the surface plate which is used for mass production.

Conventional prototypes have been made in special production lines. However, mass production lines are now used for prototyping so that predictable problems in mass production can be solved at the prototyping stage, and production can start quickly. By using mass production lines, prototypes, except for cores, can be cast in the same condition as mass production.

The endurance test for each prototype sample is repeated several times. However, with improvements in the accuracy of the casting simulation, the number of runs required to test the quality of the samples is greatly reduced. The sample which has proved its performance and durability in the test is then prepared for production. It is impossible to cast and check the quality of the sample at the same time. As a result a new permanent mould can only be produced only after quality assurance. However, the design of the permanent mould is changed considerably with mass production. It is then possible to check not only the quality but also the process capability of the individual moulds within the permanent die.

Evaluation of casting performance

Solidification simulation has been applied for over ten years as a support tool for casting runner design. Aisin Takaoka also introduced software for solidification simulation as a method of preventing casting defects. As computer performance is improving, fluid flow simulation has been extensively used and the processes from filling to solidification can be easily simulated. Initially simulation and prediction of casting conditions by computer was difficult and the combination of filling behaviour and solidification in cavities could not be simulated satisfactorily. Internal observation of moulds such as green sand and permanent moulds is extremely difficult. At present, although it is possible to see through the inside of the mould by using methods such as X-raying green sand or replacing a portion of a permanent mould with transparent glass, the applied pressure and thermal resistance make it very difficult to observe the actual internal condition of moulds.

Development of simulation and measurement technologies enables simulation of filling behaviour and solidification behaviour and prediction of defects such as cold shut, misrun, gas shrinkage and porosity. These analyses are thus being used for designing product shapes which can be cast easily. In addition, as the accuracy of fluid flow solidification simulation is improving and the need for reduction in development costs and lead time is increasing, these analyses are being used widely to specify casting conditions. Particularly in the case of die casting moulds, by specifying the position of cooling rods in advance, the number of prototypes made can be reduced and production preparation time can also be saved.

In recent years lightweight materials such as aluminium and magnesium have been utilized in response to the demand for weight reduction of car parts. When using light materials, it is necessary to design a product shape after first examining the actual strength of the materials. For aluminum, the actual strength of cast products can be estimated by incorporating the correlation between strength and dendrite arm spacing into the solidification simulation.

Computer aided engineering developments

Wide application of simulation technologies makes it possible to reduce design costs and assure quality and strength of castings. Composite and integrated complex shapes can be developed in order to respond to customer needs for further weight reduction and reduction in development lead time. In the area of computer aided engineering, the development of innovative structural simulation methods which are based on application of casting simulation to composite materials, the most suitable composition of alloys, simulation of residual strain, actual strength and black skin of casting surface as well as improvements in accuracy will be required in the future.

Improved accuracy of computer simulation and reduction in calculation time will enable the design of lighter and higher quality products. Improvements in the accuracy of computer simulation will achieve an enhancement in the use of cellular automaton methods and analysis at the molecular level. Reduction in calculation time will be achieved with the constant progress of hardware and software, leading to even lighter and higher quality products.

Summary

The key points can be summarized as follows:

- The automotive industry is living in a world of change. New global alliances are being forged between companies every day in an environment of economic uncertainty.

- A car development strategy which satisfies the demands of customers remains the first priority of producers.
- As we enter an era of environmentally friendly vehicles, the use of lightweight materials and flexibility in design will become more important. Optimization of the three key factors, shape design, material and processing method, will be essential for each component.
- The use of finite element analysis by computer aided engineering ensures that customers can be provided with a product of the highest quality and performance.

How we deal with these problems and opportunities is the key to our future success.

Chapter 5

Cast Al–Si piston alloys

John Woodthorpe, R Thomson, C Daykin and P Reed

Introduction

Federal-Mogul is a well-established manufacturer of engine components, and is consequently under constant pressure from its customers to supply lighter, better and cheaper components. The evolution of engine design has always resulted in demands for a reduction in reciprocating mass, traditionally to obtain increased performance but now equally important in improving engine refinement by reducing the noise, vibration and harshness. The piston designer's task is to cope with a series of conflicting demands. Performance improvements are achieved by an increase in engine speed or brake mean effective pressure, requiring a more robust piston, yet higher speeds demand a lower reciprocating mass in order to contain bearing loads and skirt side forces. Engine designers require pistons which are shorter, either in skirt length or in compression height (the distance between the crown and the pin centre-line), to reduce the length of the connecting rod and the height of the cylinder block. A shorter skirt, however, poses further problems for the piston designer as guidance in the cylinder bore can deteriorate with adverse effects on piston noise and ring pack performance.

The customer may sometimes also influence the piston design, requiring a particular style or component or insisting on a more conservative approach to mass reduction. Any reduction in reciprocating mass is quickly appreciated by the bearing designer. With maximum piston acceleration frequently exceeding 2000 times that due to gravity, every gramme saved in the piston translates to a 2 kg reduction in maximum bearing load. During the past decade, a typical 30% reduction in the mass of gasoline engine pistons (amounting to 100 g in many cases) has considerably enhanced bearing performance. In addition, a lower mass piston reduces thermal inertia, enabling a faster warm-up following a cold start, thus improving emissions and reducing noise during this critical period.

Design aspects

The simple cylindrical skirt pistons of the 1950s and 1960s gave way to a more imaginative design during the 1970s. For example, transverse slots were often included at the top of the skirt to give it flexibility whilst allowing closer running clearances, resulting in reduced piston slap. As the mass became a more important factor the recessed side panel or window began to appear, whilst retaining a full circular open end. Some designers reduced the skirt to the minimum, favouring the slipper configuration in order to save weight. A further benefit of all recessed panel or slipper designs is that the overall pin length tends to reduce, saving additional weight.

It became clear at the end of the 1980s that engine ratings were increasing and piston designs were required to respond accordingly. Solid skirt designs were being adopted, as stresses around transverse slots were becoming critical. Stress concentrations at the skirt/panel intersection resulted in a move away from the slipper type, giving an elliptical open end configuration as a compromise between the slipper and the full open end. The elliptical open end has now gained acceptance and has become the favoured style for current Federal-Mogul gasoline pistons.

As engine manufacturers are required to meet stringent emission standards, they demand a smaller top land height (the distance between the crown of the piston and the first piston ring) to maintain the lowest possible crevice volume. This lowering of the crown causes an increase in temperature, both within the ring grooves and around the pin hole. Furthermore,

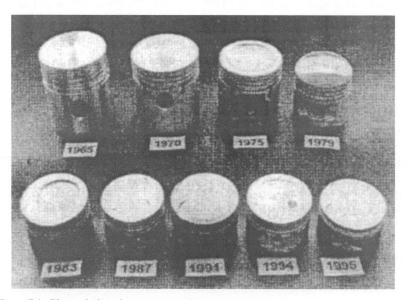

Figure 5.1. Piston design changes over a 30-year span.

modern engines are required to run close to detonation, further exposing the piston to risk, particularly in the ring land region. To assess the potential for further mass reduction, Federal-Mogul embarked on a programme involving the design and manufacture of a lightweight piston for a modern 1.6 litre engine. After completing a survey of pistons already in production in 1993 a benchmark for piston mass was set, compared with which the candidate designs showed mass reductions of between 25 and 50%. Design analysis concluded that a 25% reduction was attainable and the piston is now in volume production for the Rover K Series engine. Figure 5.1 demonstrates the progress made over the past 30 years in reducing the dimensions of a typical automotive piston, culminating in the K Series piston, which is still regarded as one of the most advanced designs. This is almost at the limit of the changes that can be made by basic design modifications, as there is very little excess material that can be removed without reducing the performance of the piston. The main ways to proceed are by looking at the piston materials and the operation of the cylinder as a whole.

Materials aspects

Despite the virtual monopoly of gravity die-cast aluminium alloys for volume production throughout the world, Federal-Mogul has continued to develop and refine both material composition and manufacturing techniques. For example, greater fatigue strength has been gained through the development of materials with a high copper content, making possible many of the piston designs discussed above. Combined with this, the application of solidification modelling has enabled consistent manufacturing quality to be achieved by improving the feeding of molten metal to the component. Further benefits have followed from this, allowing the use of reduced section thickness to lower the mass further for example.

Alternatives to gravity die-cast aluminium alloys have shown mixed benefits and have yet to prove themselves commercially. Despite their attractive strength to weight ratios, wear problems have been experienced with both metal matrix composites and magnesium alloy pistons. Some success has been achieved with aluminium-based powder metallurgy materials (e.g. Dispal S250), although cost remains a serious problem in respect of volume production. Similarly, forged pistons remain the first choice in the competition field, but price again tends to exclude their use in normal road vehicles.

Metallic alloy developments

Recent alloy development work has concentrated on strengthening the understanding of the links between component design, alloy composition and processing methods. Obviously the processing route has a major influence on the integrity of the material, and hence on the final properties of components, and other work within Federal-Mogul is developing this further. The focus of this is the development of microstructure and its influence on mechanical properties within an automotive component, and to what extent these can be modelled from a knowledge of the composition and the processing route. The greater the level of understanding of each of the links between the different stages shown in figure 5.2, the more control that can be exerted over the performance of the final component. In turn this provides an opportunity to reduce section thickness and hence mass, to improve the component's performance, or to increase its lifetime. In particular, such developments may allow lighter aluminium-based alloys to be used instead of steel in a wider range of diesel pistons in which engine temperatures are significantly higher.

In 1995, Federal-Mogul initiated a collaborative research programme with the Universities of Loughborough and Cambridge to model the development of microstructure in a range of Al–Si piston alloys. This work expanded to include contributions from the University of Southampton and the National Physical Laboratory in the UK. The research focuses on linking mechanical properties (especially fatigue behaviour) to detailed microstructure modelling. Some of the scientific aspects are dealt with below, but it is also appropriate to cover some of the industrial benefits of this work. The initial research work at Loughborough and Cambridge was closely involved in the alloy development programme within Federal-Mogul, and became an integral part of the in-house activities. The development

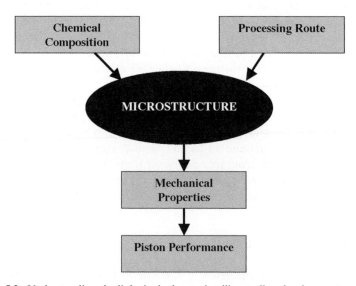

Figure 5.2. Understanding the links is the key to intelligent alloy development.

programme involved Federal-Mogul staff throughout the piston business, both in the US and in Europe, and the university researchers acted as consultants to much of that work.

The initial university research programme included:

- Detailed characterization of experimental and commercial alloys.
- Determination of phase composition.
- Modelling of phase development, both composition and amount.
- Advising the alloy development team on the effect of variations in alloy composition.
- Participating in the planning of experimental trials.

The Federal-Mogul alloy development team hit all of their targets, including those related to improvements in mechanical properties and the timing of the commercial launch of the new alloy. Without the contribution of the university work, there is no doubt that the alloy would not have been launched on time. This level of close interaction between university and industrial work is not without precedent, but neither is it commonplace. The experience within the company of this way of working has been a very positive one, and has established the value of well-chosen university research.

Microstructural characterization

Despite their widespread use, commercial near-eutectic multi-component Al–Si alloys have not been characterized in detail in the past. Most of the attention has been paid to the silicon distribution rather than the various intermetallic phases which form within the microstructure. Many of the intermetallic phases are non-stoichiometric and contain three, and usually more, different elements, making their identification in a particular multi-component commercial alloy non-trivial. It is essential to know how much of each element a particular phase can dissolve in order to be able to make accurate predictions of the volume fractions and compositions of all the different phases. To this end, a detailed characterization of the phases has been carried out (including complex ternaries, quaternaries and quinaries such as $Al_5Cu_2Mg_8Si_6$ and $Al_8FeMg_3Si_6$) contained in Al–11 wt% Si alloys with additions of iron, manganese, magnesium, copper and nickel in both model and commercial alloys using a variety of analytical techniques, including electron microscopy and X-ray diffraction.

Thermodynamic modelling

Commercial software packages (for example, MTDATA and Thermocalc) are available which determine the chemical equilibrium in a system by employing numerical routines which minimize the total Gibbs free energy with respect to the amount and composition of the phases present in an alloy of specified composition as a function of temperature and pressure. It is essential to have good quality thermodynamic data which contain interaction terms between all elements in all possible phases in order for useful and accurate information to be obtained. Thermodynamic databases for commercial aluminium alloys have been improved in recent years, although they are still some way from being able to make completely accurate predictions for all of the elements likely to be present in multi-component piston alloys. Extensive testing and data generation activities have been carried out to assist with database development to improve the quality of such predictions. A Scheil approach has also been used to predict non-equilibrium segregation and phase stability as a function of cooling rate within castings. This has been shown to be extremely useful for the prediction of phases which may only occur in segregated regions of the microstructures.

Thermodynamic calculations can provide very useful insights into the volume fraction of phases which would be present at equilibrium in a system. However, used in isolation they cannot hope to predict the distribution and morphology of such phases within a microstructure.

Modelling microstructural development on solidification

There are a number of models, developed over many years, which address the problem of modelling alloy solidification. Chapters 6, 9 and 10 discuss solidification and casting modelling developments in considerably more detail, concentrating on mould filling, defects and microstructures respectively. Deterministic models based on a classical nucleation and growth approach are well established and have been verified experimentally. However, such models are difficult to extend for complex multi-component systems. Stochastic models, based on Monte Carlo techniques, have met with considerable success in the prediction of grain structures and have been coupled with finite element heat flow modelling. These models also face difficulties in terms of prediction of second phases and in accounting for all segregation effects. An alternative approach which has emerged in recent years is the computational technique of phase field modelling. Phase field models are fundamentally different from other models in that they do not assume that the interface between the solid and liquid is sharp and do not follow the motion of the interface as in classical solidification models. They are based on the concept of a thermodynamic order parameter (the phase field) which varies between 0 and 1 depending on whether the system is in the liquid or solid phase at a particular point. This means that it is possible to follow the evolution of the phase field numerically, rather than track the motion of the solid/liquid interface. The mathematical formalism for the evolution of the phase field has its roots in statistical mechanics and the theory of phase transitions, coupled with macroscopic continuity equations for heat and/or solute transfer.

A number of phase field models for mainly pure and binary alloys have been developed for relatively simple systems to date and there are now several groups working in this area world-wide. The potential of the technique for application to Al–Si alloys has been demonstrated, and phase field models have been linked directly to thermodynamic databases. This code is currently being further developed to model microstructural development in multi-component, multi-phase Al–Si casting alloys.

Mechanical property requirements for piston alloys

Good high temperature strength and fatigue performance are key future piston requirements, and a specific strength comparable with or better than the existing Al–Si alloys is also required. Although wear properties are also important, they are a complex function of liner/ring materials, and are usually determined through simulated or real engine testing. Hence, the focus of current and future research is to develop a deeper fundamental understanding of how the microstructure critically affects both strength and fatigue resistance.

Previous studies of fatigue failure in cast Al–Si alloys have been mostly qualitative, identifying for example the role of porosity in fatigue initiation. In better quality castings, both the eutectic silicon and iron-based intermetallics have been observed to initiate fatigue crack growth. Refined microstructures are generally considered to offer the best fatigue resistance. Fatigue crack propagation studies have found that refined microstructures lead to deflected crack paths and hence increased closure/shielding effects and so offer improved fatigue propagation resistance. In general, the role of intermetallics in near-eutectic Al–Si alloys in fatigue crack initiation and growth has been identified, but there has been little direct quantification of the effects of intermetallic spatial distribution, volume fraction, and local clustering on fatigue initiation, crack growth and coalescence. The contribution to each of these processes needs to be isolated as they have differing effects on the overall fatigue resistance dependent upon the service condition. Although the detailed identification of a number of intermetallics in these alloys has been carried out, their role in the fatigue process has not yet been fully explored. Most fatigue testing has been carried out at room temperature although there has been some work on the creep behaviour of Al–Si alloys at service temperatures. To date, little consideration has been given to high temperature fatigue or creep–fatigue interactions. More detailed examination of the initiation and early crack growth of fatigue cracks at service temperatures is now being undertaken. Quantitative statistical measures of the microstructural features and local environments giving rise to crack initiation and controlling crack propagation are expected to enhance our understanding of these multi-component Al–Si alloys.

Summary

The goal of understanding all the links between the composition, processing conditions, microstructure development and final properties is an ambitious one. Nevertheless, the commercial benefits in being able to use this work to design new piston alloys are enormous and provide a strong incentive.

SECTION 2

MODELLING AND SIMULATION

The impact of modelling and simulation methods has been enormous in developing and interpreting advances in solidification theory, as well as in investigating the implications of our understanding for casting design and operation. Fluid flow, heat transfer, solidification mechanisms and microstructure development have all seen substantial theoretical advances. Fundamental understanding can often be described analytically, but numerical computational techniques play a key role in exploring the full range of theoretical predictions as well as in applications to the complex environments encountered in real casting processes. This section covers a variety of developments in underlying theory and use of computer modelling to simulate behaviour during casting.

Chapter 6 describes the use of computer modelling to understand and improve the important pressure die-casting process. Chapter 7 shows how carefully applied fundamental theory can be codified into working rules for ensuring sound casting practice. Chapters 8 and 9 discuss different aspects of casting microstructures, the first giving an example of how to control grain orientations in single crystal superalloy aeroengine components, and the second showing how to prevent defects in aluminium alloy castings. Chapter 10 discusses the theory of dendritic and eutectic solidification, and chapter 11 discusses the theory of peritectic solidification.

Chapter 6

Mould filling simulation of die-casting

Koichi Anzai, Eisuke Niyama, Shinji Sannakanishi and Isamu Takahashi

Introduction

A casting computer aided engineering (CAE) system named Stefan3D has been developed by an industry–university cooperative project. Stefan3D is specially designed for daily use by industrial casting engineers. Ease of use is one of the main features of Stefan3D. Only a few operations are necessary to carry out solidification and mould filling simulation of a casting. Fast and robust simulations can be performed with a highly tuned calculation algorithm. While Stefan3D is applicable to many casting processes, high performance mould filling analysis of die-casting is one of the main tasks of the industry–university project. In this chapter, the main features of Stefan3D and its application to die-casting mould filling simulations are described.

Background

In recent years, die-casting processes have been widely used for the manufacturing of complex shaped metallic components because of advantages that include low manufacturing costs, near net-shape and mass productivity. Most of the die-castings have been used as automobile parts. However, applications in household appliances, such as video cameras, CD players and so on, have been increasing as an alternative to plastic components. Automotive castings are discussed in chapter 4. An important aspect of this development is the need for reusability of the components.

It is very important to know the mechanical and thermal behaviour of the casting material during die-casting in order to obtain sound castings, but it is very difficult to investigate this behaviour experimentally because of the high pressure, temperature and speed of the process. The die-casting engineer's intuition and experience are still essential for good manufacturing, but numerical research on the optimization of die-casting is also necessary to overcome these difficulties.

There have been a number of commercial software packages that are applicable to die-casting optimization problems. However, die-casting engineers have greatly desired to have a more powerful and high performance casting CAE system. Stefan3D is one of the general purpose casting CAE system programs developed by the Stefan project at Tohoku University, organized by Professor Niyama under the support of over 30 Japanese industrial companies since 1992. Stefan3D runs on a wide variety of UNIX systems including PC UNIX. Tohoku University has developed Stefan3D from scratch, and therefore no commercial code is included in it. All the source codes have been distributed to the project members at every version upgrade time. Original pre- and post-processors are specially designed for practical use in industry. Only a few operations are necessary to build up the input data for the simulation. All the solvers (solidification, mould filling, etc.) have been evaluated by experiments and improved for quick and robust calculations.

Simulation methods

The conventional SOLA-VOF base algorithm is employed in Stefan3D; however, Stefan3D has been powered up by high performance tuning techniques for practical use in industry. The main features of Stefan3D are summarized as follows:

- Huge scale calculations (over 100 000 000 mesh points are available).
- Extremely fast and robust calculations.
- Time saving and easy to use system.
- Runs on small PC (less memory, less hard disk space).
- Optimum die cooling design tool, called cyclic steady-state heat balance method (CSM), included.

Practical simulations can be performed with Stefan3D on standard business PCs including laptops. The maximum number of meshes that Stefan3D can handle may vary with the memory that the PC has. Generally speaking, Stefan3D can handle over 2 000 000 mesh points on a 128 MB RAM PC. Also, the calculation speed may vary with the MPU in the PC. One hour is a typical calculation time for a one million mesh mould filling simulation on a Pentium II (400 MHz) PC.

Figure 6.1 shows the main structure of the Stefan3D software system. Casting engineers can construct three-dimensional geometric casting objects easily using a built-in simplified solid modeller named TCB. A single TCB model expresses only simple solid polyhedra, but complex shaped three-dimensional objects can be constructed by combinations of

Simulation methods 81

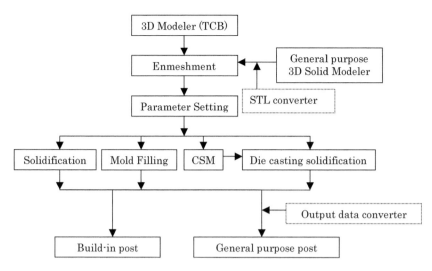

Figure 6.1. Structure of Stefan3D.

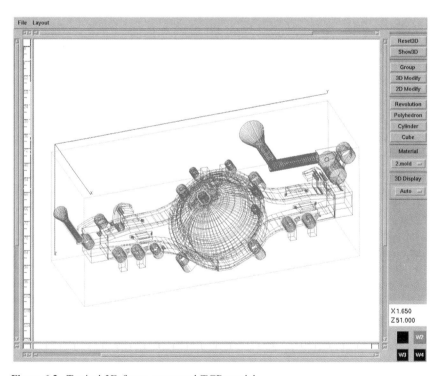

Figure 6.2. Typical 3D figure generated TCB model.

various TCB models. Figure 6.2 shows a typical casting shape generated by TCB. In this case, it took 16 h to construct the complex three-dimensional figure using TCB. If the casting designer needs to construct a more precise three-dimensional model, then a general purpose three-dimensional modeller must be utilized. Data conversion from three-dimensional computer aided design (CAD) data to mesh data for Stefan3D is prepared for the convenience of easy mesh data generation. Also, general purpose post-processors on the market can be used to show the calculated results as sophisticated three-dimensional images with the help of a built in output data converter, while the original post-processor of Stefan3D has enough functions to evaluate the simulated result.

CSM is a unique idea proposed by Professor Niyama and co-workers to narrow down the range of possible die-casting process parameters at the trial stage. With CMS, the temperature distribution of die-casting dies in cyclic steady-state conditions can be estimated quickly. CSM works well for optimization of die-casting conditions and/or preliminary estimation of die temperature before full-scale cyclic die-casting solidification simulation that can be accelerated with the help of the initial die temperature condition estimated by CSM.

Applications

Figure 6.3 shows a typical output of a die-casting mould filling simulation generated by Stefan3D. In this case, 14 million mesh points are used to represent a casting cavity and moulds. The calculation time is around 24 h on a 450 MHz Pentium II PC until 95% of the casting cavity is filled up. Plunger motion is taken into account in the die-casting mould filling simulation by introducing velocity boundary conditions, set in the middle of the runner or gate to save calculation time and cost. This assumption is reasonable and may work well in practical simulations, but the unexpected flow pattern in the neglected runner and the disturbance of flow in the plunger occasionally spoil the accuracy of the assumption. Figure 6.4 shows mould filling patterns generated by Stefan3D in a multi-cavity die-casting problem. It is clear that the length of vertical runner that is taken into account in the simulation has a strong effect on the filling pattern in the casting cavities. Casting engineers must pay attention to this kind of problem in the simulation. Simulation errors caused by ill-conditioned boundary conditions will mislead counter-measures against runner design. Defects in castings are discussed in more detail in chapter 9, and how to prevent defects by controlling melt flow and other casting parameters is discussed in chapter 7.

Plunger speed is one of the major control parameters for the manufacturing of sound die-castings, but direct simulation of the plunger

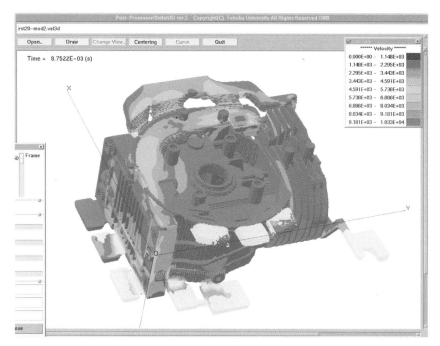

Figure 6.3. Typical output of Stefan3D (14 000 000 meshes are used).

movement is not taken into account in the current mould filling simulation. This is because the modelling of the plunger movement is quite a difficult task. Sannakanishi and co-workers have proposed a practical algorithm named the quasi-fluid model to take the effect of plunger movement into account in the die-casting mould filling simulation. The quasi-fluid is a special fluid that comes into the plunger region from the boundary velocity that is fixed on the outer surface of the calculation region. In every time step, the quasi-fluid comes into the plunger region in accordance with the actual plunger movement and pushes the preset region in the preset fluid in the plunger. The only difference between the quasi-fluid and the actual fluid is the velocity correction calculation at the end of each step. The velocity vectors of quasi-fluid cells are always maintained parallel to the direction where the plunger tip advances. As a result, the quasi-fluid acts like a plunger rod that pushes the preset fluid in the plunger. Some minor modifications were needed to the source codes of Stefan3D to implement the model.

Figure 6.5 shows a comparison of filling patterns between (a) a conventional model and (b) the quasi-fluid model by Stefan3D. Both of the filling sequences are similar. This indicates that the conventional assumption of plunger movement in the simulation works well, in this case.

84 *Mould filling simulation of die-casting*

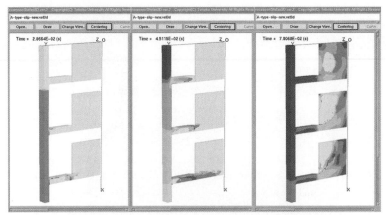

(a) Case of straight vertical runner

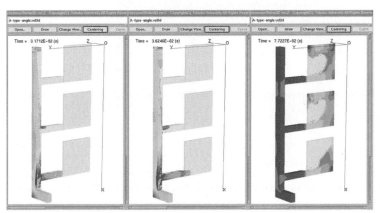

(b) Case of L-shaped short vertical runner

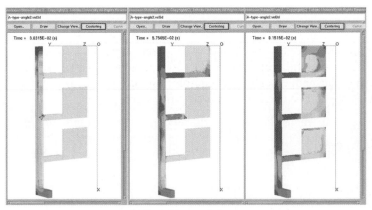

(c) Case of long vertical runner

Figure 6.4. Mould filling patterns of multicavity problem by Stefan3D.

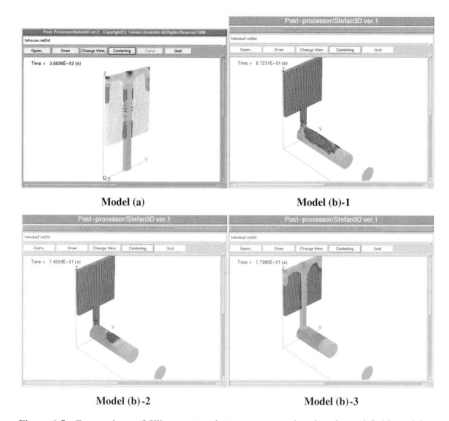

Figure 6.5. Comparison of filling pattern between conventional and quasi-fluid model.

Summary

A casting computer aided engineering system named Stefan3D has been developed by an industry–university cooperative project in Japan. Stefan3D has been successfully utilized in over 30 companies for their routine work. In particular, huge scale and complex shaped die-casting mould filling simulations can be carried out quickly and stably.

References

[1] Hao S, Anzai K and Niyama E 1995 'Numerical simulation of fluid flow as applied to casting design' *AFS Trans.* **17** 41–46
[2] Anzai K, Niyama E and Hao S 1995 'A casting CAE system with flow and solidification simulation' *Proceedings, Modelling of Casting and Solidification Processes 1995* pp 279–286

[3] Itamura M, Yamamoto N, Niyama E and Anzai K 1996 'Application of the flow visualisation technique and flow simulation to diecasting flow analysis' *Int. J. Cast Metal Res.* 71–81
[4] Xiao L, Anzai K, Niyama E, Kimura T and Kubo H 1998 'Reducing "cold shut" defects in the "H" process aided by computer simulation' *Int. J. Cast Metals Res.* 71–81
[5] Hiratsuka H, Niyama E, Horie H, Kowata T, Anzai K and Nakamura M 1998 'Computer simulation of fluidity and microstructure of thin-sectioned cast iron' *Int. J. Cast Metals Res.* 201–205
[6] Anzai K, Oda K, Kubo H and Niyama E 1998 'Die cooling design by cyclic steady state heat balance method' in *Advances in Aluminium Casting Technology* (TMS) pp 183–189

Chapter 7

The ten casting rules

John Campbell

Introduction

The past few years have witnessed an unprecedented increase in our understanding of the casting process. Thus as understanding has increased, the author has steadily added to the list of casting requirements as they have become known. From an initial list of four rules, now ten rules have been identified which incorporate the latest technology for the production of reliable castings. These are just the start. It is already known that additional rules exist, but these remain to be further researched and clarified.

The ten rules listed here are proposed as necessary, but not, of course, sufficient, for the manufacture of reliable castings. It is proposed that they are used in addition to existing conventional technical specifications such as alloy type, strength, and traceability via ISO 9000, and other well known and well understood conventional foundry controls such as casting temperature. Although not yet tested on all cast materials, there are fundamental reasons for believing that the ten rules have general validity, and thus are applicable to all types of metal and alloy including those based on aluminium, zinc, magnesium, cast irons, steels, air- and vacuum-cast nickel and cobalt, and titanium. Nevertheless, of course, although all materials will probably benefit from the application of the rules, some will benefit almost out of recognition, whereas others will be less affected.

The rules as they stand constitute a draft process specification, which a buyer of castings would demand if he wished to be assured that he was buying the best possible casting quality. If he were to specify the adherence to these rules by the casting producer, he would ensure that the quality and reliability of the castings was higher than could be achieved by any amount of expensive quality control on the finished product. Conversely, of course, the rules are intended to assist the casting industry. They will speed up the process of producing the casting right first time, and should contribute in a major way to the reduction of scrap when the casting goes into production. In this way the casting industry will be able to raise standards, without any significant increase in costs. By adherence to the rules, quality will be raised to the point at which casting reliability equal to that of forgings can be offered with confidence. Only in this way will castings be accepted by the engineering profession as reliable, engineered products, and assure the future prosperity of the casting industry and its customers.

The ten rules

The ten casting rules are listed sequentially as follows:

Rule 1. Good quality melt

Immediately prior to casting, the melt should be prepared and treated, if necessary, using the best current practice.

Naturally, it is of no use to put in the best designs of filling and feeding systems if the original melt is of such poor quality, or perhaps already damaged, that a good casting cannot be made from it. It is a requirement that either the process for the production and treatment of the melt shall have been shown to produce good quality liquid, or the melt should be demonstrated to be of good quality. A good quality liquid is one which is substantially free from non-metallic inclusions. It should be noted that such melts are not to be assumed, and, without proper treatment, are probably rare. (Additional requirements, not part of this specification, may be also placed on the melt; for example, low gas content, or low values of particular solute impurities.)

An interesting possibility for future specifications for aluminium alloy castings (where, fortunately, residual gas in supersaturated solution does not appear to be harmful) is that a double requirement may be made for the content of dissolved gas in the melt to be high, but the percentage of gas porosity in the casting to be low. The meeting of this double requirement will ensure to the customer that double oxide films are not present. This is because these damaging but undetectable defects will, if present, effectively be labelled and made plainly visible on X-ray radiographs and polished sections by the precipitation of dissolved gas.

Rule 2. Liquid front damage

The liquid metal front (the meniscus) should not go too fast.

Maximum meniscus velocity is in the range 0.4 to $0.6\,\mathrm{m\,s^{-1}}$ for most liquid metals. This maximum velocity may be raised in constrained running systems or thin-section castings. This requirement also implies that the liquid metal must not be allowed to fall more than the critical height which corresponds to the height of a sessile drop of the liquid metal.

The maximum velocity requirement

Recent research has demonstrated that if the liquid velocity exceeds a critical velocity there is a danger that the surface of the liquid metal may be folded over by surface turbulence. There is therefore a chance that the surface oxide film may be folded into the bulk of the liquid if the speed of advance of the liquid front exceeds this critical velocity. The folded-in films constitute initiation sites for (1) gas precipitation, (2) shrinkage cavities and (3) hot tears, and after being frozen into the casting become (4) effective cracks, lowering the strength and fatigue resistance. The folded films may also (5) create leak paths, causing leakage failures. Thus the entrainment of the liquid surface by surface turbulence leads to a catalogue of problems which beset the foundryman.

The folding-in of the oxide is a random process, leading to scatter and unreliability in the properties and performance of the casting on a casting to casting, day to day and month to month basis during a production run. The critical velocity is close to $0.4\,\mathrm{m\,s^{-1}}$ for dense alloys such as irons, steels and bronzes; it is $0.5\,\mathrm{m\,s^{-1}}$ for liquid aluminium alloys, and $0.6\,\mathrm{m\,s^{-1}}$ for liquid magnesium alloys. The maximum velocity condition effectively forbids top gating of castings. This is because liquid aluminium reaches its critical velocity of about $0.5\,\mathrm{m\,s^{-1}}$ after falling only 12.5 mm under gravity. The critical velocity of liquid iron or steel is exceeded after a fall of only about 8 mm. Such short fall distances are of course always exceeded in practice in top gated castings, leading to much incorporation of the surface oxide films, and consequent leakage and crack defects.

Castings which are made where velocities everywhere in the mould never exceed the critical velocity are consistently strong, with high fatigue resistance, and are leak-tight (if properly fed, of course, so as to be free from shrinkage porosity). Experiments on the casting of aluminium have demonstrated that the strength of castings may be reduced by as much as 90% if the critical velocity is exceeded. The corresponding defects in the castings are not always detected by conventional non-destructive testing such as X-ray radiography or dye penetrant, since, despite their large area, the folded oxide films are very thin, and do not necessarily give rise to any significant surface indications.

The speed requirement automatically excludes conventional pressure die-castings, since the filling speeds are over an order of magnitude in excess of the critical velocity. Some recent special developments of high pressure technology are capable of meeting this requirement, however. These include the vertical injection squeeze casting machine by UBE, and the Shot Control (SC) technique by Buhler, both of which can, in principle, be operated to fill the cavity through large gates at low speeds, and without the entrainment of air into the liquid metal. Such castings require to be sawn, rather than broken, from their filling systems of course.

Other uphill filling techniques such as the Cosworth Process and low-pressure die-casting systems are capable of meeting rule 2. Even so, many low-pressure die-casting machines are in fact so poorly controlled on flow rate that the speed of entry into the die greatly exceeds the critical velocity, thus negating one of the most important potential benefits of the low-pressure system. In addition, of course, the critical velocity is practically always exceeded during the filling of the low-pressure furnace itself because of the severe fall of the metal as it is transferred into the pressure vessel, so that the metal is damaged even prior to casting.

No-fall requirement

It is quickly shown that if liquid aluminium is allowed to fall more than 12.5 mm then it exceeds the critical $0.5\,\mathrm{m\,s^{-1}}$. Similar critical velocities and critical fall heights can be defined for other liquid metals. The critical fall heights for all liquid metals are in the range 3 to 15 mm. Thus top gating of castings almost without exception will lead to a violation of the critical velocity requirement. Many forms of gating which enter the mould cavity at the mould joint, if any significant part of the cavity is below the joint, will also violate this requirement.

In fact, for conventional sand and gravity die-casting, the requirement effectively excludes any form of gating other than bottom gating into the mould cavity. Also excluded are any filling methods which cause waterfall effects in the mould cavity. This dictates the siting of a separate ingate at every isolated low point on the casting. The initial fall down the sprue in gravity-filled systems does necessarily introduce some oxide damage into the metal. However, it seems that for some alloys much of the oxide introduced in this way is attached to the walls of the sprue, and, in the main, does not appear to find its way through into the mould cavity. This surprising effect is clearly seen in many top gated castings, where most of the oxide damage (and particularly any random leakage problem) is confined to the area of the casting under the point of pouring, where the metal is falling. Although extensive damage does not seem to extend into those regions of the casting where the speed of the metal front decreases, and where the front travels uphill, there appears to be the carry-over of some concentration of defects. Thus the provision of a filter immediately after the completion of the fall is valuable. It is to be noted, however, that some damage will still be expected to pass through the filter.

The requirement for no fall of the liquid within the running system (after the metal leaves the base of the sprue) and mould cavity implies that metal must be designed to go only uphill after it leaves the base of the sprue.

Thus the runner must be in the drag and the gates must be in the cope for a horizontal single jointed mould (if the runner is in the cope, then the gates fill prematurely, before the runner itself is filled, thus air bubbles are likely to enter the gates). The no-fall requirement may also exclude some of those filling methods in which the metal slides down a face inside the mould cavity, such as Durville casting type processes. This effect is discussed in more detail in the following section (rule 3).

It is noteworthy that these precautions to avoid the entrainment of oxide films also apply to casting in inert gas or even in vacuum. This is because the oxides of aluminium and magnesium (as in aluminium alloys, ductile irons, or high-temperature nickel-base alloys, for instance) form so readily that they effectively 'getter' the residual oxygen in any conventional industrial vacuum, and form strong films on the surface of the liquid.

Metals which behave in the same way are suggested to be the Zn–Al alloys and ductile irons. Carbon steels and some stainless steels are thought to be similar, although in some of these systems the entrained oxides agglomerate because they are partially molten, and thus float to form surface imperfections in the form of slag macro-inclusions on the surface. Duplex stainless steels exhibit a tough surface oxide film on the melt which gives special problems if surface entrainment is permitted. For a few materials, particularly alloys based on the aluminium and manganese bronzes (Cu–10Al types) the critical velocities were thought to be much lower, in the region of $0.075\,\mathrm{m\,s^{-1}}$. However, this seems from recent work at Birmingham to be a mistake probably resulting from the confusion caused by bubbles entrained in the early part of the filling system. With well-designed filling systems, the aluminium bronzes fulfil the theoretically predicted $0.4\,\mathrm{m\,s^{-1}}$ value for a critical ingate velocity.

Rule 2 applies to normal castings with walls of thickness over 3 or 4 mm. For very thin-walled castings, of section thickness less than 2 mm, the effect of surface tension in controlling filling becomes predominant, the meniscus being confined between walls which are so much closer than the equilibrium curvature that the meniscus is effectively compressed, and requires the application of pressure to force it into such narrow gaps. The liquid surface is now so constrained that it is not easy to break the surface, i.e. there is no room for splashing or droplet formation. Thus the critical velocity is higher, and metal speeds can be raised somewhat without danger of exceeding this higher limit. In fact, tight curvature of the meniscus becomes so important in very thin-walled castings, with walls less than 2 mm thickness, that filling can sometimes be without regard to gravity (i.e. can be uphill or downhill) since the effect of gravity is swamped by the effect of surface tension.

The dominance of surface tension also makes the uphill filling of such thin sections problematic, because the effective surface tension exceeds the effect of gravity, so that instabilities can occur, whereby the moving parts of the meniscus continue to move because of the reduced thickness of the oxide skin at that point whereas other parts of the meniscus which drag back are further suppressed in their advance by the thickening oxide, so that a run-away instability condition occurs. This dendritic advance of the liquid front is no longer controlled by gravity in very thin castings, making the filling of extensive horizontal sections a major problem. The problem is relieved by the presence of regularly spaced ribs or other geometrical features which help to organize the distribution of liquid, and thus avoid the starving of areas which the flow happens to avoid because of random meandering. Such chance avoidance, if prolonged, leads to the development of strong oxide films, or even freezing of the liquid front. Thus the final advance of the liquid to fill such regions is hindered or prevented altogether.

Rule 3. Liquid front stop

The liquid metal front should not go too slowly and, more exactly, should not stop at any point on the front.

The advancing liquid metal meniscus must be kept alive (i.e. moving) and therefore free from thick oxide film which may be incorporated into the casting. This is achieved by the liquid front being designed to progress only uphill in a continuous uninterrupted upward advance (i.e. in the case of gravity-poured casting processes, from the base of the sprue onwards). This implies:

- only bottom gating is permissible,
- no stopping due to arrest of pouring,
- no downhill flow (either falling or sliding), and
- no horizontal flow (of any significant extent).

The liquid metal front must continue to advance at all points on its surface. If it stops, a thick surface film has a chance to form, and thus resist the re-starting of the flow at some later moment. The incorporation of this film in the casting as metal flows over it from other regions then forms a major defect, often spanning the casting from wall to wall. This problem can occur in several ways.

If the liquid metal is allowed either to fall or to spread horizontally, the unstable propagation of the front in the form of jets or streams bounded by an oxide flow tube leads to a situation where the oxide flow tube grows to a considerable thickness, and is finally sealed into the casting as the liquid metal envelops it. A double oxide film defect is created, with a highly asymmetric double layer, in which one oxide is thick and the other thin. However, since dry side is arranged against dry side as usual, the defect is once again a crack. Such cracks are extensive and serious. They are different from the fragmented and chaotic double films introduced by surface turbulence. They have predictable geometric shapes such as planes, cylinders, or riverine forms. In common with the irregular and unpredictable forms produced by surface turbulence, they are normally invisible to X-rays and to dye penetrant testing. They are only revealed by gross mechanical deformation which opens the crack.

The deleterious flow tube structure which forms when filling downwards or horizontally is usually eliminated when filling vertically upwards. However, even in this favourable mode of filling, a related defect can still occur if the advance of the meniscus is stopped at any time. This effect is discussed below.

Continuous uphill advance of the meniscus

The uphill motion of the liquid front, if it can be arranged in a mould cavity, helps to keep all parts of the front alive, i.e. the meniscus does not become pinned in place with oxide, and thus become dead. As explained earlier, the front becomes pinned if the rate of advance of the metal front is too slow, or if it stops. If this happens, the surface oxide has a chance to thicken, becoming so strong that there is a danger that any further advance of the front will be prevented. The breaking through of the meniscus at a weak point will then flood over the fixed, thickened oxide, sealing it into the casting as a major lap defect.

One of the reasons for not allowing any fall of the metal in the mould is not only to avoid exceeding the critical velocity, and thus impairing the metal which falls, but also to avoid the period when the rise of the metal is interrupted, causing an oxide to form on the stationary front (whilst the metal level is fixed during the period of the overflow), and on the tube of flow surrounding the fall itself, which is, of course, also effectively stationary.

The no-fall requirement may also exclude some of those filling methods in which the metal slides down a face inside the mould cavity, such as Durville casting type processes. These processes can be used to good advantage if used correctly. In such cases the liquid metal front progresses steadily at all points, expanding and breaking its surface oxide, folding it out against the mould wall as the front progresses. In this way the oxide never enters the bulk of the metal: it only continues to augment the oxide already present on the surface of the casting. The production of ingots by the controlled tilting of the mould can be seen to maintain the live nature of the front at all times, as the flow expands into the ingot mould.

When a Durville-type pour is carried out badly, the metal can be seen to flow down into the mould as a stream. Its boundaries, i.e. its outer surface, are of course stationary, being formed as a tube of oxide through which the metal flows while it is running down one of the faces of the mould. This oxide flow tube then becomes sealed into the finished casting, creating a serious defect. The problem is analogous to the oxide tube defect which forms around the falling metal during any waterfall effect during the filling of the casting. These flow tubes are a common defect seen in a wide variety of castings which exhibit either extensive horizontal or downhill sections.

Similar defects which have the appearance of laps also occur on the side walls of flows as they meander aimlessly across large horizontal parts of the mould. The rat-tail defect is a familiar reminder of such wanderings of the metal stream, although in this case the rat-tail line is usually formed by the failure of the mould. However, a corresponding line of oxide defect in the metal casting may also be present as an additional problem. The avoidance of extensive horizontal sections in moulds is therefore essential for reproducible and defect-free castings. Any horizontal sections should be avoided by the designer, or by the caster by tilting the mould. (This is easily provided by some casting techniques such as the Cosworth Process, where the mould is held in a rotatable fixture during casting.) The flow across such inclined planes is therefore progressive, if slow, but the continuous advance of the front at all points assists the aim of keeping the meniscus alive.

Rule 4. Bubble damage

No bubbles of air entrained by the filling system should pass through the liquid metal in the mould cavity.

This may be achieved by:

- the use of a properly designed pouring basin;
- preferred use of a stopper in small castings;
- the use of a sprue and runner designed to fill in one pass;
- avoidance of the use of wells or other volume-increasing features of filling systems;
- possible use of ceramic foam filter close to sprue/runner junction;
- possible use of bubble traps;
- no interruptions to pouring.

The passage of a single bubble through an oxidizable melt results in the creation of a bubble trail as a double oxide crack in the liquid. Thus even though the bubble may be lost by bursting at the liquid surface, the trail remains as permanent damage in the casting. Poor designs of filling systems can result in the entrainment of much air into the liquid stream during its traverse through the filling basin, during its fall down the sprue, during its travel along the runner, and during its entry into the mould cavity if the melt velocity is still too high at that point, or if the melt is allowed to fall inside the mould cavity. The repeated passage of many bubbles through the liquid metal leads to an accumulation of a tangle of bubble trails and, if the density of trails is sufficiently great, residual fragments of entrapped bubbles wedged in among the trails. This mixture is collectively christened bubble damage. Bubble damage is probably the most common defect in castings, but is almost universally unrecognized.

The problem is commonly observed just inside and above the first ingate from the runner, and is often mistaken for shrinkage porosity as a result of its irregular form. However, it is easily recognized on a plan view by X-ray from the outlines of residual entrapped bubbles, and from close optical examination: in an aluminium alloy casting, instead of shiny dendrite tips characteristic of shrinkage porosity, mainly a series of dark, non-reflective oxidized surfaces interleaved as of crumpled pages of sepia-coloured paper will be found. (Some associated shrinkage porosity with its dendrites will almost always be present of course, but should not be allowed to cause the major non-reflective regions to be overlooked. It is easy to fall into this trap because under the microscope, in contrast to the eye-catching regular crystallographic order of the dendrites, the oxidized regions do not attract attention.) In some stainless steels the phenomenon is seen under the microscope as a mixture of bubbles and cracks (a remarkable combination which would normally be difficult to explain as a metallurgical phenomenon!) because the high cooling strain leads to high stresses in such strong material, and thus opens up the double oxide bubble trails.

Gravity-filled running systems

In gravity-filled running systems the requirement to reduce bubbles in the liquid stream during the filling of the casting calls for offset stepped basins, or other advanced basin designs. The conventional conical pouring basin is not suitable.

The requirement also demands properly engineered and manufactured sprues. The normal sprue is required to have a calculated taper, in which the choke of the running system (i.e. the point at which the rate of flow through the running system is controlled) is constituted preferably by the base of the sprue, or immediately adjacent at the entrance to the runner. Parallel or reversed taper sprues are not permitted. It is mandatory that the taper of the sprue contains no perturbations to upset the smooth fall of the liquid metal. Thus it must be well-fitting with the pouring basin, or other mould or die joints; no steps, ledges or abrupt changes in direction are permissible. Also no branching or joining of other ducts, runners, gates or sprues is allowable. All such features (especially common in investment castings) have the potential for the introduction of air into the stream, or the uncontrolled escape of liquid into other parts of the mould cavity.

At one time it was mandatory that each sprue had a sprue well at its base. The well was thought to facilitate the turn of the metal through the right-angle bend into the runner with minimum turbulence. Previous work on well development had been based on water models. However, more recent work at Birmingham using liquid metals has demonstrated that, at best, the well is no better than no well at all, and at worst causes considerable extra turbulence. Thus the new designs of filling systems incorporate no well at the base of the sprue.

It is mandatory that no interruption to the pour occurs which leads to the uncovering of the entrance to the sprue, so that air enters the running system. A provision should be made for the foundry to automatically reject any castings which have suffered such an interrupted pour.

Pumped and low-pressure filling systems

Pumped systems such as the Cosworth Process, or low-pressure filling systems into sand moulds or dies, are highly favoured as having the potential to avoid the entrainment of bubbles if the processes are carried out under proper control. (Good control of a *potentially* good process should not be assumed; it is required to be demonstrated. In addition, bubbles can be released erratically from the interior wall of a tube launder system, as well as from the underside of a badly designed distribution plate, or from a poorly maintained and cleaned launder tube, if one is used.)

Although low-pressure filling systems can in principle satisfy the requirement for the complete avoidance of bubbles in the metal, a leaking riser tube in a low-pressure casting machine can lead to serious problems. The stream of bubbles from a leak in the riser tube will then rise up the tube and directly enter the casting. Thus regular checks of such leakage, and the rejection of castings subjected to such consequent bubble damage, are required.

The other major problem with conventional low-pressure delivery systems where the melt is contained within a pressurized vessel is that the quality of the melt is usually damaged by the filling of the pressure vessel. This is an uncontrolled fall, first from the foundry transfer ladle, then down a chute, and finally into the melt. This unsatisfactory transfer process introduces much oxide into the metal, only part of which, of course, is subsequently removed by a filter at the entrance to the mould cavity.

Rule 5. Core blows

- No bubbles from the outgassing of cores should pass through the liquid metal in the mould cavity. Cores must be shown to be of sufficiently low gas content and/or adequately vented to prevent bubbles from core blows.
- No clay-based core or mould repair paste must be used (unless demonstrated to be free from the risk of creating bubbles).

Serious oxide films, which can cause leaks and mechanical cracking, can be caused by the outgassing of sand cores through the melt, causing bubbles to rise through the metal, and leaving an oxidized bubble trail in its wake. A succession of such bubbles from a core is very damaging to the upper parts of castings. The bubble from the core contains a variety of gases, including water vapour, which are highly oxidizing to metals such as aluminium and higher melting point metals. Bubble trails from core blows are usually particularly noteworthy for their characteristically thick and leathery double oxide skin, which is probably why core blows result in such efficient leak defects through the upper sections of castings.

Considerable volumes of water vapour are known to be driven into the melt from clay-based core repair and mould repair pastes (this is because the clay contains water, and is impermeable, preventing the escape of the water into the core or mould, and thus forcing the escape path to be through the melt). The use of such repair pastes is to be avoided (unless they can be demonstrated to avoid the generation of blows in the melt).

To demonstrate that a core, or assembly of cores, does not produce blows may require a procedure such as the removal of all or part of the cope or overlying cores, and taking a video recording of the filling of the mould. If there are any such problems, the eruption of core gases will be clearly observable, and will be seen to result in the creation of a froth of surface dross (which would of course normally be entrapped inside the upper walls of the casting). A series of video recordings might be found to be necessary, showing the steady development of solutions to a core blowing problem, and recording how individual remedies resulted in progressive elimination of the problem. The video recording shall be retained by the foundry for inspection by the customer for the life of the component. Any change to the filling rate of the casting, or core design, or the core repair procedure, will necessitate a repeat of this exercise.

For castings with a vertical joint where a cope cannot be conveniently lifted clear to provide such a view, a special sand mould may be required to carry out the demonstration that the core assembly does not cause blows from the cores at any point. This will have to be constructed as part of the tooling to commission the casting. This will have to be seen as an investment in quality assurance.

Rule 6. Shrinkage damage

- There should be no feeding uphill because of (i) unreliable pressure gradient and (ii) complications introduced by convection.
- Good feeding design should be demonstrated, following all ten casting rules, by an approved computer solidification model, and test castings.
- It is essential to control (i) the level of flash at mould and core joints, (ii) mould coat thickness and (iii) temperatures of metal and mould.

Where computer modelling is not carried out, the observance of the casting rules is strongly recommended. Even when computer simulation is available, the rules will be found to be good guidelines. In addition, all five mechanisms for feeding (as opposed to only liquid feeding) should be used to advantage. For instance solid feeding (a kind of self-feeding by the solidified casting) will be found to be especially useful when attempting to achieve soundness in an isolated boss or heavy section where the provision of feed metal by conventional techniques is impossible.

Feeding with gravity

As opposed to filling uphill (which is of course quite correct) feeding should only be carried out downhill (using the assistance of gravity). Attempts to feed uphill, although possible in principle, can be unreliable in practice, and may lead to randomly occurring defects which have all the appearance of extensive shrinkage porosity. These occur because of the difficulties caused by two main problems: (i) adverse pressure gradient and (ii) adverse density gradient leading to convection. Reason (i) is dealt with below; reason (ii) is dealt with in Rule 7.

The atmosphere is capable of holding up several metres of head of metal. For liquid mercury the height is approximately 760 mm of course (the height of the barometer column). There are similar equivalent heights for other liquid metals, allowing for the density difference; thus the atmosphere will hold up about 4 m of liquid aluminium for instance. (Whilst no pore exists, the tensile strength of the liquid will in fact allow the metal, in principle, to greatly exceed this height since in the absence of defects it can withstand tensile stresses of thousands of atmospheres. However, the random initiation of a single minute pore will instantly cause the liquid to break, causing such feeding to stop and go into reverse. The casting will then empty down to the level at which atmospheric pressure can support the liquid.)

However, if there is a leak path to atmosphere, allowing atmospheric pressure to be applied in the liquid metal inside the mould cavity, the melt will then fall further, attempting to equalize levels in the mould and feeder. Thus the residual liquid will drain from the casting if the feeder is sited below the casting. This is an efficient way to cast porous castings and sound feeders. Clearly, the initiation of a leak path to atmosphere (via a double oxide film, or via a liquid region in contact with the surface at a hot spot) is rather easy in many castings, making the whole principle of uphill feeding so risky that it should not be attempted where porosity cannot be tolerated. The recent practice of active feeding, whereby small feeders placed low on the casting are pressurized to cause the casting to feed uphill is almost certainly allowable in small castings. For larger castings the technique will be subject to the problems of convective flow, and is expected to become unreliable.

Computer modelling of feeding

Computer modelling has demonstrated its usefulness in being able to predict shrinkage porosity with accuracy. It should now be specified as a prior requirement to be carried out before work is started on making the tooling. This minor delay will have considerable benefits in shortening the overall development time of a new casting, and will greatly increase the chance of being right first time. An example of computer modelling of die-casting is discussed in chapter 6. A second example applying computer modelling to an automotive part is given in chapter 9.

It has to be recognized, however, that the computer will not necessarily be capable of any design contribution. Thus filling and feeding will be required as inputs. Additionally, of course, many computer simulations do not simulate the important effects due to convection, and some do not include conduction either. Some neglect, or only crudely allow for, the effect of the heating of the mould by the flow of metal during filling. The belief in the results from such models is therefore required to be tempered at least with caution, if not scepticism.

Random perturbations to feeding patterns from casting to casting

Flash of approximately 1 mm thickness and only 10 mm wide has been demonstrated to have a powerful effect on the cooling of local thin sections up to 10 mm thick, speeding up local solidification rates by up to ten times. Thus flash has to be controlled, or used deliberately, since it has the potential to cut off feeding to more distant sections. The erratic appearance of flash in a production run may therefore introduce uncertainty in the reproducibility of feeding, and the consequent reproducibility of the soundness of the casting. Moderate flash on thicker sections is usually less serious because convection in the solidifying casting conveys the local cooling away, effectively spreading the cooling effect over other parts of the casting, so that an averaging effect over large areas of the casting is created. In general, however, it is desirable that these uncertainties are reduced by good control over mould and core dimensions.

The other known major variable affecting casting soundness in sand and investment castings is the ability of the mould to resist deformation. This effect is well established in the case of cast irons, where high mould hardness leading to good mould rigidity is needed for soundness. However, there is evidence that such a problem exists in castings of copper-based alloys and steels. A standard system such as statistical process control (SPC), or other technique, is required to be demonstrated to be in place to monitor and facilitate control of such changes. (Permanent moulds such as metal or graphite dies are relatively free from such problems.)

The solidification pattern of castings produced from permanent moulds such as gravity dies and low-pressure dies may be considerably affected by the thickness and type of the die coat which is applied. A system to monitor and control such thickness on an SPC system is required to be demonstrated to be in place. Alternatively, for each casting on a case by case basis, the thickness of the die coat is required to be demonstrated to be of no consequence.

For some permanent moulds, pressure die and some types of squeeze casting the feeding pattern is particularly sensitive to mould cooling. Changes to cooling channels in the die, or to the cooling spray during die opening, will have to be checked to ensure that corresponding deleterious changes have not been imposed on the casting.

Rule 7. Convection damage

The freezing time should be assessed in relation to the time for convection to cause damage.

Thin and thick section castings automatically avoid convection problems. For intermediate sections either (i) reduce the problem by avoiding convective loops in the geometry of the casting and rigging, or (ii) eliminate convection by roll-over after filling. Convection enhances the difficulty of uphill feeding in medium section castings, making them extremely resistant to solution by increasing the (uphill) feeding. In fact, increasing the amount of feeding, e.g. by increasing the diameter of the feeder neck, makes the problem worse by increasing the driving force for convection. Many of the current problems of low-pressure casting systems derive from this source. In contrast, having feeders with hot metal at the top of the casting, and feeding downwards under gravity, is completely stable and predictable, and gives reliable results.

Convection damage and casting section thickness

Damage to the micro- and macro-structure of the casting can occur if the solidification time of the casting is commensurate with the time taken for convection to become established and start extensive remelting. This incubation time appears to be in the region of 2 min. In 3 min or more, convection can cause extensive remelting, the development of flow channels, and the redistribution of heat in castings. Castings which freeze either quickly or slowly are free from convection problems as indicated below.

Thin-section castings can be fed uphill simply because the thin section gives (i) a viscous constraint, reducing flow and (ii) more rapid freezing thus allowing convection less time to develop and wreak damage in the casting. Thus instability is both suppressed and given insufficient time, so that satisfactory castings can be made.

Thick-section castings are also relatively free from convection problems, because the long time available before freezing allows the metal time to convect, reorganizing itself so that the hot metal floats gently into the feeders at the top of the casting, and the cold metal slips to the bottom. Thus, once again, the system reaches a stable condition before any substantial freezing has occurred, and castings are predictable.

The convection problem arises in the wide range of intermediate section castings, such as automotive cylinder heads and wheels, and the larger investment cast turbine blades in nickel-based alloys etc. This is an important class of castings. Convection can explain many of the current problems with difficult and apparently intractable feeding problems with such common products. The convective flow takes about one or two minutes to gather pace and organize itself into rapidly flowing plumes. This is occurring at the same time as the casting is attempting to solidify. Thus channels through the newly solidified material will be remelted, or partially remelted and redistributed, by the action of the convecting streams. These channels will contain a coarse microstructure because of their greatly delayed solidification, and in addition may contain shrinkage porosity if unconnected to feed metal. This situation is likely if the feeders solidify before the channels, as undoubtedly happens on occasions.

For conventional gravity castings which require a lot of feed metal, such as cylinder heads and blocks, and which are bottom gated, but top fed, this will dictate large top feeders, because of their inefficiency as a result of being farthest from the ingates, and so containing cold metal, in contrast to the ingate sections at the base of the casting which will be preheated. This unfavourable temperature regime is of course unstable because of the inverted density gradient in the liquid, and thus leads to convective flow, and consequent poor predictability of the final temperature distribution and effectiveness of feeding.

The upwardly convecting liquid within the flow channels usually has a freezing time close to that of the preheated section beneath which is providing the heat to drive the flow. In the case of many low-pressure die systems, artificially heated metal supply systems lead to a constant heat input, so that the convecting streams never solidify. They therefore empty, or partially empty, when the casting (which should by now have been fully solidified) is lifted from the casting station.

Alternatively, a technology where the mould is inverted after casting effectively converts the preheated bottom ingate system into a top feeding system. Furthermore, of course, the inversion of the system to take the hot metal to the top, and the cold to the bottom, confers stability on the thermal regime. Convection is eliminated. This is a powerful and reliable system used by such operations as Cosworth and an increasing number of others at the present time. It is likely that techniques involving roll-over *immediately after* casting will become the norm for many castings in the future.

Durville-type casting processes (where the roll-over is used *during* casting—actually to effect the filling process) also satisfy the top feeding requirement. However, in practice many geometries are accompanied by waterfall effects, if only by the action of the sliding of the metal in the form of a stable, narrow stream down one side of the mould. Thus meniscus control is, unfortunately, often poor. Where the control of the meniscus can be improved to eliminate surface film problems, the Durville-type techniques would be valuable.

An alternative kind of convection, that driven by density differences due to segregation, may lead to other problems as outlined below.

Rule 8. Segregation

Segregation should be analysed and predicted to be within limits of the specification, or out-of-specification compositional regions should be agreed with the customer. Channel segregation formation should be avoided if possible.

At regions in which the local cooling rate of the casting changes, such as at a change of section or at a chill or at a feeder, it is to be expected that a change in composition of the casting will occur. One of the best understood segregations of this type is inverse segregation, which the author prefers to call simply 'dendritic segregation'. In this case the partitioned solute is segregated preferentially to the face of the mould, especially if this is a chill mould. A similar effect will occur, of course, at the junction with a thinner section which will act as a cooling fin. However, in a complex thermal field, and where the geometry of the casting is requiring a complex distribution of residual liquid to feed shrinkage, these chemical variations can be complex in distribution, and not always easily predicted, except perhaps by a sophisticated computer simulation.

Other segregations are driven by gravity, and account for the concentration of carbon and other light elements in the tops of large ferrous castings, and the concentration of heavy elements such as tungsten and molybdenum at the base of large tool-steel castings. Strong concentrations of segregated solutes and inclusions are found in channel segregates which are once again a feature of large, slowly cooled castings.

When extensive and/or intensive, such changes in composition of the casting may cause the alloy of the casting to be locally out of specification. If this is a serious deviation, the coincidence of local brittleness in a highly stressed region of the casting might threaten the serviceability of the product. The possibility of such regions therefore needs to be assessed prior to casting if possible, and demonstrated to be within acceptable limits in the cast product.

Rule 9. Residual stress

There should be no quenching of light alloy castings into water following solution treatment. (Boiling water is also not permitted, but polymer quenchant or forced air quench may be acceptable if casting stress is shown to be negligible.)

Following high-temperature heat treatment, if the castings are quenched, the cooling of the outer sections of castings may be too fast to allow time for heat to diffuse out from interior sections. In this case the internal sections will cool and contract after the outer parts of the casting have cooled to form a rigid, unyielding frame. Thus the interior sections go into tension. It is a fact, therefore, that quenching into water causes high residual stresses in large and complex castings. The stresses will be tensile in some regions, mainly in the centre of the casting volume, and compressive in others, mainly the outer walls. The use of a boiling water quench has been demonstrated to be of insignificant assistance in reducing the stresses introduced by water quenching. Furthermore, the stresses are not significantly reduced by the subsequent ageing treatment.

Immediately following the quench, the residual stress in aluminium alloy castings solution treated and quenched into water is well above the yield point of the alloy. Even after the strengthening during the ageing treatment, the stress remains usually around $50 \pm 20\%$ of the yield stress. Thus the useful strength of the alloy is reduced from its unstressed state of 100%, down to 70, 50 or even 30%. This massive loss of effective strength makes it inevitable that residual tensile stresses are a significant cause of casting failure in service.

Many national standards for heat treatment specify water quenching. This situation needs to be remedied. In the meantime such national standards are recommended to be avoided.

The reduced mechanical strength when using polymer or forced air quenching is more than compensated by the benefit of increased reliability from putting unstressed castings into service. Thus somewhat reduced mechanical strength requirements should be specified by the casting designer and/or customer. The reductions are expected to be in the range 5–10% for strength and hardness. Although the strength of the *material* will therefore be lowered by the slower quench, the strength of the *casting* (acting as a whole, as a load-bearing component in service) will effectively be increased.

Rule 10. Location points

All castings should be provided with agreed location points for dimensional reference and for pick-up for machining.

Location points have a variety of other names such as tooling points, pick-up points etc. It is proposed that location points describes their function most accurately. It is essential that every casting has defined locations which will be agreed with the machinist and all other parties who require to pick up the casting accurately. Otherwise, it is common for an accurate casting to be picked up by the machinist using what appear to be useful features, but which may be formed by a difficult-to-place core, or a part of the casting which requires extensive dressing by hand. The result is a casting which does not clean up on machining, and is thus, perhaps somewhat unjustly, declared to be dimensionally inaccurate. This rule is designed to ensure that all castings are picked up accurately, so that unnecessary scrap is avoided.

Different arrangements of location points are required for different geometries of casting. Some of the most important types are listed below:

- Six points are required to define the position of a component with orthogonal datum planes which is designed for essentially rectilinear machining, as for an automotive cylinder head. (Any fewer points are insufficient, and any more will ensure that one or more points are potentially in conflict.) Three points define plane A, two more define a second orthogonal plane B, and one more defines a third mutually orthogonal plane C.
- For cylindrical parts to be picked up in a three-jaw chuck, one axis is required to be defined as a datum reference, together with four additional tooling points, three of which define the plane orthogonal to the axis, and the last tooling point defines a clock position with respect to the axis.
- For prismatic shapes, comprising hollow, boxlike parts such as sumps (oil pans), the pick-up may be made by averaging locations defined on opposite internal or external walls. This is a more lengthy and expensive system of location often tackled by a sensitive probe on the machine tool, which then calculates the averaged datum planes of the component, and orients the cutter paths accordingly.

For maximum internal consistency between the tooling points, all should be arranged to be in one half of the mould, usually the fixed or lower half. However it is sometimes convenient and correct to have all tooling points in one half of a core (defined from one half of the core box) if the machining of the part needs to be defined in terms of the core.

Data located at the end of a long component, or at the opposite end to a critically located feature, give rise to casting rejections, or unnecessary manufacturing difficulties. It is helpful if the part can have its data for dimensions arranged to go through or near to the centre of the part. The slight variations in size of the part, from one part to the next, then remain more easily within tolerance, since any variations in length are halved. Alternatively the part can be datumed through the feature which is most critical to have in a fixed location, such as the dip-stick boss in the oil pan casting. The problem of variability of other dimensions is then much reduced.

The tooling points shall be defined on the drawing of the part, and it shall be agreed by the manufacturer of the tooling, the caster and the machinist, that all parties will work from *only* these points when checking dimensions and when picking the part up for machining. The use of the same agreed points by tool maker, caster and machinist leads to an integration of manufacturing between these parties. Disputes about dimensions then rarely occur, or if they do occur they are easily settled. Casting scrap apparently due to dimensioning faults, or faulty pick-up for machining, usually disappears.

Summary

The ten rules above are necessary but not sufficient for the manufacture of reliable castings, and constitute essentially a process specification for the casting manufacture.

Chapter 8

Grain selection in single-crystal superalloy castings

Philip Carter, David Cox, Charles-Andre Gandin and Roger Reed

Introduction

It is well known that investment casting [1] is used routinely nowadays for fabrication of nickel superalloy single-crystal turbine blading of very complex geometries, with intricate channels often incorporated so that cooler air can be forced to flow within and along the blades during operation [2, 3]. Although $\langle 001 \rangle$ is the preferred growth direction, single crystal turbine blades are often employed in a condition that has the centrifugal loading away from this exact orientation, perhaps by up to 15° [4]. Practical considerations and cost implications necessitate that this is the case, and for this reason a trade-off between orientation and performance must always be made. This situation demands that proper attention be paid to the factors influencing the orientation of the blading conferred during casting and the grain selection process in particular. This is one of the aims of this work. A process model is described, which is based upon a thermal analysis of the heat transfer occurring in a commercial casting furnace, and a cellular-automaton model for competitive dendrite growth [5, 6]. For the purposes of model validation, a number of castings have been instrumented and characterized via several analytical techniques. Particular attention is paid to the choice of parameters which are required as input into the model. The model is used to study the geometrical factors influencing competitive growth and the efficiency of a common design of grain selector.

Casting details

A fully instrumented industrial directional casting furnace was used at Rolls-Royce, Bristol, UK [7]. The data reported pertain to the CMSX-4 single

Figure 8.1. Photograph of the wax assembly prior to coating with the ceramic slurry. The running system can be seen clearly.

crystal superalloy. Wax replicas consisting of a feed arrangement, platen, starter, selector and test bar sections were assembled. A single casting consisted of eight such test bars arranged on a circular carousel, as shown in figure 8.1. This was bottom fed via a central riser which was connected to the starter blocks by a series of horizontal runners, as described by Higginbotham [8]. Specifically for the present project, a vacuum port was installed on the ram, beneath the copper chill, which allowed up to ten thermocouples to be placed in the mould. During casting, the temperatures within the copper chill were recorded, as they were at various positions in the starter blocks, as shown in figure 8.2.

The orientation of each test piece was determined by the automatic indexing of back-reflection Laue patterns using the SCORPIO method [7]. Quantitative wavelength dispersive spectroscopy in the electron microprobe analyser was used to characterize the microsegregation present after casting. Further analysis was accomplished using orientation imaging microscopy (OIM); electron back-scattered diffraction patterns (EBSPs) were collected using regular square grids of 10 to 50 µm edge-length, over the whole of the prepared sections. On horizontal sections, typically 3000 patterns were taken, while on vertical sections some 60 000 patterns were collected; these were subsequently indexed and grain boundaries determined using the methods reported by, for example, Adams *et al* [9]. The grain size/density

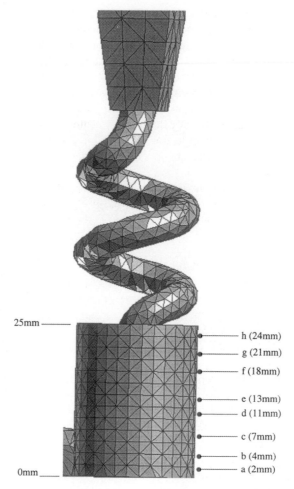

Figure 8.2. Illustration of the positions of the thermocouples, with respect to the copper chill, starter block and grain selector.

relationships determined in this way were compared with those obtained using traditional optical metallography.

Heat transfer in the starter block: modelling and measurements

The soaking and withdrawal processes were modelled with a three-dimensional, finite element analysis using the ProCAST software (UES Inc, Ohio, USA) with temperature/time boundary conditions and temperature dependent materials properties. Mould filling was not considered. The geometries of the components were created individually using purpose-built software tools in the following way. First, the mould was created by extracting the major surfaces of the casting, which then formed the inner surface of the mould. After surface meshing, this was offset by an amount equal to the mould thickness, to obtain a mesh of the outer surface of the mould. These surface meshes were then combined with those of the chill and ram, and a solid tetrahedral mesh created in the enclosed volume. A solid mesh of the furnace was then created by sweeping its two-dimensional axisymmetrical description through an appropriate angle, which in this case was 45°. Finally, the models of the moving assembly and furnace were combined, as shown in figure 8.3.

It was necessary to deduce some of the heat transfer coefficients pertinent to the alloy/chill and water/chill interfaces, and this was done by comparing the measured thermal cycles with those predicted by the model.

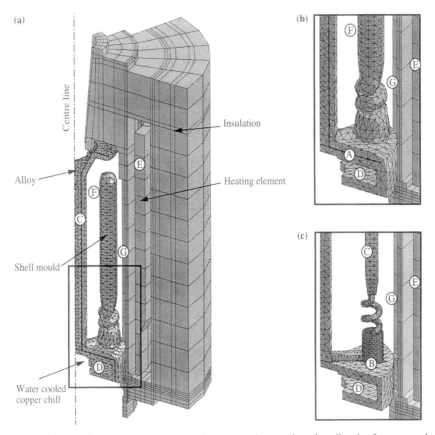

Figure 8.3. (a) Illustration of the finite element meshes used to describe the furnace and casting, and more detailed views in the vicinity of (b) the grain selector and (c) alloy chill.

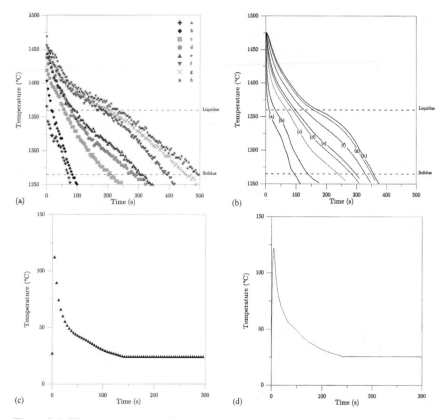

Figure 8.4. The temperature readings measured from the thermocouples after pouring. (a) Measurements from the starter block region (the labels correspond to the positions in figure 8.3), (b) prediction from the process model, (c) measurements from the copper chill and (d) model predictions.

Once this was done the model was able to reproduce the measured thermal cycles with a very reasonable degree of accuracy, as shown by comparing the experimental results and predictions.

Modelling of competitive dendritic growth

Grain structure and texture evolution in the castings were simulated using a three-dimensional version [10, 11] of the stochastic microstructural model detailed by Gandin *et al* [5, 6]. The thermal model described previously was used to calculate the thermal history and the microstructure model was then applied subsequently in a post-processing operation. This way, the Scheil approximation was used by the microstructural model to calculate the evolution of latent heat, via the fraction of solid versus temperature relationship. For use in this mode, a law for the dendrite-tip velocity–undercooling relationship was deduced from consideration of Bobadilla et al's [12] multicomponent extension to the Kurz, Giovanola and Trivedi (KGT) model [13].

Comparison of observed and predicted grain distributions

In figure 8.5, the predicted and measured grain densities are compared as a function of distance from the copper chill. The experimental data are presented in the rawest possible form, i.e. as the mean linear density, N_L (the number of grain boundaries intersected per unit length on a line normal to the axis of the furnace). The analysis shows that the experimental data are reproduced by the model. Moreover, it appears that under the

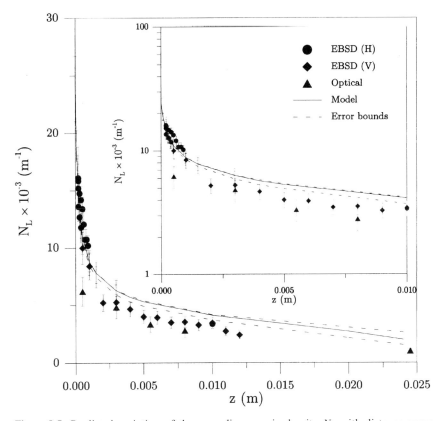

Figure 8.5. Predicted variation of the mean linear grain density N_L with distance away from chill. Also shown are experimental results obtained from traditional optical microscopy and the indexing of electron back-scattered diffraction patterns (EBSPs)

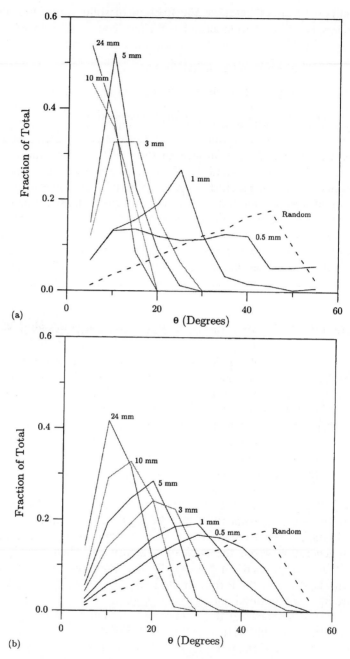

Figure 8.6. (a) Measured and (b) predicted variations of θ, the shortest angular rotation between the casting $\langle 001 \rangle$ and the normal to the chill surface, at several locations in the starter block.

imposed conditions the grain-size distribution is relatively insensitive to uncertainties in the growth kinetics.

Comparison of observed and predicted texture

The measured and predicted orientation distributions are compared in figure 8.6. The comparison is based on the angle θ, defined as the shortest angular rotation between a $\langle 001 \rangle$ direction in the casting and a reference direction; here, this is taken to be the geometrical axis of the casting which is nominally equivalent to the axis of the furnace, or the direction of withdrawal. Also plotted is a theoretically random distribution for the same number of measurements. It can be seen that the majority of primary selection occurs within 5 mm. On this basis it would appear that the length of the starter block could be reduced to approximately 5 mm, particularly as the thermal field of the runner does not seem to influence primary selection. Numerical trials have in fact shown it is possible to use shorter starters; however, higher withdrawal rates are required to prevent remelting of the alloy within the selector, prior to the casting being withdrawn from the furnace.

The efficacy of the selection of primary orientation using a helix

For the fabrication of single-crystal components, commercial practice involves the use of a grain selector to ensure that only a single grain enters the cavity of the casting. Typically, this takes the form of a narrow constriction, a semi-circular arc or alternatively a helix as employed here. The model developed is particularly suited to the simulation of the efficacy of this grain selection process and the geometrical factors which govern whether it is successful or not. It is appropriate to examine whether the results from the model are consistent with experience gained in the foundry.

One way of doing this is to determine the angle between the $\langle 001 \rangle$ crystallographic axis and the long axis of the casting, denoted θ, and to compare this with experimental data. During the numerical computations, the stochastic nature of the simulations was maintained by using a random number generator within the nucleation algorithm. Then, provided that sufficient runs are carried out, statistically meaningful data can be acquired and then displayed in the form of histograms of frequency versus angle θ. In figure 8.7 the measured and predicted θ angles are compared. Since the number of castings fabricated was rather small, data are also reported for a larger dataset which relates to CMSX-4 high-pressure turbine blading produced in the commercial foundry with an identical helical grain selector. The production data span a range of θ values from 2 to 18° and yield an average θ value of approximately 8°. The predicted θ values fall within the range of the production data, and span 11° with a slightly higher average

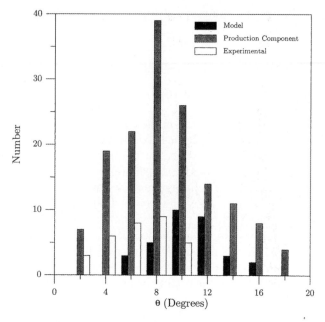

Figure 8.7. Histograms illustrating the variation of the θ angle, defined as the shortest angular rotation between the casting $\langle 001 \rangle$ and the normal to the chill surface. Experimental data correspond to the casting trials reported here, and foundry data corresponding to a CMSX-4 high-pressure turbine blade.

of approximately $10°$. It appears that the experimental trials, with which the thermal model was validated, are consistent with the production data and this gives some confidence that they are representative of foundry practice.

The numerical treatment of microsegregation

The applicability of the Scheil assumption is open to question and, for this reason, further experiments and analyses were carried out. The variation of the fraction solid with temperature, as determined by quantitative differential calorimetry, lies between that expected from the lever and Scheil rules, as shown in figure 8.8; this confirms that back-diffusion during the growth of the solid phase does indeed occur. This conclusion is further supported by the results from a statistical treatment [14] of the microprobe data, as shown in figure 8.9. Consistent with the behaviour of other single-crystal alloys, in CMSX-4 rhenium and tungsten segregate to the dendrite cores and tantalum to the interdendritic regions. A theoretical model based upon the principles described by Maton et al [15] and data from Saunders [16] can be used to

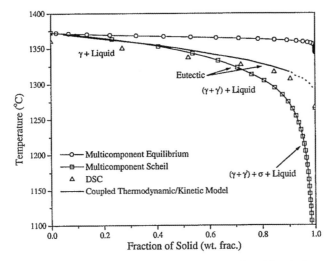

Figure 8.8. Variation of the volume fraction of solid as predicted by the lever and Scheil rules. The experimental data deduced from differential scanning calorimetry are also shown; these can be reproduced with a coupled thermodynamic-kinetic model [15] which accounts for back diffusion in the solid.

explain these results, although at the present time more work needs to be done to simulate in a robust manner the eutectic reactions which occur in the later stages of solidification.

Traditionally, the superalloy community has evaluated the resistance to freckling, which is driven by differences in the density of the solid and liquid regions in the mushy zone, by appealing to empirical expressions of the type

$$\frac{Ta}{W + Re} \quad \text{or} \quad \frac{Ta + 1.5Hf + 0.5Mo - 0.5Ti}{W + 1.2Re}$$

with the exact form varying from user to user. These expressions acknowledge in a qualitative sense the known partitioning behaviour of the elements of interest. It should be noted that a combination of the thermodynamic models and accurate estimates of the densities of the phases represents a rational explanation for these expressions. However, it is emphasized that more work on the prediction of freckling phenomena in single-crystal alloys is required.

Future directions

Process models for the investment casting of nickel-base superalloy blading are now at a very advanced stage. It is possible to take an engineering

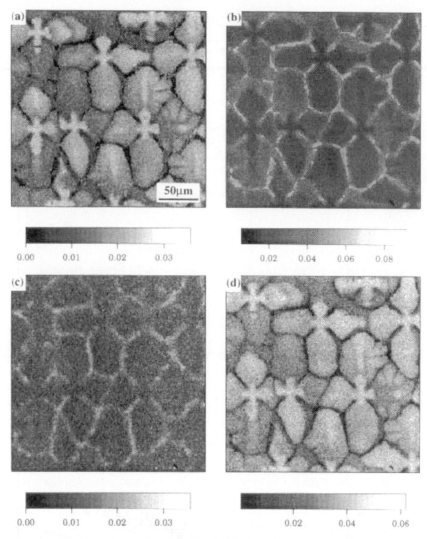

Figure 8.9. Electron microprobe maps illustrating the microsegregation of (a) rhenium, (b) tantalum, (c) titanium and (d) tungsten in CMSX-4.

definition of the surface geometry of the blading—including interior cooling passages—and to generate the meshes necessary to perform very realistic calculations in three dimensions. Simulations of heat flow have been possible for some time but predictions of microstructure evolution are now becoming possible. However, it is emphasized that there is a continuing need for properly targeted processing trials which are aimed at calibrating, verifying and testing the models.

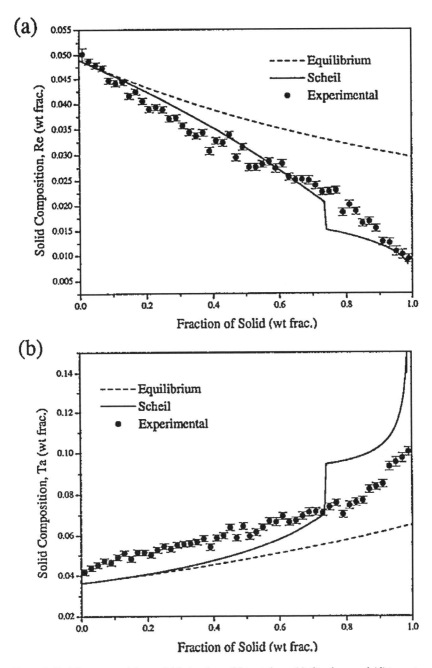

Figure 8.10. The compositions of (a) rhenium, (b) tantalum, (c) titanium and (d) tungsten in the microprobe dataset, after statistical treatment following ref. [14], as a function of fraction solid. Comparison is made with the lever and Scheil rules using data from ref. [16].

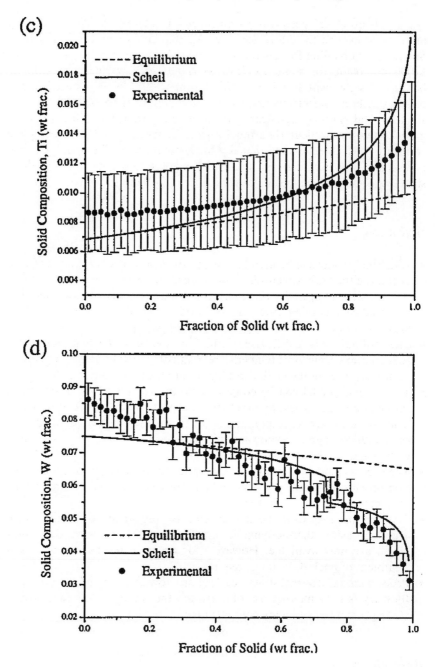

Figure 8.10. *(Continued)*

It is, however, pertinent to consider the factors which are driving industry to expend the effort necessary for this kind of process simulation. Single-crystal blading for gas turbine applications is expensive—a figure of £1000 per blade for some applications is not unreasonable—and this is because a significant fraction of the castings which are produced in the foundry exhibit defects which lead to non-conformance and waste. For single-crystal blading, significant causes of scrap are (i) freckling, (ii) stray grain formation particularly around re-entrant features and (iii) inadequate control of wall thickness, particularly for cooled blades. From an industrial perspective, it is expected that the use of process modelling will contribute significantly to the cost reductions which are required nowadays.

Summary

- A thermal model can be used to describe the heat transfer occurring in the vicinity of the grain selector in a commercial single-crystal casting furnace. It is possible to optimize the interface heat transfer coefficients such that the predicted cooling rates are in good agreement with experimental measurements.
- The cooling rate is of the order of 2 K s^{-1} at the alloy/chill interface, where solidification occurs in a temperature gradient of $20\,000 \text{ K m}^{-1}$. A distance 25 mm away from the chill, these figures are 0.2 K s^{-1} and 2000 K m^{-1}.
- The thermal model can be coupled with an algorithm for the stochastic simulation of grain and texture evolution. Reasonable predictions can be made from a microstructural model for (i) the evolution of grain size, (ii) the evolution of grain texture and (iii) the distribution of casting orientation with respect to the long axis of the casting.
- The process model is suitable for assessing the efficiency of the grain selection process. As such, it is very suitable for assessing and optimizing the geometry of the grain selector.
- The microsegregration behaviour can be predicted to within useful accuracy using thermodynamic models and the assessed databanks which are now available. However, for accurate results the numerical treatment of back-diffusion is required.
- Process models for investment casting are now at a very advanced stage. Industry is now making use of them for the rapid prototyping of new blade geometries and alloy compositions.

References

[1] Beeley P R and Smart R F 1995 *Investment Casting* (London: The Institute of Materials)

[2] Betteridge W and Shaw S W K 1987 *Mater. Sci. Technol.* **3** 682
[3] *The Jet Engine* 1992 fourth edition, The Technical Publications Department, Rolls-Royce plc, Derby, UK
[4] McLean M 1983 *Directionally Solidified Materials for High Temperature Service* (London: The Metals Society)
[5] Gandin C-A, Rappaz M and Tintillier R 1993 *Metall. Trans.* **24A** 467
[6] Gandin C-A, Rappaz M and Tintihier R 1994 *Metall. Trans.* **25A** 629
[7] Goulette M, Spilling P D and Arthey R P 1984 'Cost effective single crystals' in *Superalloys 1984* ed M Gell, C S Kortovich *et al* (The Metallurgical Society of AIME, Warrendale, PN, USA) pp 167–176
[8] Higginbotham G J S 1986 *Mater. Sci. Technol.* **2** 442
[9] Adams B L, Wright S I and Kunze K 1993 *Metall. Trans.* **24A** 819
[10] Gandin C-A and Rappaz M I 1997 *Acta Mater.* **45** 2187
[11] Calcom S A, Parc Scientifique PSE-EPFL, Lausanne, Switzerland
[12] Bobadilla M, Lacaze J and Lesoult G 1988 *J. Crystal Growth* **89** 531
[13] Kurz W, Giovanola B and Trivedi T R 1986 *Acta Metall.* **34** 823
[14] Gungor M N 1989 *Metall. Trans.* **20A** 2529
[15] Matan N, Winand H M A, Carter P, Bogdanoff P D and Reed R C 1998 'A coupled thermodynamic/kinetic model for diffusional processes in superalloys' *Acta Mater.* **46** 4587
[16] Saunders N 1995 *Phil. Trans. R. Soc. Lond.* **A351** 543

Chapter 9

Defects in aluminium shape casting

Peter Lee

Introduction

Application of aluminium castings in the automotive and aerospace industries has risen significantly in recent years. Casting provides reduced costs over processes such as forging because of the reduced number of production steps required. However, the lack of thermomechanical processing means that the defects formed during solidification are present in the final product. These defects limit the mechanical properties of aluminium castings. Numerical simulation of the macroscopic heat and fluid flow during solidification is a commonly used analysis tool which helps the casting designer avoid defects such as hot spots and cold shuts. This chapter both reviews the application of macroscopic modelling for defect prediction and illustrates how these macroscopic models can be combined with mesoscopic models for the prediction of the solidification microstructure. Linkage of the macro- and meso-scales extends our understanding of how solidification conditions affect the final properties of aluminium shape castings.

The automotive and aerospace industries are becoming increasingly competitive. Therefore, they are seeking methods for producing components with better mechanical properties at reduced cost. To achieve this goal, numerical simulations of both the production and operating conditions of these components are used to optimize the parts' service life. For parts produced by shape casting, the simulation of the filling and heat transfer during production has now become a powerful tool for the optimization of casting processes [1, 2]. This is particularly true for the prediction of macroscopic defects in die casting processes where simulation can reduce high tooling development costs. As applications become even more demanding, finer-scale defects must be eliminated, requiring prediction of the microstructural features and their relationship to the mechanical properties. For example, in as-cast structures microporosity can be one of the most detrimental defects formed during solidification and is often the limiting factor controlling fatigue and elongation properties [3].

A number of mesoscopic models are described in this chapter, including simple empirical functions for dendrite arm spacing, deterministic models for grain size, and continuum-stochastic models for the formation of microporosity. This chapter will first review the development of the models for the prediction of defects. The theory behind the macroscopic and mesoscopic models applied in the chapter are then detailed. The application of macroscopic modelling for the prediction of overall heat transfer and filling-related defects is then given, followed by the prediction of the solidification microstructure and defects related to the microstructural development. The macro- and mesoscopic modelling is illustrated by application to the design of an automotive tyre mount hinge component using an A356 type alloy. This component requires reliable mechanical properties for both strength and toughness, together with the cosmetic requirement of very smooth surfaces. Chapter 4 gives an overall discussion of castings in the automotive industry and the importance of modelling. Chapter 7 discusses how to avoid defects during casting.

Background

Many different types of micro/meso-model have been developed for the prediction of solidification structures as reviewed by Rappaz [4] and by Stefanescu [5]. The models have ranged from empirically based functions (e.g. the Niyama criteria function for porosity [6]) to continuum models solving for the structure using a microscopic discretization (e.g. phase field method for predicting dendrite shapes [7]). Some empirical models have limited application because they do not encompass the basic governing physics and hence are not appropriate for extrapolation to processes with significantly different conditions (i.e. the Niyama criterion was developed for steel sand castings and its application to directional solidification of aluminium is not appropriate). Other mesomodels involving a very fine discretization often include all the relevant physics, but are currently too computationally intensive for application to large complex automotive castings.

The microstructural properties required for a hinge part are a fine secondary dendrite arm spacing (DAS), a fine grain size, and very low porosity. A brief overview of the methods for modelling these microstructural features is given below.

Flemings illustrated that the dendrite arm spacing is controlled by coarsening [8]. Therefore, it can be determined by an empirical law for most conditions encountered in various shape casting processes. The prediction of grain size has recently been modelled by deterministic [9], cellular automata (CA) methods [10], and phase field calculations [11]. The deterministic methods benefit from computational simplicity but only track average values. The cellular automata methods implicitly include effects such as impingement and track a population of grains, generating both mean and distributions of morphologies, but these methods are more computationally expensive. The phase field technique allows the most realistic incorporation of thermodynamic and interfacial energy, but is prohibitively expensive computationally when applied to three-dimensional shape cast structures.

The third critical property, percentage and size of porosity, has been the subject of many experimental (reviewed by Tynelius [3]) and modelling studies. Many different models of pore formation and growth have been proposed; these methods are categorized below into five different groups. Each has its own benefits and drawbacks as far as industrial application is concerned.

1. Analytic models.
2. Criterion function methods, based on empirical functions.
3. Numerical solutions of general Stokes flow (Darcy's law), coupled with energy, mass and momentum conservation and continuity equations.
4. Models coupling hydrogen diffusion-controlled growth with the interacting gas–solid system.
5. Models using a stochastic approach to nucleation of pores and grains in combination with continuum solutions for diffusion, taking into consideration the pore–microstructure interactions.

The simulation techniques encapsulated in each of these categories are applicable to the prediction of many different microstructural problems, and are therefore reviewed below focusing on the example of the simulation of microporosity formation to elaborate on their features. The reader is also referred to Raabe [12], who describes the application of the common techniques to materials microstructures and properties in general.

Porosity models

Analytical models range from exact mathematical solutions to approximate asymptotic analytical solutions, all of which share the common feature of simplifying assumptions that make an analytical solution of the resulting equations tenable. Analytical models of microporosity formation are limited in their applicability, and have historically focused on shrinkage-driven pore growth, in the limited case of a constant thermal gradient and solidification velocity, which leads to a constant liquid feeding velocity. With these assumptions the flow between the dendrites can be considered to be analogous to flow down a bundle of pipes. The latter is governed by the Hagen–Poiseuille equation if the flow is laminar. This allows the pressure to be calculated in the mushy zone and if a correlation between pressure and percentage porosity is assumed, the latter can be calculated. Walther *et al* [9] first used this technique to model the formation of a single centreline pore in a long tube solidifying inwards in the radial direction. Assuming a pure metal and low solidification velocity, the interface will be planar, giving an ever narrowing tube to feed metal down. If all the shrinkage associated with this rate of change of radius is fed, the pressure drop ΔP is

$$\Delta P = \frac{\rho L}{g} \frac{64 \beta c^4 L^2}{r^4} \left(\frac{1}{2} + \frac{\beta f L}{3r} \right), \qquad (9.1)$$

where L is the length of the casting, r is the radius of the liquid central cylinder, c is a constant, β is the thermal coefficient of expansion, ρ is the density, and f is the friction factor. Making two additional arbitrary assumptions—a pore will form at $\Delta P = 1$ atm; and once formed, the pore will occupy all the remaining space previously occupied by molten metal—they achieve results that give reasonable agreement with the centreline pores formed in metal solidified in Pyrex tubes. However, these types of analytic solution have not been successfully applied to the prediction of microporosity formation in castings.

Criterion functions

Criterion functions are simple rules that relate the local conditions (e.g. cooling rate, solidification velocity, thermal gradient, etc.) to the propensity to form porosity. Some criterion functions are based solely upon experimental observations, whilst others are based upon the physics of one of the driving forces (as reviewed by Murali and Sharma [13]). In the latter case, the physical basis gives the form of the equation, whilst the coefficients are experimentally fitted. One of the best-known criterion functions was proposed by Niyama [6] in 1982. Niyama used Darcy's law in cylindrical coordinates to obtain an empirical function to predict difficult-to-feed regions in castings. He proposed that there would be a larger ratio of the thermal gradient, G, to the cooling rate, R, in the regions where shrinkage porosity is most likely to form:

$$\frac{G}{\sqrt{R}} > \text{constant}, \qquad (9.2)$$

where the constant is dependent upon the alloy being cast. Niyama illustrated that his criterion for predicting porosity works reasonably well in cylindrical castings of ferrous alloys. Since its original development, the Niyama criterion has been applied to many different alloys, including aluminium alloys, but often without success [14].

Darcy law models

The third category of modelling approach, continuum numerical solutions of general Stokes flow, are based on the assumption that the casting metal is a

Newtonian fluid and flow through the inter-dendritic region is creeping (i.e. Stokes' flow). With this assumption, Darcy's law can be used to evaluate the pressure drop resulting from restricted feeding of the volumetric shrinkage. Darcy's law is given by [15]

$$\bar{v} = -\frac{K}{\mu}(\nabla P - \rho g), \qquad (9.3)$$

where K is the permeability of the porous medium, μ is the shear viscosity and $\bar{v}$ is the superficial velocity (average velocity over both liquid and solid).

Kubo and Pehlke [16] were the first to publish a comprehensive model coupling Darcy's law to the equations of continuity to calculate the restriction to fluid flow in the mushy zone. They used the continuity equation to relate the reduction in volume due to shrinkage to the liquid feeding and pore growth:

$$\left(\frac{\rho_s}{\rho_l} - 1\right)\frac{\partial f_l}{\partial t} - \frac{\partial f_l v_i}{\partial x_i} + \frac{\partial f_p}{\partial t} = 0, \qquad (9.4)$$

where f_l and f_p are the fraction liquid and porosity, respectively, and v_i is the velocity in the ith direction. By combining equations (9.3) and (9.4), they constructed a relationship between the fraction porosity and the pressure. If no porosity formed, they solved for the pressure and then checked whether sufficient supersaturation had occurred for nucleation. This was done by relating the pressure of the gas inside the pore, P_g, the pressure in the metal, P_m, and the liquid-gas interfacial energy, γ_{lg}, with the equation

$$P_g = P_m + \frac{2\gamma_{lg}}{r}, \qquad (9.5)$$

and a conservation equation for hydrogen, which assumes hydrogen partitions according to the lever rule. The additional term uses the ideal gas law (α is the gas constant) to relate the mass concentration of hydrogen in the pores to the volume porosity:

$$C_H^0 = (1 - f_l)C_s + f_l C_l + \alpha \frac{P_g C_v}{T}, \qquad (9.6)$$

where C_H^0 is the initial hydrogen concentration and the subscripts s, l and v indicate the solid, liquid and vapour concentrations, respectively.

Kubo and Pehlke first calculated the change in fraction liquid, then, using an explicit finite difference approximation, they solved equations (9.3) and (9.4) to determine the local metal pressure. Next they used equation (9.5) to determine the gas pressure. To do this they assumed that the pore will nucleate on the solid–liquid interface of a dendrite, and that the pore diameter will equal the secondary dendrite arm spacing.

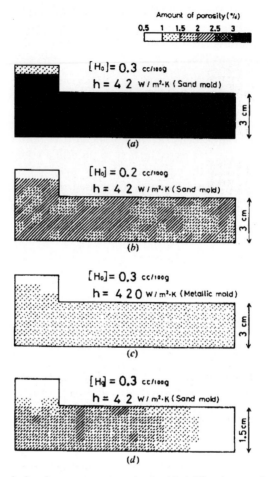

Figure 9.1. The calculated percentage porosity in Al–4.5Cu plate castings for different initial hydrogen concentrations, heat transfer coefficients, and plate thicknesses [16].

Kubo and Pehlke's results are shown in figures 9.1 and 9.2. They claimed good qualitative agreement in percentage porosity predictions with earlier experiments where they unfortunately did not measure the hydrogen content in the melt. The correlation between experimental and predicted pore size was not as good as for percentage porosity; perhaps because they set the pore size to be a function of dendrite arm spacing, and did not include any competitive nucleation and growth laws.

Another interesting feature of Kubo and Pehlke's results is that the majority of pore growth occurs during the last 20% of the solidification. This may be seen in figure 9.3 where both the pressure in the liquid metal and the amount of porosity are plotted as a function of fraction solid. The

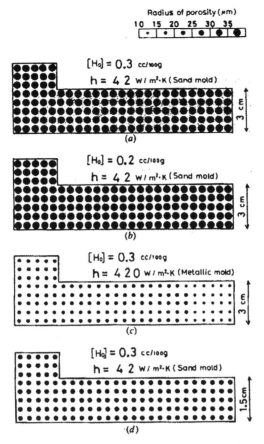

Figure 9.2. The calculated radius of porosity in Al–4.5Cu plate castings for different initial hydrogen concentrations, heat transfer coefficients, and plate thicknesses [16].

reason for this result is due to their description of the permeability:

$$K = \frac{f_l^3 \lambda_2^2}{80(1-f_l)^2}. \tag{9.7}$$

This equation produces a very small permeability and hence a high pressure drop as the fraction liquid approaches zero. In order to avoid a singularity at $f_l = 0$, they set f_l to have a minimum value of 0.01. Using their data for the Al–4.5Cu system, the fraction eutectic from the Scheil equation is approximately 9%. At this stage, a determination of the permeability becomes very difficult, and the eutectic may grow as a planar front between the dendrites unless there are impurities present [22]. *In situ* experimental

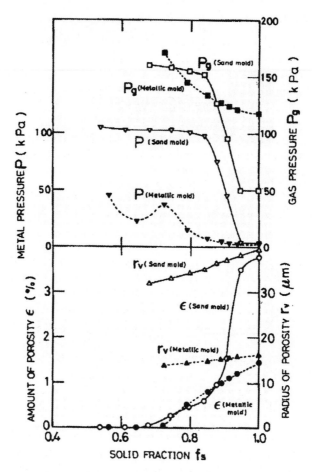

Figure 9.3. The calculated metal pressure and volume of porosity as a function of the fraction solid in Al–4.5Cu plate castings with an initial hydrogen concentration of 0.3 ml STP/100 g. The heat transfer coefficients used are 42 and 420 W m^{-2} K^{-1} for the sand and metallic mould respectively [16].

observations of the formation of porosity in the Al–10Cu system showed that pores form and grow at much lower fraction solids in many cases [17], which their model would not be able to predict. Many other authors have also developed similar models, such as Poirier *et al* [18] and Zhu and Ohnaka [19], with similar reasonable predictions for percentage porosity but not pore size. In summary, models based on the Darcy law predict the experimentally observed trends in final percentage porosity, illustrating that shrinkage is an important driving force, but the lack of correlation in predicted pore size suggests it is not the whole picture.

Hydrogen diffusion controlled and stochastic models

The final two categories of micro-porosity modelling approaches both solve for the nucleation and growth of pores controlled by hydrogen diffusion, sometimes combined with shrinkage driven growth. Some of these models combine the continuum solutions for hydrogen diffusion with a stochastic approach to nucleation of pores and grains, taking into consideration the pore–microstructure interactions [20, 21]. This chapter focuses upon the use of these types of model, therefore they are presented in the theory section which follows, together with a model for the prediction of grain nucleation and growth.

Macromodelling approach

Most of the commercially available macroscopic modelling packages consider only heat transfer and fluid flow and simulate the filling of the mould cavity by the molten metal and its subsequent solidification. The packages use either the finite element method (FEM) or the finite difference (or control volume) method (FDM) to discretize the model system. The macroscopic solidification and thermal analyses performed for this paper used CAP [22], a finite element method heat transfer solver. CAP models heat conduction, taking into account latent heat of fusion by solving the Fourier equation:

$$\rho C_p \left(\frac{\partial T}{\partial t} \right) = \left(\frac{\partial}{\partial x_i} \right) \left(k \frac{\partial T}{\partial x_i} \right) + \rho L \left(\frac{\partial f_s}{\partial t} \right), \quad (9.8)$$

where T is the temperature, ρ is the density, C_p is the specific heat, t is the time, L is the latent heat of fusion and f_s is the fraction solid. A time integration scheme which incorporates a fully implicit method at an early calculation stage is used, followed by an unconditionally stable integration method. The filling was simulated using the complementary finite element method code, WRAFTS [22], which solves the Navier–Stokes equations tracking the free surface.

Mesomodelling approach

A description of the deterministic model used to predict the grain size is given below, followed by a description of the continuum-stochastic model for porosity formation. These two models, together with an empirically fitted coarsening law for dendrite arm spacing (detailed first), provide the three required structure predictions (grain size, porosity and dendrite arm spacing), which were incorporated into the multi-scale model by using functions generated from the fundamental physics of the processes involved.

Dendrite arm spacing

As given by Flemings, the dendrite arm spacing DAS was found to be proportional to the local solidification time, t_s, and for an A356 alloy:

$$\text{DAS} \approx 7.5 t_s^{0.39}, \qquad (9.9)$$

where dendrite arm spacing is the secondary dendrite arm spacing in microns and t_s is in seconds.

Deterministic model for grain nucleation and growth

The transient grain evolution was calculated using a time step procedure similar to that of Rappaz *et al* [23]. The pertinent assumptions are as illustrated in figure 9.4: (i) no back-diffusion occurs within the solid; (ii) complete mixing takes place in the liquid within the dendritic envelope; (iii) a solutal diffusion layer of thickness δ exists in the liquid outside the envelope; and (iv) diffusion across this layer can be approximated as steady state.

The code treats the grain growth in four stages as shown schematically in figure 9.5:

(i) Initial spherical growth where the internal solid fraction is 1.
(ii) Boundary layer diffusion controlled growth before solute layer impingement.
(iii) Dendritic growth with solute layer impingement.
(iv) Grain impingement where the final solid is simulated using the Scheil equation.

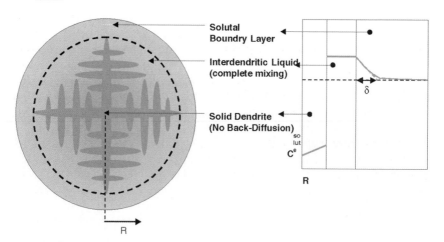

Figure 9.4. Schematic of a growing dendrite illustrating the assumptions used in the model for dendritic grain evolution.

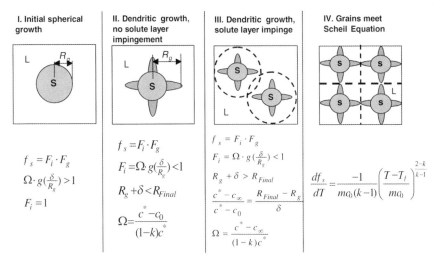

Figure 9.5. The four stages of growth implemented in the dendritic grain evolution model. f_s is the fraction solid, F_i the fraction of the grain envelope, F_g the internal solid fraction within the grain envelope, Ω the supersaturation, R_g the grain radius, R_{final} the final grain size, C^* the solute concentration in the liquid at the solid/liquid interface, the ambient solute level, C_0 the initial solute level, k the partition coefficient, m the liquidus slope and g the non-spherical correction factor.

In stages (i)–(iii), the fraction solid, f_s, is equal to the fraction of a spherical grain envelope, F_i, multiplied by the internal solid fraction within the grain envelope, F_g:

$$f_s = F_i \cdot F_g. \qquad (9.10)$$

Initially in stage (i), the dendrite grows as a solid sphere and thus the internal solid fraction has a value of one. As the term Ωg falls below one, the growth is assumed to be non-spherical. When the solutal boundary layers meet, the ambient solute level increases and the growth slows down (stage (iii)). In stage (iv), the grain envelope stays constant whilst the interdendritic-liquid regions solidify according to the Scheil equation.

Nucleation was calculated using a model of Charbon and Rappaz [10] where the number of nuclei, n, formed at a particular undercooling, ΔT, is given by

$$n(\Delta T) = d_{\max} \frac{1}{\sigma\sqrt{2\pi}} \int_{T_l}^{T_l - \Delta T} \exp\left[-\frac{1}{2}\left(\frac{T - T_n}{\sigma}\right)^2\right] d(T), \qquad (9.11)$$

where $d_{\max}$ is the maximum grain density for large undercooling, T_n is the Gaussian centre of the distribution function, and σ is the deviation.

Continuum-stochastic model

There are three primary causes of porosity in shape castings: entrapped insoluble gas bubbles, shrinkage porosity and gas porosity. Shrinkage porosity occurs because of an increase in the density of the metal as it changes phase from liquid to solid. Gas porosity occurs when dissolved gases within the liquid metal come out of solution during solidification due to a reduced solubility in the solid, and form bubbles. Shrinkage and dissolved gas often interact to provide a combined driving force for the formation of porosity. Both the shrinkage and soluble gas pressure are determined by the evolution of the developing solid. Therefore, a model of the growth of the equiaxed grains or columnar dendrites must be incorporated. For equiaxed grains, the deterministic model can be used (as described above), or a cellular automaton model can be applied (as described by Lee et al [21]). For columnar grains, the model of dendritic growth by Hunt and Lu [24] was used. These models of the developing solid provide both the fraction solid as a function of temperature and the characteristic spacing of the liquid regions in which the pores can grow. This characteristic spacing is also used to give the radius of growth of the pore, r, which is used in equation (9.5) to determine the pore pressure.

Using the macromodel to solve for the heat transfer and the grain growth model to provide the fraction solid and curvature terms, the porosity model must solve for the nucleation of the pores followed by pore growth dominated by hydrogen diffusion or the shrinkage pressure. For this paper, it was assumed that the major driving force was the diffusion of hydrogen. Shrinkage was only incorporated by altering the P_m term in equation (9.5). Developments to remove this limitation are currently being investigated by coupling the CA-based porosity model [21] directly into the macromodel. The diffusion of hydrogen was solved on a continuum level and pore nucleation was modelled stochastically using an experimentally determined distribution of nucleation supersaturations [25]. A random number generator was used to determine the nucleation potentials within the given distributions and to determine the location of each pore. Once a pore nucleated, it was coupled into the continuum model, acting as a sink or source of hydrogen.

The diffusion of hydrogen is governed by [26]

$$\frac{\partial}{\partial t}(\rho CH) + \nabla j = R_H \qquad (9.12)$$

where C_H is the concentration of hydrogen, R_H is the generation or consumption of hydrogen per unit volume and j is the diffusion or Fickian flux, given by

$$j = -\rho D \nabla C_H \qquad (9.13)$$

where D is the diffusivity of hydrogen in aluminium. Dropping the subscript H and replacing it with an s or l to denote whether C represents the hydrogen concentration in the solid or liquid respectively, equations (9.12) and (9.13) can be combined to obtain

$$\frac{\partial}{\partial t}(\rho_l C_l f_l + \rho_s C_s f_s) = \nabla(\rho_l D_e \nabla C_l) + R_H, \qquad (9.14)$$

where the subscript s or l denotes a value in the solid or liquid phase respectively and D_e is the effective diffusivity calculated using Markworth's suggested improvement on the law of mixtures [27].

$$D_e = D_l \left(\frac{D_l(1-f_s) + D_s(1+f_s)}{D_l(1+f_s) + D_s(1-f_s)} \right). \qquad (9.15)$$

Assuming that on a local (or microstructural) scale the hydrogen concentration in the solid and liquid within a unit volume can be related by the equilibrium partition coefficient for hydrogen, $C_s = kC_l$, C_l can be determined using

$$\frac{\partial}{\partial t}[C_l(\rho_l f_l + k\rho_s f_s)] = \nabla(\rho_l D_e \nabla C_l) + R_H. \qquad (9.16)$$

Equation (9.16) was solved using an explicit finite difference algorithm to capture pore nucleation with a minimal over-shoot of the nucleation supersaturation. At each time step T was determined first, followed by the calculation of the values for all material properties. Using these updated values, equation (9.16) was solved using central differences, determining C_H at each node. The pore nucleation potential and the growth of existing pores were subsequently calculated.

The deterministic grain model combined with the continuum-stochastic model for porosity were used to solve for the solidification structures over the expected parameter space encountered in real castings, and then these results were approximated by analytic equations using multilinear regression analysis. The statistically derived equations were fitted for a range of parameters of interest for both the mean values and extremes, for example, maximum pore size at three standard deviations from the mean value. These functions were then used in a decoupled manner to give an estimate of the properties by applying the fitted functions in a post-processing step.

Casting process selection

The three processes considered for the production of the automotive hinge part were sand casting, permanent mould, and squeeze casting. Starting from the computer aided design (CAD) description, all three options were meshed, including the surrounding sand mould or steel die (with appropriate

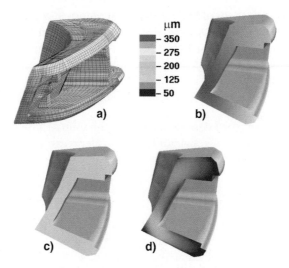

Figure 9.6. Predicted grain size shown on (a) the same cross-section of the hinge; for three different casting processes: (b) green sand casting, (c) permanent mould casting and (d) squeeze casting.

water cooling). In order to compare the structures produced, only the solidification of the part was simulated. Using the macroscopic results the grain size was calculated using the deterministic model, and the predictions are compared graphically in figure 9.6. The predicted percentage porosity is shown graphically in figure 9.7. Average values for both grain size and percentage porosity are listed in table 9.1. In addition to predicting the mean value, the mesomodel for porosity tracks a population of pores allowing a function to be fitted characterizing the distribution. Assuming a normal distribution (as experimentally supported by the work of Lee and Hunt [17]), a function was fitted to three times the standard deviation in the average pore length, generating a prediction for a 3σ maximum pore

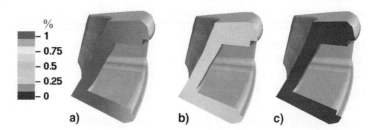

Figure 9.7. Predicted percentage porosity shown on the same section of the hinge for three different casting processes: (a) green sand casting, (b) permanent mould casting and (c) squeeze casting.

Table 9.1. Predicted values for DAS, grain size, percentage porosity and 3σ pore length averaged over the entire part. The $\pm$ values are a standard deviation in the mean over all nodes in the part.

Process	DAS (μm)	Grain size (μm)	Percentage porosity (%)	3σ pore length (μm)
Sand	51 ± 6	339 ± 46	1.0 ± 0.1	383 ± 37
Perm. mould	36 ± 4	198 ± 31	0.5 ± 0.1	276 ± 27
Squeeze	15 ± 5	76 ± 28	<0.01	5 ± 1

length (see table 9.1). The prediction of a maximum pore length is important because failure due to fatigue has been related to the largest pore size, rather than the average.

Sand casting can be eliminated as a viable process for this part because of both poor microstructural properties and surface finish. Permanent mould casting may be acceptable with respect to dendrite arm spacing and grain size, but the 3σ pore length is large. The predictions show that squeeze casting provides a fast solidification rate that produces better microstructural properties (see table 9.1), and the high pressure aids in reducing porosity that could form in the thicker sections of the casting. Therefore, in order to achieve a high quality casting for this application, squeeze casting is the preferred method of production.

Squeeze casting design optimization

Once the optimal process has been selected, the macro- and meso–macromodels can be applied to optimize the design of the dies, cooling passages, and filling speed for the squeeze casting. The macromodel, which consists of approximately 144 000 elements and 143 899 nodes, includes the aluminium casting, steel die retainers, steel die inserts and slides with respective water cooling lines, shot tip, and an outside air boundary condition. The full process cycle was modelled: (i) a dwell period (metal solidifying in the die); (ii) an open die period (exposing the die to air); (iii) a spraying period (chill coating of the die); (iv) another open die period; and (v) a closed die period (clamping dies together) prior to injecting metal again. Due to the typical warming of the dies during the first few cycles of production, 30 cycles were analysed to ensure that steady-state conditions were reached before analysing a final solidification. Water and air were modelled as boundary conditions at 30 and 50°C respectively. The material properties, initial temperatures, and heat transfer coefficients used in the analyses are listed in tables 9.2 and 9.3.

Table 9.2. Material properties and initial temperatures.

Material	ρC_p (MJ m^{-3} K^{-1})	k (W m^{-1} K^{-1})	Initial temperature (°C)
Al	2.71	100.0	700.0
Steel	3.72	25.9	250.0

ρ is the density, C_p is the specific heat, k is the thermal conductivity, T is the temperature.

Two thermal simulations will be discussed to illustrate the design optimization. These are analysis 1 which does not include water cooling, and analysis 2, which includes water cooling. Due to the comparison of these analyses, water cooling is now used in the current industrial implementation. In analysis 2 the casting (excluding runner and sprue) completely solidifies in 13.5 s, whereas in analysis 1 the casting does not solidify over 19 s. With a fast solidification rate any pores that form will remain small and scattered and are more likely to be fed by high pressure in the cavity. As shown in figure 9.8, the propensity for a hot spot forming is also reduced in analysis 2.

The filling of the cavity is simulated by varying the shot sleeve velocity profile of the metal at the gate. Figure 9.9 illustrates two fluid flow analyses. The first features a profile with a 50% reduction in the initial speed, while the second reduces the initial speed by 75%. The faster shot sleeve velocity promotes filling of the rib section prematurely and causes initial turbulence and possible oxide entrainment when the fluid fronts meet at the top of the casting. The slower shot sleeve velocity provides a quieter, directional filling from the bottom to the top of the casting and adds only 0.5 s to the total filling time.

Having selected squeeze casting as the best process, and determining an optimal design, predictions of the final properties can be made using a complete filling and cyclical analysis followed by application of the deterministic and continuum-stochastic mesomodels described earlier. The component requirements specified a fine dendrite arm spacing and grain size, along with minimal porosity. Figure 9.10 shows the predicted solidification structures for the optimized, water cooled, squeeze casting design. A

Table 9.3. Interface heat transfer coefficients.

Interface materials	Heat transfer coefficient (W m^{-2} K^{-1})
Aluminium to steel	16 000
Steel to steel (clamped)	1000
Steel to steel (retainer/insert)	5000
Steel to water cooling	6000
Steel to air	50

Squeeze casting design optimization 137

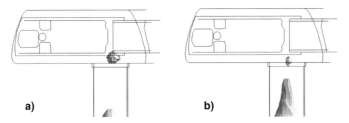

Figure 9.8. Iso-surfaces of solidification time, t_s, illustrating a potential hot-spot near the gating: (a) analysis 1 (no cooling, iso $t_s = 18$ s), (b) analysis 2 (cooling, iso $t_s = 13$ s).

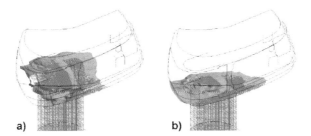

Figure 9.9. Filling pattern for (a) analysis 1 (fast) at 0.35 s and (b) analysis 2 (slow) at

very fine dendrite arm spacing (figure 9.10(a)) and grain size (figure 9.10(b)) are predicted, and almost no porosity is anticipated (figures 9.10(c) and 9.10(d)). This microstructural state is expected to give excellent mechanical properties. If the casting is not well fed, the real values for porosity will be larger than those predicted.

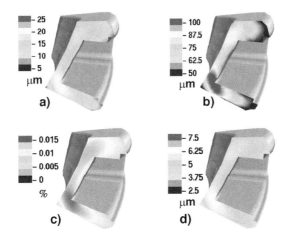

Figure 9.10. Squeeze casting microstructural and defect predictions: (a) secondary dendrite arm spacing, (b) grain size, (c) percentage porosity and (d) pore length.

Summary

Simulation of the shape casting process can be used to determine defects at both a macroscopic and mesoscopic level. Macroscopic models of the molten flow into the mould cavity allow the propensity for oxide formation to be evaluated, whilst macroscopic heat transfer calculations allow defects such as hot spots to be predicted. The use of empirical, deterministic and continuum-stochastic mesoscopic models for the prediction of the solidification microstructures allows defects related to the microstructural development to be determined.

The coupling of the length scales between mesoscopic and macroscopic models is costly, therefore the use of empirical and mesomodel derived functions for the prediction of the casting solidification structures are best incorporated into an efficient finite element method macromodel and applied to both casting process selection and optimization. Deriving constitutive equations from the mesomodels allows the fundamental physics underlying the development of the solidification structures to be incorporated, while negligibly affecting the computational times. This allows an efficient prediction of the process/structure/property relationships during the casting of aluminium alloys.

The future

With the advent of ever more powerful computers, the complexity of the physics encapsulated in the models used to predict solidification defects will continue to increase. As an example, the fast deterministic model for the prediction of grain size presented in this paper could be replaced by a much more complex phase field or cellular automaton simulation of the grain growth. The level of resolution of both the grain structure and the pore structure that is possible with such techniques is illustrated by the preliminary results of a combined CA–continuum model shown in figure 9.11. The results shown are for a few cubic millimetres of material, yet the computation took about an hour on an SGI Origin2000. Running this model for a full casting, such as the hinge, would currently take months, but may be much faster within the next ten years. The benefits of this type of modelling are the prediction of both the average values of the microstructural features and their distributions. Like actual materials processes, they have a stochastic element, and it is the reduction in the variation which is often sought. These models may be coupled to a finite element method model at the continuum scale, and also use sub-models from smaller scales to incorporate calculations of atomistic scale phenomena. The coupling of complex mesoscopic models into continuum models has already been shown to be practical by Rappaz *et al* [28], whilst Grimes *et al* [29] have

The future 139

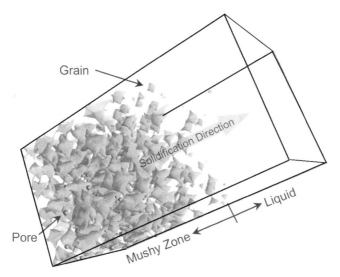

Figure 9.11. Three-dimensional visualization of a model that simulates both the grain structure and pore growth. The grains are the shaded structures with fully molten Al–7Si alloy to the right of them; the pores are rendered as spheres. (The bounding box shown is 1.25 mm × 1.25 mm × 2.50 mm.)

shown the feasibility of deriving the rules used in CA simulations from atomistic models. The range of modelling techniques used for the different length scales is illustrated in figure 9.12. Currently, models are being combined on neighbouring scales. However, it is the coupling of models

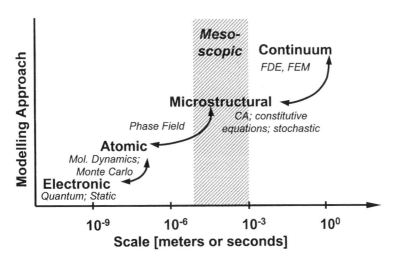

Figure 9.12. The different spatial and temporal scales of simulation techniques applied to materials modelling. (Based on private communication with D G Pettifor, 1995.)

across increasingly greater length and temporal scales that will give the greatest new insight during the next decade.

References

[1] Major J F, Makinde A, Lee P D, Chamberlain B, Scappaticci T and Richman D 1994 'Vehicle suspension system advancements' SAE International Congress and Exposition, Detroit, Michigan, 28 February–3 March, Paper No 940874

[2] Lee P D, Major J F and Hofmann L May 1991 *Numerical Simulation of Casting Solidification in Automotive Applications* ed Chongmin Kim and Chung-Whee Kim (TMS) pp 231–251

[3] Major J F, Makinde A, Lee P D, Chamberlain B, Scappaticci T and Richman D 'Vehicle Suspension System advancements' SAE International Congress and Exposition, Detroit, Michigan, 28 February–3 March, Paper No 940874 Tynelius, Major J F and Apelian D 1993 *AFS Trans.* **166** 410–413

[4] Rappaz M 1989 *Int. Mater. Rev.* **34**(3) 93–123

[5] Stefanescu D M 1995 *ISIJ* **35**(6) 637–650

[6] Niyama E, Uchida T, Anzai K and Saito S 1982 *AFS Int. Cast. Metals. J.* 52–63

[7] Boettinger W J and Warren J A 1996 *Metall. Trans.* **A27**(3) 657–669

[8] Flemings M C 1974 *Solidification Processing* (New York: McGraw-Hill)

[9] Thevoz P, Desbiolles J L and Rappaz M 1989 *Metall. Trans.* **A20**(2) 311–322

[10] Charbon Ch and Rappaz M 1993 *Mod. Simul. Mater. Sci. Eng.* **1** 455–466

[11] Boettinger W J, Coriell S R, Greer A L, Karma A, Kurz W, Rappaz M and Trivedi R 2000 *Acta Mater.* **48**(1) 43–70

[12] Raabe D 1998 *Computational Materials Science: The Simulation of Materials Microstructures and Properties* (Weinheim: Wiley–VCH)

[13] Murali A V and Sharma G R 1987 *Trans. Indian Inst. Met.* **40**(2) 101–108

[14] Spittle J A, Almeshhedani M and Brown S G R 1994 *Cast Met.* **7** 51

[15] Darcy H 1856 *Les Fontaines Publiques de la Ville de Dijon* (Paris: Delmont)

[16] Kubo K and Pelke R D 1985 *Metall. Trans.* **16B** 359–366

[17] Lee P D and Hunt J D 1997 *Acta Mater.* **45**(10) 4155–4169

[18] Poirier D R, Yeum K and Maples A L 1987 *Metall. Trans.* **18A** 1979

[19] Zhu J D and Ohnaka I 1991 *Modeling of Welding and Advanced Sol. Proceedings V 1991* ed Rappaz *et al* (TMS) p 435

[20] Lee P D and Hunt J D 1995 *Modeling of Casting, Welding and Advanced Sol. Proceedings VII* ed M Cross and J Campbell (TMS) pp 585–592

[21] Lee P D, See D and Atwood R C 1999 *Cutting Edge of Computer Simulation of Solidification and Casting, Osaka, Japan, 14–16 November* ed I Ohnaka and H Yasuda (ISIJ) pp 97–111

[22] CAP and WRAFTS, registered by EKK Inc, 2065 West Maple, Walled Lake, MI 48390, USA

[23] Rappaz M and Thevoz Ph 1987 *Acta Metall.* **35** 1487

[24] Hunt J D and Lu S-Z 1996 *Metall. Trans.* **27A** 611–623

[25] Atwood R C, Sridhar S and Lee P D 1999 *Scr. Mater.* **41** 1255

[26] Bird R B, Stewart W E and Lightfoot E N 1960 *Transport Phenomena* (New York: Wiley)

[27] Markworth A J 1993 *J. Mater. Sci. Lett.* **12** 1487–1489
[28] Rappaz M, Gandin C-A, Desbiolles J L and Thevoz P H 1996 *Metall. Trans.* **27A**(3) 695–705
[29] Grimes R W, Atkinson K J W, Zacate M O, Lee P D, Jackson A G and LeClair S R 1999 *JOM* **51**(8) 8

Chapter 10

Pattern formation during solidification

John Hunt

Introduction

Regular patterns form in many solidification processes. Examples occur during the growth of lamellar and rodlike eutectics, the growth of dendrites and when cells are formed in faceted and non-faceted materials. The scale and regularity of the microstructure can determine the properties of the cast materials and thus understanding the processes is important practically. The present chapter points out the common features that occur in many pattern forming processes.

Analysis of steady-state growth indicates that a wide range of possible spacings could occur during eutectic, cellular or dendritic growth. The degree of freedom can be removed by considering the mechanism determining the minimum and maximum spacing on a specimen. It is found that the minimum spacing occurs when the array first becomes stable for a lamellar or rodlike eutectic, for cell growth and for some dendrites. The maximum spacing for eutectics and for cells is determined by tip splitting. The maximum spacing for dendrites occurs when a tertiary arm becomes a new primary. Good agreement is obtained between theory and experiment using this approach to predict spacing limits. The average spacing on a specimen can approach either limit depending on past history. The two extreme spacings are found to span the spacing of the minimum undercooling for eutectic and cellular growth and this allows an average spacing to be estimated using a single condition.

It is concluded that three conditions are necessary to form regular structures. A mechanism must exist to eliminate members of the array when the spacing is too small. A mechanism must exist to form new members of the array when the spacing is too wide. The structure must be stable to fluctuations in the range of spacing between the two limits.

Background

Regular patterns are formed throughout nature. In the physical sciences the details of pattern formation are not well understood. Regular structures form in a number of solidification processes. Lamellar or rodlike structures can be produced in eutectics. Regular cell or dendrite arrays form during single-phase growth. In directionally grown peritectics, the pattern and scale of the microstructure is usually determined by the cells or dendrites of the high-temperature phase. In each of these examples the scale of the microstructure can control later reactions and the eventual properties of the material. This chapter concentrates on pattern formation during dendrite, cellular and eutectic solidification. The more complex and less well understood peritectic solidification process is discussed in chapter 11.

The first observation to notice is that solidification structures are never perfectly regular. On a directionally grown lamellar eutectic, a small range of spacings will always be present. Regular patterns could not be produced if only one spacing were permitted. Any slight change in growth conditions would need an instantaneous change in spacing over the whole specimen.

Lamellar or rodlike eutectics

As a lamellar eutectic grows, different amounts of solute are deposited in the two solid phases. At steady state, concentration gradients build up to transport the solute across the lamella. The concentration gradient in the liquid near the solid means that if the effect of surface energy is neglected the interface could not be at local equilibrium at more than one point. By considering the effect of curvature on the melting temperature and adjusting the interface shape, it is possible to satisfy local equilibrium at all points on the solid–liquid interface. Figure 10.1 shows experimental and calculated interface shapes for a lamellar eutectic; note the small radii of curvature near the alpha-beta liquid groove which is needed to compensate for the near eutectic composition in these regions.

An approximate solution to the lamellar eutectic problem can be obtained without difficulty. The earliest models by Zener [1] or Tiller [2] produce the correct type of solution at small spacings. The eutectic undercooling is shown to be of the form

$$\Delta T = AV\lambda + \frac{B}{\lambda} \qquad (10.1)$$

where V is the velocity, λ is the spacing, and A and B are constants. The equation is plotted in figure 10.2 as the solid line at small spacings then continued as a dotted line.

The analytic solution for equation (10.1) by Jackson and Hunt [3] was obtained by approximating the diffusion problem to that of a two-phase planar front, and the effect of surface energy was included as an average curvature undercooling. When surface energy is included in this way it seems as if the spacing can be increased indefinitely. At the time of the

144   *Pattern formation during solidification*

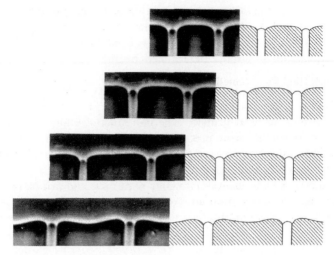

Figure 10.1. Calculated and observed interface shapes in carbon tetrabromide–hexachloroethane eutectic [3]. Magnification ×1200.

early work it was pointed out that there was a maximum steady-state spacing and an attempt was made to calculate its value. A fully self-consistent solution shows [4] that, as the spacing increases, the tip of the widest phase becomes flat, then a hollow appears and eventually a steady-state solution is not possible [3, 4]. The widest growth spacing is probably near the widest steady-state spacing. The widest spacing is typically 2–3 times the

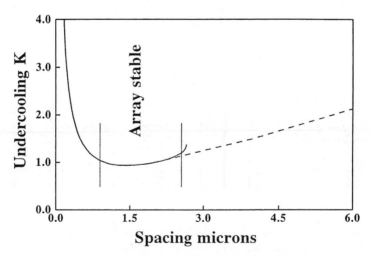

Figure 10.2. Plot of undercooling against spacing. The solid line shows the limits of a self-consistent solution; the dashed line shows equation (10.1) at large spacings.

spacing of the minimum undercooling. The important point is that a large range of solutions are possible and these are shown by the solid line on figure 10.2. The question is what determines the growth spacing?

Spacing selection for a lamellar eutectic

Zener [1], Tiller [2] and later workers got rid of the degree of freedom by assuming arbitrarily that growth occurred at the minimum undercooling. This was called the extremum spacing. Although the assumption predicts the experimental average spacing rather well there is no real justification for its use.

By examining the growth process using transparent materials, it was shown [3] that there was a minimum spacing below which one lamella was overgrown by the surrounding lamellae. It was argued [3] that this should occur just before the minimum undercooling. The argument is illustrated in figure 10.3. For a spacing wider than the extremum spacing, a lamella with a slightly narrower spacing will tend to grow at a higher temperature (smaller undercooling). This will mean it will grow ahead and thus grow wider (see figure 10.3(a)). For a spacing narrower than the extremum spacing, a narrower lamella will grow at a lower temperature, and will grow behind and thus be overgrown (figure 10.3(b)). This intuitive approach was justified by later workers [5, 6]. The minimum spacing is the smallest spacing which is stable to perturbations in spacing. This is referred to in the present chapter as the array stability limit. It is concluded that lamellar eutectic structures grow somewhere between minimum and maximum spacing. For a lamellar eutectic, the assumption that growth occurs at the minimum undercooling is approximately valid because it predicts a spacing somewhere between the minimum and maximum possible spacings. Although the discussion has only considered lamellar growth, similar considerations apply to rodlike eutectic growth.

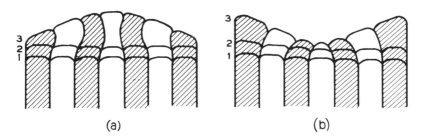

Figure 10.3. The lamellae in (a) are wider than the extremum spacing, (b) narrower than the extremum spacing. Interface position 1 shows the initial position, position 2 the formation of a narrower lamellae, position 3 the change in interface shape due to the change in local undercooling. (a) Leads to restabilization of the array, (b) leads to overgrowth.

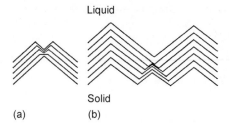

Figure 10.4. A schematic diagram showing (a) the cell tip splitting and (b) the loss of a cell by overgrowth.

Faceted cellular growth

Faceted cellular growth is one of the problems associated with the growth of thin-film silicon. Because of the relative simplicity of the interface shape it has been possible to model [7] numerically a large number of cells. Facet growth occurs because atomic steps can spread easily over a facet but it is difficult to produce new steps. The rate of growth of the facet is determined by the maximum kinetic undercooling on the facet. This in turn is determined by the size of the facet and the flow of heat and solute. A new cell is produced by tip splitting and it is argued [7] that this occurs when the facet grows sufficiently fast so that the kinetic undercooling at the tip drops to zero. This produces a small non-faceted region which becomes unstable and forms two cell tips, as shown in figure 10.4(a). Cells are lost when overgrowth of one cell by its neighbours occurs, as shown in figure 10.4(b). This takes place when the maximum kinetic undercooling on adjacent faces of a groove are different.

Examples of calculated interface positions are shown in figure 10.5. The initial spacing gets larger on the left hand and middle but gets narrower on

Figure 10.5. Examples of the calculated time evolution of the interface with different initial spacings. The growth condition is given in reference [7].

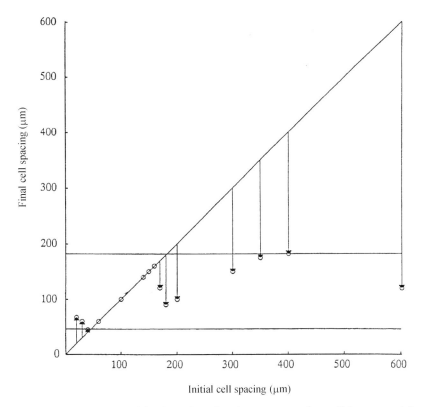

Figure 10.6. The initial and final spacings for the same growth condition as used for figure 10.5.

the right hand example. Figure 10.6 summarizes a number of runs with different initial spacings. The vertical lines show how the spacing changes with time. The important feature is that the spacing changes only when the initial spacing is outside a stable band. Once inside the stable region, spacing changes almost cease, as shown in figure 10.5.

Using this approach it is possible to explain the structures found in thin-film silicon and it is found that the ideas can also be used for non-faceted growth.

Non-faceted cellular and dendritic growth

When growth occurs in a positive temperature gradient, a planar interface is stable at a low enough and high enough velocity [8]. Between these velocities arrays of cells or dendrites are formed. As the cell or dendrite grows, solute is rejected in front and to the sides of the tip. Depending on the spacing and tip

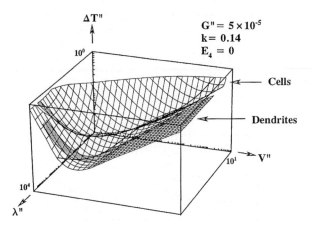

Figure 10.7. Plot of dimensionless undercooling $\Delta T''$, velocity V'' and spacing λ''. Dimensionless terms defined in reference [9].

radius, the diffusion fields overlap and an interaction occurs between neighbours. To model non-faceted cellular growth, a smooth steady-state interface shape is needed which satisfies curvature undercooling, solute and heat flow at all points on the interface. Numerical solutions have been obtained for smooth axi-symmetric shapes [9, 10]. The axi-symmetric cylinder is meant to approximate one member of a hexagonal array. An example of the calculated results for a fixed temperature gradient is shown in figure 10.7. The figure is a plot of undercooling against spacing and velocity. The smallest velocity shown is approximately the constitutional undercooling velocity and the largest the absolute stability limit [8]; outside these limits growth should occur with a planar front.

A feature of the numerical work is that smooth cell and dendrite-like shapes are found on different regions of figure 10.7. The smooth dendrite shapes are taken to be a representation of the averaged fraction solid. This in fact was the assumption made in all analytical models of dendritic growth. Traditionally cells are considered to be dendrites without arms. A conclusion from the numerical work is that it is better to differentiate them by considering the interface shape. For a dendrite, the interface near the tip is a paraboloid of revolution and the tip has the smallest radius of curvature, as shown in figure 10.8(b). For a cell the interface is almost spherical but tends to have the largest radius of curvature at the tip, as shown in figure 10.8(a).

At a constant velocity, cells occur at small spacings and dendrites at large spacings. A plot of undercooling against spacing for a particular velocity is shown in figure 10.9. There is a separate line for cells and dendrites and these appear as different surfaces on figure 10.7. In the same way as for the lamellar eutectic there is a maximum spacing for cells; as this spacing is

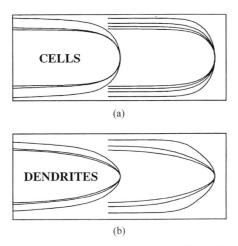

Figure 10.8. Calculated interface shapes for (a) cells and (b) dendrites.

approached the cell becomes flat at the tip and eventually a steady-state solution cannot be obtained. There appears to be no maximum spacing for dendrites.

Spacing selection for non-faceted cells and dendrites

From figure 10.9 it is apparent that, as for eutectics, a condition is necessary to fix the spacing, or conditions are necessary to fix a range of spacings. Experimental studies of the space change mechanisms have been carried out. These are illustrated schematically in figure 10.10. When the spacing is too wide for a cell, a hollow appears in the centre of a cell and the tip splits giving an additional member of the array. For dendrites a different process occurs. When the spacing is too wide, a tertiary arm catches up the

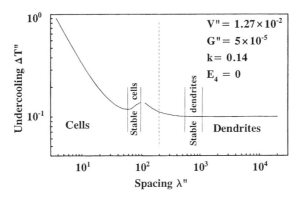

Figure 10.9. Constant velocity section from figure 10.7, showing cell and dendrite regions and stable bands.

Pattern formation during solidification

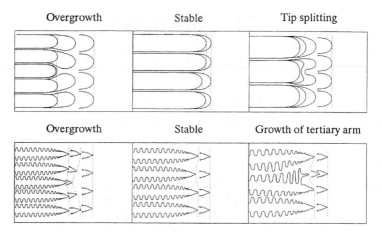

Figure 10.10. Schematic representation of spacing adjustment mechanism for cells and dendrites.

front and becomes a new primary arm. When the spacing is too narrow both for cells and dendrites, a slightly smaller member of the array gets smaller and becomes overgrown by its neighbours.

Maximum spacing

The maximum spacing for cells can be treated by examining the time-dependent growth of a single cell [9, 10]. It is found that the maximum stable spacing for cells is close to the maximum steady-state spacing. The maximum spacing for dendrites cannot be modelled by the smooth interface model under discussion. It would be expected to depend on the competitive growth between the tertiary and the next secondary arm. It is clear that the maximum spacing must be greater than twice the minimum spacing otherwise the new dendrite cannot catch up. As a crude first approximation the maximum dendrite spacing is taken to be twice the minimum spacing.

Minimum spacing

When the spacing is narrow, overgrowth occurs until one of two things happen:

- The array becomes stable. When the array is stable any small local variation in spacing tends to be reversed and the spacing of the array becomes more uniform. The spacing when this first takes place is the array stability limit. It will later be argued that the presence of a stable array is an essential feature of the formation of regular structures. Both cells and dendrites can form stable arrays.

- If the array does not become stable, the spacing eventually becomes so large that the dendrites no longer interact with one another. The spacing when interaction first becomes appreciable has been termed the interaction limit; the arbitrary amount of overlap is defined by Hunt and Lu [10].

If an array stability limit exists it must occur at a smaller spacing than the interaction limit, since interaction is necessary for a stable array.

Array stability limit

In earlier work [9, 10] a multi-cell/dendrite model was used to examine stability. The model was set up with a slightly smaller central cell surrounded by six others. The numerical problem was started by using compositions obtained for steady-state single cells. When the wall composition near the tip is initially such that solute flows from a wider to the narrower cell, then the narrower cell slows down and is overgrown (see figure 10.11(a)). Conversely, when solute initially flows from the narrower to the wider dendrite the narrower cell grows more easily and a stable array is formed (see figure 10.11(b)). This is an intuitively reasonable condition and might have been anticipated. The condition was applied to the simpler single-cell model to predict the array stability limit.

A search is made for the spacing when the wall composition adjacent to the tip of a narrower dendrite first becomes larger than that of a slightly wider dendrite. The search is illustrated in figure 10.12. The lines on the figure show calculated dendrite wall compositions plotted against distance for five different spacings. The position of the tip is shown as a point on each line. The spacing increases in the order 1, 2, 3, 4 and 5 and for dendrites the tip undercooling decreases with increasing spacing so that the tip position moves from left to right of figure 10.12 as the spacing increases. Considering the wall composition near the tips for dendrites 1 and 2, the line for the wider spacing dendrite 2 is above that for 1 and thus the array is unstable (figure 10.11(a)). On the other hand, for dendrites 4 and 5 the wall composition for the wider spacing dendrite 5 is lower than that for spacing 4 and thus

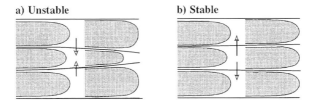

Figure 10.11. Schematic illustration of the direction of solute flow at the cell walls for (a) overgrowth and (b) array stability.

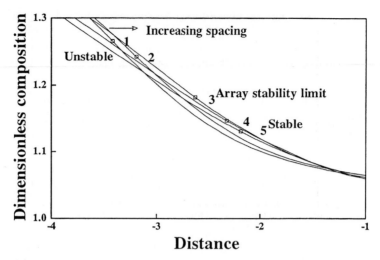

Figure 10.12. Calculated wall compositions plotted against distance for five dendrites. The squares show the position of the tip. The spacing increases 1–5. Line 2 is above 1, thus overgrowth occurs. Line 4 is above 5 so the array is stable.

the array is stable (figure 10.11(b)). The transition occurs somewhere between spacings 2 and 3.

The transition was searched for numerically and typical results together with maximum spacing predictions are shown on figures 10.7 and 10.8.

Minimum spacings limited by the interaction limit

Complications occur in calculating the array stability limit for dendrites at high velocities or low temperature gradients. An example of the calculated minimum spacing as a function of velocity for dendrites is shown in figure 10.13. Interaction of the diffusion fields occurs for any spacing below and to the left of the interaction limit, straight line (b). The interaction leads to a loss of a dendrite unless the array is stable. There are three regions shown on figure 10.13: in region A, the dendrites are so far apart that no interaction takes place and thus the spacing remains unchanged; in region B overgrowth occurs leading to an increase in the average spacing; in region C, the array is stable and no loss of dendrites occurs.

If the dendrite array moves into or starts in region B, overgrowth occurs and the average spacing gets larger. The minimum spacing is limited by either the array stability limit, line (a), or the interaction limit, line (b). Regular structures form when the minimum spacing is determined by the array stability limit. Much less regular structures are formed when the minimum spacing is determined by the interaction limit and a very wide range of spacings exist on a specimen [11].

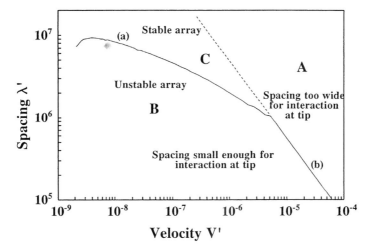

Figure 10.13. Plot of calculated spacing for the array stability limit, line (a), and the interaction limit line (b). In region A the spacing is too wide for interaction near the tip, in region B interaction occurs and dendrites are lost by overgrowth, in region C interaction occurs but here the array is stable.

Comparison with experiment

Correlation of experiment with theory is shown in figures 10.14–10.16. For cells and dendrites figure 10.14(a) shows a comparison of the predicted stable band with the results from Eshelman *et al* [12] for succinonitrile–acetone and figure 10.14(b) shows a comparison with the balanced pseudo-binary system of Al–Si–Mg [13]. Figure 10.14(c) shows the array stability limit for dendrites (solid line) compared with the results of Somboonsuk and Trivedi [14]. The average spacing is expected to be greater than the array stability limit. Correlation with the temperature gradient [14] is shown by the solid lines in Figure 10.14(d) and with composition [15] in figure 10.14(e).

Huang *et al* [16] investigated dendrite spacing limits experimentally by using stepped increments or decrements in velocity. The work is shown in figure 10.15. Overgrowth occurs for the spacings shown by the open circles. This should be compared with the array stability limit (lower line). Tertiary arms grow between dendrites for the open triangles. This appears to be larger than the factor of two times the array stability limit suggested by earlier work.

The minimum dendrite spacing for figures 10.14 and 10.15 was determined by the array stability limit. A correlation of the predicted minimum spacing where the spacing is determined by the array stability limit at low velocities and the interaction limit at high velocities is shown in figure 10.16. The agreement is good both for the transition velocity and for the

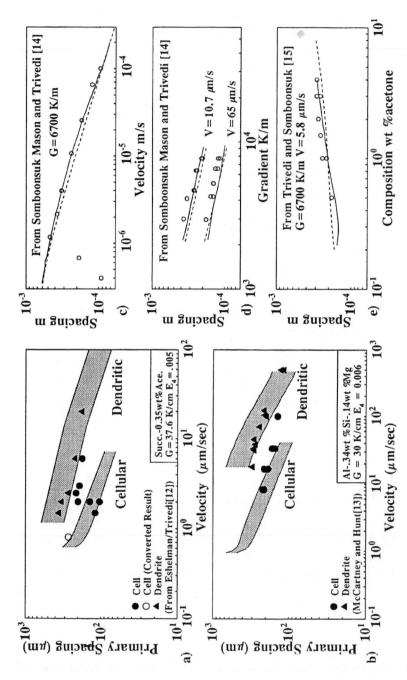

Figure 10.14. (a) and (b). Comparison of measured cell and dendrite regions with theory; (c)–(e), comparison of experiment with the predicted array stability limit; the dashed line shows the analytic expression.

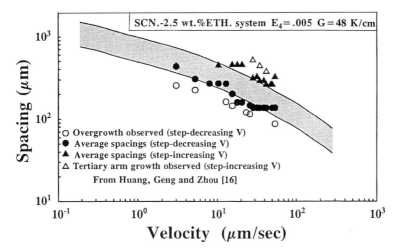

Figure 10.15. A comparison of theory and experiment [16]. The experiments were performed with stepped increments or decrements in velocity. The filled and open circles might be expected to agree with the array stability limit (lower line).

different dependence of spacing on velocity [11]. As expected for velocities below the transition velocity the spacings were relatively uniform but for higher velocities the spacing was almost random. The high-velocity effect can be equally well regarded as a low-temperature-gradient effect [11]. This is the reason why a regular spacing is not generally found in castings.

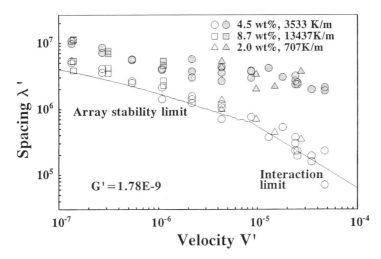

Figure 10.16. A comparison of theory and experiment [11]. At low velocities the minimum spacing should be determined by the array stability limit, and at high velocities by the interaction limit.

The numerical results have been fitted by approximate analytical expressions. These may be used to compare experiment with theory without carrying out numerical calculations. The expressions are given in the appendix, and are shown on figure 10.14 as dashed lines.

Summary

The factors controlling lamellar eutectic, faceted cells, and non-faceted cell and dendritic growth are very similar. To predict spacing it is necessary to consider and model the mechanisms that limit the minimum and maximum spacing. Between the two limits the regularity of the structure depends on whether the array is stable. For a stable array any local variation in spacing tends to decay leading to a more uniform spacing.

The maximum spacing occurs when a mechanism leads to the formation of a new member of the array. For a lamellar eutectic and for cells, the interface for a single member of the array becomes unstable and the tip splits. For dendrites a tertiary arm moves forwards and becomes a new primary. This occurs when a tertiary arm competes successfully with the next secondary arm. For lamellar eutectics, faceted and non-faceted cells and some dendrites the minimum spacing is determined by the spacing when the array first becomes stable. When this occurs, regular structures are formed. For dendrites grown at low gradients and high velocity, the array does not become stable and dendrites are lost until the tips are far enough apart for no interaction to occur. Any spacing up to the maximum can then exist on the specimen.

The average spacing on a specimen probably has little real meaning. If the growth velocity is slowly decreasing the average spacing is found to migrate towards the minimum spacing. If the growth rate is gradually increasing the average spacing moves towards the maximum spacing limit. On a specimen grown at steady state, local regions will lose members while new members of the array are formed at other regions. The local changes may occur at grain boundaries or for eutectics at fault lines. It is likely that the average spacing will depend on the presence of these irregularities.

In the past it has been suggested that a single condition can remove the degree of freedom present during array growth. It does not seem likely that a regular structure could have been formed if only a single spacing were permitted. A single condition can, however, estimate the spacing if it predicts a spacing between the minimum and maximum spacing. For eutectics, the spacing of the minimum undercooling is valid for this reason. The minimum undercooling could be used for cells in the same way. Use of this condition in the early analytical models [17] cannot be justified because the diffusion solution was not accurate enough to determine where the minimum occurred. Minimum undercooling cannot be used as an estimate for dendrites because there is no minimum in undercooling with change in spacing. The marginal stability condition was used in early analytical analyses of dendritic growth. The condition cannot be used as an additional condition when a self-consistent solution is obtained. It cannot be used in the numerical work because all dendrites of whatever spacing grow near marginal stability. The relevance of marginal stability is discussed by Hunt and Lu [9, 10].

It is tempting in the light of the examples of solidification structure considered to suggest that regular structures can be formed when three conditions are met. A mechanism must exist to eliminate members of the array when the spacing is too small. A mechanism must exist to form new members of the array when the spacing is too wide. The structure must be stable to fluctuations in spacing in the range between the two limiting spacings.

Appendix

In the numerical work on the array growth of cells and dendrites [9, 10] it was found that the dimensionless undercooling $\Delta T' = \Delta T/\Delta T_0$ could be calculated as a function of four parameters. The four dimensionless parameters were chosen to be

$$G' = \frac{G\Gamma k}{\Delta T_0^2}, \qquad V' = \frac{V\Gamma k}{D\Delta T_0}, \qquad \lambda' = \frac{\lambda \Delta T_0}{\Gamma k}, \qquad k = \frac{C_S}{C_L}$$

where G, V, λ, D, m, C_0 and k are respectively the temperature gradient, tip velocity, Gibbs–Thomson coefficient, liquid diffusion coefficient, liquidus slope, bulk composition and the distribution coefficient, and where $\Delta T_0 = mC_0(k-1)/k$. These parameters were chosen because $\Delta T_0 = 1$ for a planar front, $V' = 1$ at the absolute stability limit and $V' = G'$ at the constitutional undercooling limit. The form of λ' was chosen so that $V'\lambda' = V\lambda/D$, which is a spacing Peclet number.

The predicted undercooling and minimum spacing obtained using the numerical model [9, 10] have been fitted using analytical expressions [10, 18]. These expressions allow experimental results to be compared with theory without using the numerical model. An expression was given for dendrites in reference [10] but a better expression was obtained in reference [18], the latter behaving more realistically as $k \to 0$. The dimensionless dendrite undercooling can be represented by

$$\Delta T' = \frac{G}{V} + \frac{kI(p)}{1 - I(p)(1-k)} + 0.33(V' - G')^{0.5}, \qquad (10.1A)$$

where $p = V'^{0.5}/(0.33k(1 - G'/V'))$ and $I(p) = p\exp(p)E_1(p)$ is the Ivantsov function [19] and can be evaluated using rational or polynomial expansions [20].

The minimum primary spacing or array stability limit is given by the smaller of

$$\lambda' = 2.5 V'^{-b}\left[1 - \frac{G'}{V'}\right]^{0.5} G'^{-2(1-b)/3} \quad (10.2A)$$

$$\lambda' = 12 V'^{-1} \quad (10.3A)$$

where $b = 0.3 + 1.9 G'^{0.18}$.

It was found that the results for cells could be fitted [10] by

$$\Delta T' = \frac{G'}{V'} + a + (1-a)V'^{0.45} - \frac{G'}{V'}(a + (1-a)V'^{0.45})$$
$$+ b(V' - G')^{0.55}(1 - V')^{1.5} \quad (10.4A)$$

where

$$a = 5.273 \times 10^{-3} + 0.5519k - 0.1865k^2$$

and

$$b = 0.5582 - 0.2267 \log_{10} k + 0.234 \log_{10} k^2.$$

The numerical results for spacing may be fitted by an expression of the form

$$\lambda' = 4.09 k^{-0.485} V'^{-0.29} (V' - G')^{-0.3} \Delta T'_S{-0.3}(1 - V')^{-1.4}. \quad (10.5A)$$

References

[1] Zener C 1946 *AIME Trans.* **167** 550
[2] W A Tiller 1958 *Liquid Metals and Solidification* (Cleveland, OH: ASM) p 276
[3] Jackson K A and Hunt J D 1966 *Trans. Met. Soc. AIME* **236** 1129
[4] Liu J and Elliott R 1995 *Acta Metall. Mater.* **43**(9) 3111
[5] Strässler S and Schneider W R 1974 *Phys. Condens. Matter* **17** 153
[6] Datye V and Langer J S 1981 *Phys. Rev.* **B24**(8) 4155
[7] Shangguan D K and Hunt J D 1989 *J. Cryst. Growth* **96** 856
[8] Mullins W W and Sekerka R F 1964 *J. Appl. Phys.* **34** 444
[9] Lu S-Z and Hunt J D 1992 *J. Crystal Growth* **123** 17
[10] Hunt J D and Lu S-Z 1996 *Metall. Mater. Trans.* **27A** 611
[11] Wan X, Han Q and Hunt J D 1997 *Acta Mater.* **45**(10) 3975
[12] Eshelman M A, Seetharaman V and Trivedi R 1988 *Acta Metall.* **36** 1165
[13] McCartney D G and Hunt J D 1981 *Acta Metall.* **29** 1851
[14] Somboonsuk K, Mason J T and Trivedi R 1984 *Metall. Trans.* **A15** 967
[15] Trivedi R and Somboonsuk K 1984 *Mater. Sci. Eng.* **65** 65
[16] Huang W, Geng X-G and Zhou Y 1993 *J. Cryst. Growth* **134** 105
[17] Hunt J D 1979 *Solidification and Casting of Metals* (London: The Metals Society) p 3

[18] J D Hunt and R W Thomas 1997 *SP97 Proceedings of the 4th Decennial International Conference on Solidification Processing* ed J Beech and H Jones (Sheffield: University of Sheffield) p 350
[19] Kurz W and Fisher D J 1984 *Fundamentals of Solidification* (Switzerland: Trans Tech) p 204
[20] Abromowitz M and Stegun I A 1972 *Handbook of Mathematical Functions* (New York: Dover) p 233

Chapter 11

Peritectic solidification

Hideyuki Yasuda, Itsuo Ohnaka, Kentaro Tokieda and Naohiro Notake

Peritectic microstructures

Peritectic and eutectic reactions are both well known in metallic and non-metallic alloy systems. During unidirectional solidification of peritectic alloys, two constituent solid phases grow competitively, while in unidirectional solidification of eutectic alloys two constituent solid phases grow cooperatively. The competitive growth causes various kinds of growth morphology for the peritectic systems. In conventional casting or solidification processes, solidification starts by precipitation of the primary phase and the secondary phase tends to grow along the primary phase/liquid interface below the peritectic temperature. Two kinds of growth mechanism have been reported for peritectic alloys [1]. One is called the peritectic transformation, in which growth of the secondary phase is controlled by solute diffusion in the primary phase [2, 3]. The other is called the peritectic reaction, in which growth of the secondary phase is controlled by solute diffusion in the liquid phase at the three-phase junction [2].

A third peritectic growth morphology has been reported for several systems, in which the secondary phase does not tend to grow along the primary phase, and instead the primary phase dissolves in front of the independent growing interface of the secondary phase. *In situ* observation of this kind of peritectic growth has been performed for an organic alloy, the salicylic acid–acetamide system [4]. At low growth rates, the secondary phase has a faceted interface and is not likely to grow along the primary phase. The peritectic reaction proceeds by dissolution of the primary phase and independent growth of the secondary phase [2]. A similar peritectic growth mode has also been reported for the yttrium–barium–copper–oxygen superconducting system [5–8]. At higher growth rates, growth of the secondary phase takes place along the primary phase [4], and the growth morphology again becomes similar to the peritectic transformation described above [3]. The growth morphology of the secondary phase in peritectic systems clearly depends on the alloy system and on the growth condition.

A fourth peritectic growth morphology is the banded structure, where the two constituent solid phases grow alternately perpendicular to the growth direction at high values of G/V, where G is temperature gradient and V is growth rate. The banded structure has been observed for lead–bismuth [9–12], tin–cadmium [13–15], tin–antimony [16], copper–zinc [16] and silver–zinc [17] alloys, although the formation conditions have not yet been clearly identified. At high values of G/V, a coupled growth eutectic-like structure has also been suggested [18]. Trivedi [19] reported a simple and suggestive model for the formation of the banded structure. This model indicates that the banded structure can form in the hypoperitectic alloys. The band spacing is proportional to the reciprocal of the growth velocity if undercoolings to nucleate both phases are constant.

A fifth peritectic growth morphology is the fluctuated structure in which the secondary solid phase does not completely cover the primary phase, and the three-phase junction fluctuates perpendicularly to the growth direction during unidirectional solidification [14, 20]. The period of such fluctuations shows a clear dependence on the growth rate [15].

The banded and fluctuated structures observed in peritectic alloys can be regarded as periodic structures formed during solidification. Formation of periodic structures contains several interesting points. One interesting point is that non-steady-state growth occurs even under carefully controlled unidirectional solidification conditions, with the temperature of the furnace, pulling and growth rate, and other external experimental parameters kept constant [15]. Most solidification models and theories have been based on steady state. Another interesting point is that competition between the two constituent solid phases occurs in peritectic alloys. Morphological maps have been reported for peritectic alloys [21] and criteria are used to explain the phase and morphology selection, i.e. maximum interface temperature and/or maximum growth rate. In the case of periodic structures, these criteria cannot apply directly [22], and new criteria need to be established.

In this chapter, periodic structures such as fluctuated and banded structures are discussed for lead–bismuth alloys and tin–cadmium metallic alloys. The origin of the periodic structures are discussed by considering mass transfer at the solidifying interface. An unsteady-state growth model is explained, based on tracing the interface temperature and the fraction of the primary phase. Periodic dendritic cellular and eutectic structures are discussed in chapter 10, and peritectic non-metallic oxide solidification is discussed in chapters 23 and 24.

Lead–bismuth system

Lead–25 at% bismuth and lead–33 at% bismuth alloys were made from 99.99% lead and 99.99% bismuth in an argon atmosphere. According to

162 Peritectic solidification

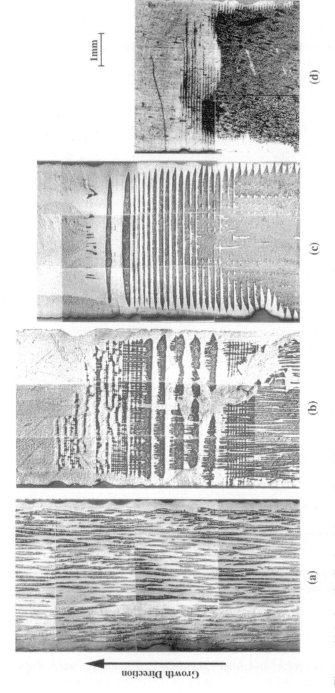

Figure 11.1. Longitudinal sections of unidirectionally solidified lead–33 at% bismuth alloys (100 mm in initial melt length) at $G = 2.7 \times 10^4\,\mathrm{K\,m^{-1}}$. (a) $V = 1.4\,\mu\mathrm{m\,s^{-1}}$, (b) $V = 0.83\,\mu\mathrm{m\,s^{-1}}$, (c) $V = 0.56\,\mu\mathrm{m\,s^{-1}}$ and (d) $V = 0.28\,\mu\mathrm{m\,s^{-1}}$. The black part is the α phase and the white part is the β phase.

the phase diagram [23], the lead–25 at% bismuth alloy is a hypoperitectic composition, while the lead–33 at% bismuth alloy is hyperperitectic. Specimen sizes were 4 mm in diameter and 100 mm in length. Details of unidirectional solidification equipment and experimental conditions are given in reference [12].

Figures 11.1(a)–(d) show longitudinal microstructures of the lead–33 at% bismuth alloys grown at the same temperature gradient of $2.7 \times 10^4 \, \text{K m}^{-1}$ but at different growth rates. The cellular primary α grew as the leading phase, and the secondary β phase was located between the cells at a growth rate of $1.4 \, \mu\text{m s}^{-1}$, as shown in figure 11.1(a). The spacing of the α cells was 130 μm. A partially banded structure was observed in a transient region between the α and β phases at the slower growth rate of $0.83 \, \mu\text{m s}^{-1}$, with a meshlike structure also observed before formation of the banded structure, as shown in figure 11.1(b). In the banded structure, the α phase showed a cellular interface and the β phase showed a planar interface. The α phase did not completely cover the β phase, and the banded structure persisted over a length of 3 mm. The band spacing decreased monotonically as solidification proceeded, from 400 μm at the beginning to 200 μm at the end of the 3 mm length.

A more stable banded structure consisting of planar α and planar β phases was formed at even slower growth rates, from 0.28 to $0.56 \, \mu\text{m s}^{-1}$, as shown in figures 11.1(c) and (d). The banded structure was again formed in a transient region between the α and β phases. Although formation of the banded structure was reproducible, the length of the banded structure was not consistent and varied from 1 mm to 30 mm. In the most stable case, as shown in figure 11.2, a banded structure with 40 layers was formed over a length of more than 30 mm.

Figure 11.3 summarizes the growth morphologies of lead–33 at% bismuth alloys as a function of temperature gradient and growth rate. The banded structure was observed in the region where the value of G/V was higher than $3.3 \times 10^{10} \, \text{K s m}^{-2}$. In this region, the α phase grew with a cellular or planar interface, while the β phase always grew with a planar interface. The region where the banded structure can form has been discussed [11,

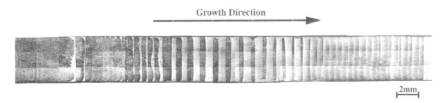

Figure 11.2. Banded structure in lead–33 at% bismuth (100 mm in initial melt length) grown at $G = 2.7 \times 10^4 \, \text{K m}^{-1}$ and $V = 0.56 \, \mu\text{m s}^{-1}$. The black part is the α phase and the white part is the β phase.

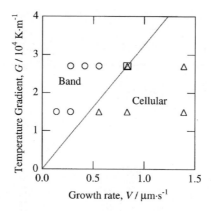

Figure 11.3. Growth morphologies of unidirectionally solidified lead–33 at% bismuth alloys (100 mm in initial melt length). ○: band (planar α + planar β), □: band (cellular α + planar β), △: cellular α + intercellular β.

12, 22]. The condition to form the banded structure is that the interface shape of the α phase is cellular or planar, while that of the β phase is always planar. As shown in figure 11.3, the present experimental result is consistent with this requirement.

Tin–cadmium system

Tin–cadmium alloys were made from 99.9% tin and 99.99% cadmium in a silica crucible under an argon atmosphere. The tin–cadmium alloy compositions were 1.0, 1.5 and 2.0 at% cadmium. Since the composition of the secondary phase precipitating at the peritectic temperature is 1.0 at% cadmium [11], the tin–1.0 at% cadmium alloy is peritectic and the tin–1.5 and 2.0 at% cadmium alloys are hyperperitectic. Specimens were 2 mm and 4 mm in diameter, and 250 mm in length. Details of unidirectional solidification equipment and experimental conditions are again given in reference [12].

Figures 11.4(a) and (b) show longitudinal microstructures of tin–1.0 at% cadmium alloys of 2 mm diameter grown at the same temperature gradient of $5.3 \times 10^4 \, \text{K} \, \text{m}^{-1}$, but at different growth rates. The banded structure was formed at a growth rate of $0.83 \, \mu\text{m s}^{-1}$, after the single primary α phase grew with a planar interface. In the banded structure, bands of the secondary β phase were not completely isolated by bands of the primary α phase and were interconnected in some places. Figures 11.4(c) and (d) show the longitudinal microstructures of tin–1.0 at% cadmium alloys of 4 mm diameter grown at the same temperature gradient of $2.7 \times 10^4 \, \text{K} \, \text{m}^{-1}$, but at different growth rates. The primary α phase did not cover the secondary β phase, and the banded structure was only partially observed. No full

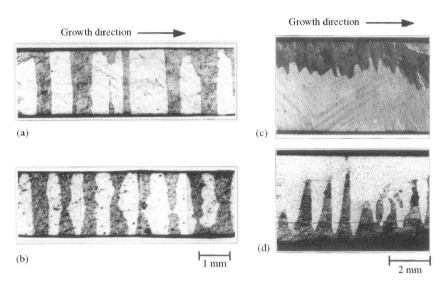

Figure 11.4. Longitudinal sections of unidirectionally solidified tin–1.0 at% cadmium. (a) $V = 0.83\,\mu m\,s^{-1}$ ($G = 5.3 \times 10^4\,K\,m^{-1}$, 2 mm in diameter), (b) $V = 5.6\,\mu m\,s^{-1}$ ($G = 5.3 \times 10^4\,K\,m^{-1}$, 2 mm in diameter), (c) $V = 2.8\,\mu m\,s^{-1}$ ($G = 2.7 \times 10^4\,K\,m^{-1}$, 4 mm in diameter), (d) $V = 5.6\,\mu m\,s^{-1}$ ($G = 2.7 \times 104\,K\,m^{-1}$, 4 mm in diameter).

banded structure in the specimens of 4 mm diameter was observed, even at G/V of $3 \times 10^{10}\,K\,s\,m^{-2}$ where the banded structure was clearly observed in the specimens of 2 mm diameter.

Figure 11.5 summarizes the growth morphologies of the tin–cadmium alloys as a function of temperature gradient, growth rate and composition. The microstructures can be roughly classified into five kinds of morphology, as shown in figure 11.5(a). In the band structure, each phase covered the other phase and complete banding occurred, while in the incomplete band structure, the primary phase did not completely but almost covered the secondary phase. In the competitive structure, each of the two phases did not cover the other phase and the three-phase junction fluctuated perpendicularly to the growth direction during solidification. In the periodic cell structure, the volume fraction of the primary phase fluctuated along the growth direction, although the morphology consisted of cellular primary phase and intercellular secondary phase. In the cell structure, the secondary phase existed between the cellular primary phase, as generally observed for metallic alloys. The complete banded structure was formed only in specimens of 2 mm diameter. The incomplete banded structure and/or competitive structure were mainly formed in specimens of 4 mm diameter. Formation of the banded structure depended significantly on the specimen radius.

Figure 11.6 shows histograms of the fluctuation period for the tin–1.0 at% cadmium alloy. At a growth rate of $0.83\,\mu m\,s^{-1}$, the average spacing

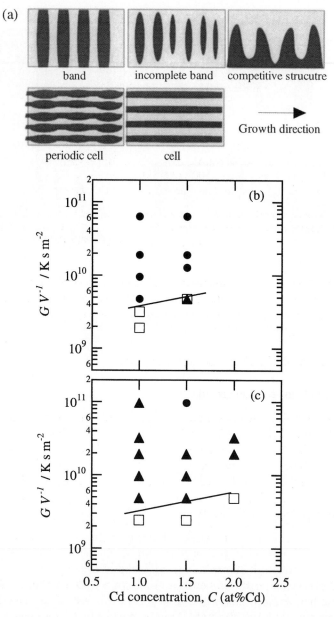

Figure 11.5. Growth morphologies of unidirectionally solidified tin–cadmium alloys. (a) Schematic morphology and classification, (b) morphological map for 2 mm diameter specimens ($G = 5.3 \times 10^4 \, \text{K m}^{-1}$), and (c) morphological map for 4 mm diameter specimens ($G = 2.7 \times 10^4 \, \text{K m}^{-1}$). ●: banded structure (discrete); ▲: incomplete band and competitive structure; □: periodic cell and cell. The lines indicate transition from cellular to planar growth.

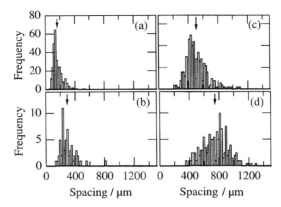

Figure 11.6. Spacing of the competitive microstructure, defined by figure 11.5(a), observed in the tin–1.0 at% cadmium with 4 mm diameter. (a) $V = 0.83\,\mu\text{m s}^{-1}$, (b) $V = 1.4\,\mu\text{m s}^{-1}$, (c) $V = 2.8\,\mu\text{m s}^{-1}$, and (d) $V = 5.6\,\mu\text{m s}^{-1}$. The arrows indicate the average spacing.

was about 200 μm, and dispersion of the spacing was relatively small. The average spacing and the dispersion of the spacing both increased with increasing growth rate, the former almost proportionally.

Mechanism of forming the periodic structures

Instability of side-by-side growth

In order to consider steady state growth of the two constituent phases, solute transfer around the interface was considered [22]. Figure 11.7 shows a schematic picture of the solidifying front of the primary phase, α, and the secondary phase, β. The average interface temperature and solute concentration in the liquid at the interface are respectively T_α and C_α, for the primary α phase, and T_β and C_β for the secondary β phase. The interface temperature and solute concentration satisfy

$$T_\alpha = T_\text{p} - m_\alpha(C_\alpha - C_\text{p}) \qquad (11.1)$$

$$T_\beta = T_\text{p} - m_\beta(C_\beta - C_\text{p}) \qquad (11.2)$$

where T_p is the peritectic temperature, C_p is the liquidus concentration at the peritectic temperature, and m_α and m_β are the slopes of the liquidus line of the α and the β phases (defined as having positive values).

Solute fluxes in the growth direction at the solid/liquid interface ahead of the α and β phases are J_α^1 and J_β^1; solute fluxes in the growth direction far from the solid/liquid interface ahead of the α and the β phases are J_α^2 and J_β^2; and the lateral solute flux from the α and β phase in the liquid just ahead of the solid/liquid interface is $J_{\alpha\beta}$. These solute fluxes are shown in

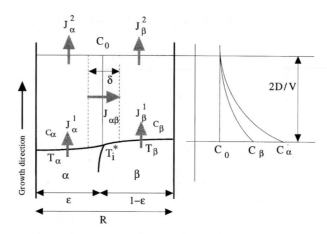

Figure 11.7. Schematic diagram of solute transfer around the solidifying interface when the two constituent peritectic phases α and β grow side by side like a coupled eutectic structure. (a) Morphology and (b) solute concentration profile in the liquid phase.

figure 11.7, and can be expressed as

$$J_\alpha^1 = (1 - k_\alpha)C_\alpha(V + u_\alpha)(\varepsilon R) \tag{11.3}$$

$$J_\alpha^2 = -\frac{C_0 - C_\alpha}{D/(V + u_\alpha)}(\varepsilon R) \tag{11.4}$$

$$J_\beta^1 = (1 - k_\beta)C_\beta(V + u_\beta)(1 - \varepsilon)R \tag{11.5}$$

$$J_\beta^2 = -\frac{C_0 - C_\beta}{D/(V + u_\beta)}(1 - \varepsilon)R \tag{11.6}$$

$$J_{\alpha\beta} = -D\frac{C_\beta - C_\alpha}{\delta}\left(\frac{1}{2}\frac{2D}{V}\right) \tag{11.7}$$

where k_α and k_β are α and β equilibrium partition coefficients, V is the average growth rate, $(V + u_\alpha)$ and $(V + u_\beta)$ are the growth rates of the α and β phases respectively, ε is the fraction of the primary phase at the solidification front, R is the specimen radius, and D is the solute diffusivity in the liquid. The thickness of diffusion layer is assumed to be a constant value of $2D/V$ for both phases, while the growth rates of the two phases are defined independently by using u_α and u_β which represent α and β deviations in growth rate from the average value V. The relation between T_α, C_α and u_α and between T_β, C_β and u_β can be obtained from equation (11.1) assuming a linear temperature gradient G in the growth direction:

$$dT_\alpha = -m_\alpha \, dC_\alpha = G \cdot u_\alpha \, dt \tag{11.8}$$

$$dT_\beta = -m_\beta \, dC_\beta = G \cdot u_\beta \, dt. \tag{11.9}$$

The above equations can be used to obtain the variation of solute concentration in the liquid ahead of the α and the β phases:

$$\frac{dC_\alpha}{dt} = \frac{\dfrac{V^2}{D}(C_0 - k_\alpha C_\alpha) - \dfrac{D}{\varepsilon R \delta}(C_\alpha - C_\beta)}{1 + \dfrac{m_\alpha V}{GD}(C_0 - k_\alpha C_\alpha)} \tag{11.10}$$

$$\frac{dC_\beta}{dt} = \frac{\dfrac{V^2}{D}(C_0 - k_\beta C_\beta) + \dfrac{D}{(1-\varepsilon) R \delta}(C_\alpha - C_\beta)}{1 + \dfrac{m_\beta V}{GD}(C_0 - k_\beta C_\beta)} \tag{11.11}$$

where R and δ are defined in figure 11.7. As shown in equations (11.10) and (11.11), the variations of solute concentration in the liquid are functions of the α phase fraction, ε. Under steady-state growth conditions, the variations in solute concentration in equations (11.10) and (11.11) should be zero. Therefore, solute concentrations which satisfy the steady-state condition are obtained as functions of the α phase fraction.

To make clear the α phase fraction dependence on liquid composition, it is assumed that the temperature gradient is sufficiently large that the second term in the dominator of equations (11.10) and (11.11) is negligible. The liquid composition can be obtained then as follows:

$$C_\alpha = \frac{C_0\left(k_\beta + \dfrac{A}{\varepsilon} + \dfrac{A}{1-\varepsilon}\right)}{k_\alpha k_\beta + A\left(\dfrac{k_\alpha}{1-\varepsilon} + \dfrac{k_\beta}{\varepsilon}\right)} \tag{11.12}$$

$$C_\beta = \frac{C_0\left(k_\alpha + \dfrac{A}{\varepsilon} + \dfrac{A}{1-\varepsilon}\right)}{k_\alpha k_\beta + A\left(\dfrac{k_\alpha}{1-\varepsilon} + \dfrac{k_\beta}{\varepsilon}\right)} \tag{11.13}$$

$$A = \frac{D^2}{V^2 R \delta}. \tag{11.14}$$

As expressed in equations (11.12)–(11.14), the solute concentration C_α in the liquid at the α phase interface always differs from the solute concentration C_β in the liquid at the β phase interface.

Figure 11.8 shows the relation between the liquid compositions at the α and β interfaces C_α and C_β for the lead–33 at% bismuth alloys with specimens 4 mm in width at an average growth rate, $V = 1\,\text{mm}\,\text{h}^{-1}$. In the lead–bismuth system, k_α and k_β are 0.58 and 0.76, respectively, δ is assumed to be 2 mm, D is $1 \times 10^{-9}\,\text{m}^2\,\text{s}^{-1}$, and the slopes of the α and the β phase liquiduses are 455 and 271 K respectively.

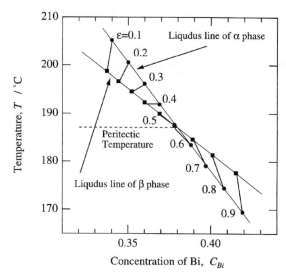

Figure 11.8. Relation between average interface temperature and solute concentrations at the interface which satisfy equation (11.11). Physical properties for the lead–33 at% bismuth alloy were used for the calculation.

In the region where ε values are small, the interface temperatures of both the α and β phases are above the peritectic temperature, indicating that the liquid ahead of the β phase is undercooled relative to the α phase. In the region where ε values are large, the β phase can grow and cover the α phase. At $\varepsilon = 0.5$–0.6, it is possible to have a three-phase junction at the peritectic temperature with the liquid concentrations ahead of both phases equal. However, the α interface is undercooled relative to the β phase, and, therefore, the solidifying structure does not have a stable condition, although the solute balance is satisfied.

Overall, therefore, a growth morphology in which the two constituent phases grow side by side is not stable. An unstable morphology of this type for peritectic alloys is sensitive to external disturbances. Park and Trivedi [19] and Karma *et al* [25] have reported that convection results in an oscillatory structure. In the tin–cadmium system, lateral temperature and solute gradients cause oscillatory convection which contributes to the formation of this oscillatory structure [19] and, as suggested by Trivedi [19], distortions.

Trace of interface temperature and phase fractions

Changes in liquid composition at the interface are derived from the diffusive model of peritectic growth [10, 19]. The banded structure is basically explained by this model. However, there is still uncertainty about the

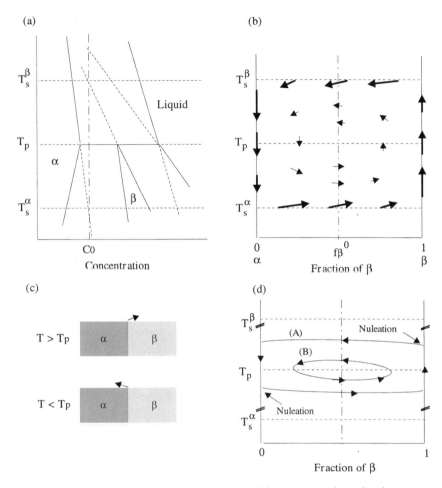

Figure 11.9. Model for the banded and the competitive structures in peritectic systems. (a) Schematic phase diagram, (b) motion of interface temperature and β phase fraction, (c) motion of the three phase junction, and (d) traces of the temperature and the β phase fraction for the banded structure (line A) and the competitive structure (line B). C_0 is the average concentration and f_β^0 is the average fraction of the α phase. T_s^α, T_s^β are solidus temperatures of the α and β phases respectively for the average concentration C_0.

growth mechanism of the periodic structures, such as the incomplete band and competitive structure shown in figure 11.5.

It is valuable to try to have a consistent explanation of periodic structures including the complete band, incomplete band and competitive structures. Here, we consider tracing the interface temperature and the volume fraction of the constituent phases in order to explain all these periodic structures [15]. We consider the schematic phase diagram shown in figure 11.9(a), with two-phase unidirectional peritectic solidification taking place. The assumed conditions are as follows:

1. growth is two dimensional,
2. both phases grow with a planar interface,
3. the two-phase region has the average composition,
4. macrosegregation in the growth direction can be neglected due to an infinite specimen length.

In the case of growth of only the primary α phase, the interface temperature of the primary phase decreases towards its solidus temperature T_s^α, which is below the peritectic temperature T_p, in order to achieve steady-state growth with a planar interface. On the other hand, the interface temperature of the secondary β phase increases towards its solidus temperature T_s^β, which is above the peritectic temperature T_p. These changes in interface temperature are indicated by arrows on the α and β phase sides in figure 11.9(b). The arrows indicating change of interface temperature and fraction of the β phase exhibit counterclockwise movement.

Next, we consider movement of the three-phase junction as a function of interface temperature. As shown in figure 11.9(c), overgrowth of the α phase occurs if the interface temperature is above the peritectic temperature, since the liquid ahead of the β phase is undercooled relative to the α phase. Overgrowth of the β phase occurs if the temperature is below the peritectic temperature since then the situation is reversed, and the liquid ahead of the α phase is undercooled relative to the β phase. The counterclockwise movement of interface temperature and β phase fraction during unidirectional solidification occurs rather stably, and consequently causes periodic structures, such as the banded, incomplete banded and competitive structures. On the other hand, the cellular α and β phases of a eutectic-like structure will only be observed if the interface temperature and the β phase fraction converge to constant values.

A sudden increase in the interface temperature leads to a sharp decrease in the β phase fraction, and the interface temperature and β phase fraction do not return linearly to a stable point. This instability of the interface shape is in contrast to the stable coupled growth for eutectic alloys.

The banded structure in small-diameter specimens and the competitive structure in large-diameter specimens can be understood in a similar manner. The discrete banded structure occurs if the diameter of the specimen is sufficiently small in comparison with the width of the β phase fluctuation, as shown in figure 11.9(d). The interface temperature and the β phase fraction tracing line A in figure 11.9(d) result in a complete banded structure. In the complete banded structure, the nucleation temperatures determine the oscillating cycle and the band spacing. As indicated by line B, the competitive structure occurs if the specimen diameter is larger than the width of the fluctuation. No nucleation is needed and both phases grow continuously. The periodic spacing of the competitive structure is proportional to the growth rate as shown in figure 11.6. This clear dependence may be attributable to the counterclockwise movement in figure 11.9(b), although further study is still required in order to arrive at a quantitative explanation.

The banded and competitive structures can be explained qualitatively, based on the proposed model. The periodic structure observed in the cadmium–tin alloys is intrinsic when the two constituent phases grow competitively in peritectic alloys.

Summary

Unidirectional solidification of lead–bismuth and tin–cadmium alloys can be carried out under high-temperature gradients where the planar interface or the cellular interface is stable. The growth mechanism of periodic structures including the banded and competitive structures have been discussed.

The banded and the competitive structures are both observed in lead–bismuth alloys and tin–cadmium alloys. The banded structure can consist of a cellular or planar α phase and a planar β phase. The morphology of the banded structure is explained by considering the interface temperature during steady growth for the constituent α and β phases.

Tin–cadmium specimens of 2 mm diameter show complete banding, while 4 mm diameter tin–cadmium specimens show incomplete banding or a competitive structure. In other words, the detailed formation criteria for banded structures depend on the specimen diameter.

Growth morphologies in which the two peritectic phases grow side by side are unstable and convection can be a trigger to disturb the solidification process and lead to periodic structures. Formation of the banded and the competitive structures can be explained by a model of the trace of the interface temperature and the β phase fraction during unidirectional solidification. The periodic structures observed in peritectic alloys, including the banded structure, the incomplete banded structure, and the competitive structure, are thus considered to be intrinsic for peritectic alloys in the region of planar front growth.

References

[1] Kerr H W, Cisse J and Bolling G F 1974 *Acta Metall.* **22** 677
[2] Hillert M 1979 *Solidification and Casting of Metals* (London: The Metals Society) p 81
[3] St John D H and Hogan L M 1987 *Acta Metall.* **35** 171
[4] Yasuda H, Ohnaka I, Matunaga Y and Shiohara Y 1996 *J. Cryst. Growth* **158** 128
[5] Nakamura Y, Izumi T, Shiohara Y and Tanaka S 1992 *J. Japan. Inst. Met.* **56** 810

[6] Cima M J, Flemings M C, Figueredo A M, Nakade M, Ishii H, Brody H D and Haggerty J S 1992 *J. Appl. Phys.* **72** 179
[7] Izumi T, Nakayama Y and Shiohara Y 1993 *J. Cryst. Growth* **128** 757
[8] Mori N, Hata H and Ogi K 1992 *J. Japan. Inst. Met.* **56** 648
[9] Barker N J W and Hellawell A 1974 *Met. Sci.* **8** 353
[10] Kurz W and Trivedi R 1996 *Metall. Trans.* **27A** 625
[11] Yasuda H, Ohnaka I and Tokieda K 1997 *Proceedings of the International Conference on Solidification and Casting* (London: Institute of Metals) p 448
[12] Tokieda K, Yasuda H and Ohnaka I 1999 *Mater. Sci. Eng.* **A262** 238
[13] Boettinger W J 1974 *Metall. Trans.* **5** 2023
[14] Brody H D and David S A 1977 *Proceedings of the International Conference on Solidification and Casting*, Sheffield, UK, p 144
[15] Yasuda H, Notake N, Tokieda K and Ohnaka I *J. Cryst. Growth* (in press)
[16] Tichener A P and Spittle J A 1975 *Acta Metall.* **23** 497
[17] Ostowski A and Langer E W 1977 *Proceedings of the International Conference on Solidification and Casting*, Sheffield, UK, p 139
[18] Chalmers B 1959 *Physical Metallurgy* (New York: Wiley) p 271
[19] Trivedi R 1995 *Metall. Trans.* **26A** 1583
[20] Park J-S and Trivedi R 1998 *J. Crystal Growth* **187** 511
[21] Ha H P and Hunt J D 1977 *Proceedings of the International Conference on Solidification and Casting*, Sheffield, UK, p 444
[22] Yasuda H, Ohnaka I, Tokieda K and Notake N 1999 *Mater. Trans. JIM* **40** 373
[23] Massalski T B, Okamoto H, Subramanian P R and Kacprzak L (eds) 1996 *Binary Alloy Phase Diagrams* (Metals Park, Ohio: ASM International) 2nd edition
[24] Massalski T B, Okamoto H, Subramanian P R and Kacprzak L (eds) 1996 *Binary Alloy Phase Diagrams* (Metals Park, Ohio: ASM International) 2nd edition p 1028
[25] Karma A, Rappel W-J, Fuh B C and Trivedi R 1998 *Metall. Trans.* **29A** 1457

SECTION 3

STRUCTURE AND DEFECTS

There is currently a great deal of interest in controlling all aspects of solidification microstructures during industrial processing due to significant improvements in castability and reduction in defects, both of which reduce reject rates, improve mechanical and surface properties, and increase component integrity and reliability. Additionally, control of structure can result in improved consistency of properties throughout a material due to reduced segregation and more uniform grain size distributions. Hence there is significant scope for control of structure and defects leading to improved yields and higher quality materials at reduced cost. This section concentrates on how these benefits can be achieved.

Chapters 12, 13 and 14 describe the control of the primary grain size and shape during casting via the use of grain refiner additions, control of nucleation and growth processes and by a novel step processing technique. Chapter 15 discusses the importance of controlling secondary phases during solidification. Chapter 16 compares the structures obtained during powder processing and more conventional casting, and the resultant effects on final properties. Chapter 17 discusses the control of structure using externally applied electromagnetic fields during the casting process. Chapter 18 uses real time observations to show how structures develop during solidification and Chapter 19 describes specific examples of the effects of solidification parameters on solidified structures.

Chapter 12

Heterogeneous nucleation in aluminium alloys

Peter Schumacher

Introduction

Controlled solidification of aluminium alloys can be achieved by grain refinement of primary aluminium leading to uniform and small grain sizes, thereby facilitating better intergranular feeding, faster casting, reduced porosity and hot tearing, and better surface finish during casting [1]. During subsequent eutectic solidification insoluble intermetallics can be formed, such as those containing iron, which may act as crack initiation sites when sufficiently large. Finer aluminium grain sizes disperse these intermetallics along grain boundaries and interdendritic spaces, thereby reducing brittle intermetallics in size and distributing them on a finer scale, leading to improved mechanical properties. Interestingly, particles added to improve the grain refinement of primary aluminium, which for statistical reasons may not have nucleated aluminium, will be pushed to grain boundaries in similar locations to intermetallics and may even contribute to their nucleation. Control of the as-cast microstructure becomes increasingly important when the final microstructure cannot be sufficiently influenced by subsequent thermo-mechanical processes, such as in twin roller casting.

Grain refinement of aluminium alloys is controlled by the addition of particles and growth-restricting solute, leading to the formation of heterogeneous nuclei and a decreased growth rate of both nucleated equiaxed crystals and competing columnar grains. Essentially, a grain refinement addition facilitates a columnar to equiaxed transition (CET) for a given alloy composition and temperature gradient in a specific casting process [2, 3]. In casting practice this is achieved by addition of grain refiner particles, such as TiB_2 or TiC particles, remaining relatively stable within the aluminium melt, and particles such as titanium aluminide, which can readily dissolve in aluminium melts, to provide growth restriction by titanium in solution in the aluminium melt. Such carbide and boride particles have shown good lattice matching to aluminium and are believed to be good nucleation sites. The titanium aluminide also shows good lattice matching but is thermodynamically not stable for typical titanium concentrations in aluminium alloys of 0.01–0.05 wt% titanium. The dissolution of the titanium aluminide particles and stability of the carbide and boride particles is dependent on thermodynamic and kinetic considerations, which are often unknown in commercial casting practice.

Within commercial aluminium casting, a wealth of empirical observations on grain refinement have been made, often affecting nucleation and growth restriction simultaneously [3]. Moreover, a detailed understanding of nucleation mechanisms for highly potent nucleation sites is lacking, hindering further grain refiner development and employment. The emphasis of this chapter is on the nucleation mechanisms of highly potent heterogeneous sites in aluminium castings, separating the potential influences of nucleation and growth restriction. Chapter 13 concentrates on overall grain refinement processes and mechanisms in solidification and castings, and chapter 14 gives a specific example of a new grain refinement technique. Chapter 15 concentrates on phase selection effects of nucleation.

Nucleation and growth restriction

Unlike grain refinement during containerless solidification [4], in conventional casting a columnar dendrite front will grow from the container walls which act as nucleation sites and locations of largest undercooling. Qualitatively, to achieve grain refinement equiaxed grains have to be nucleated ahead of the columnar dendrite front and result in the solidification of the majority of the volume fraction, thus achieving a columnar to equiaxed transition. For a given temperature gradient imposed by the casting process, the columnar to equiaxed transition is favoured by a high number of heterogeneous nucleation sites, high potencies (i.e. low activation undercoolings necessary for heterogeneous nucleation) and slow growth velocities of the columnar dendritic front caused by solute growth restriction. For a given alloy system and casting process, the columnar to equiaxed transition can be controlled by addition of further solute, affecting growth restriction, and the number and potency of added heterogeneous sites. The columnar to equiaxed transition is dependent on the temperature gradient, which is different for different processes. Hence, for example, direct chill (DC) casting requires different addition levels of heterogeneous sites and growth restricting elements than sand casting [2, 3]. Temperature gradients, similar to those observed in DC casting, are mimicked in the 'TP1' grain refiner test [3] where grain sizes are measured on the cross sections of conical casts. In contrast no standard test for cooling conditions in sand or permanent mould castings has been established. In wrought alloys small additions of a strong growth restricting element, such as titanium, can have a large affect on observed grain sizes, whereas in casting alloys further alloying may form additional solid phases resulting in a decrease in free solute and reducing the growth restriction [5, 6]. Grain refiner additions to aluminium alloys often contain titanium in the form of particles such as TiB_2, TiC and titanium aluminide, as well as titanium in solution in the melt acting as a strong growth restrictor. This gives titanium a unique and complex role affecting growth restriction and heterogeneous nucleation simultaneously.

Casting practice

Cibula, as early as 1947 [7], qualitatively described the effects of carbide and boride additions on grain refinement, in the presence of titanium in solution in the melt. Since then, grain refinement practice has developed on an empirical basis. In general, refinement practice in wrought alloys differs from that in casting alloys where refiner additions remain in contact longer with the melt and contain higher titanium concentrations and experience lower temperature gradients. Further complexities can arise in hypoeutectic casting alloys when grain refiner additions for primary aluminium crystals and modifier agents such as strontium for the remaining eutectic are used simultaneously.

Wrought aluminium alloys

Titanium diboride, TiB_2, and titanium aluminide, $TiAl_3$, particles are added to the melt in the form of aluminium-rich rods or waffles. Microstructural features of the refiner rod can be directly related to its closely guarded manufacturing secrets but less so to its refining behaviour [3]. The microstructure of refiner rods has been extensively studied without clarifying a nucleation mechanism. However, faster dissolution times for the titanium aluminide particles can be achieved when smaller particle size distributions are produced during manufacture of the refiner rod. For that reason, grain refiner rods are often mechanically worked to break up particles and the resulting microstructures in the rod are more a reflection of the deformation processes than that of the need to facilitate nucleation. Overall, the behaviour of refiner addition in contact with the melt is time dependent [8]. First, dissolution of solute (mainly from the titanium aluminide particles) and distribution of the boride particles takes place. Simultaneously, particles may alter their nucleation potency in contact with the melt. Empirical evaluations of a given refiner have established a necessary contact time for the refiner in the melt to achieve an ultimate grain size (typically <0.18 mm), after which longer holding times result in fading of the refiner efficacy

[8, 9]. Overall a fine microstructure of titanium aluminide and boride particles embedded in aluminium can give fast dissolution times for the titanium aluminide and less agglomeration of boride particles in the final cast products [8, 3].

The key feature for grain refiner performance is the nucleation mechanisms of the added particles. Various investigations of as-cast cross section have attempted to elucidate the nucleation mechanism. The first difficulty with this approach is to section through nucleating particles in statistically significant numbers. Secondly, particles found in grain centres were described as efficient, whereas particles in grain boundaries were described as inefficient, thereby neglecting potential statistical considerations during the nucleation and growth of particles [10]. Furthermore, theoretical arguments based on possible crystallographic orientation relationships were made to show potential nucleation potency [11], although no actual orientation relationships were shown in conventional refiner tests for the boride particles. Some recent work on the nucleation behaviour of TiC found a cube-to-cube orientation relationship in cross sections using electron backscatter diffraction (EBSD) [12]. Nevertheless, by this route it became apparent that at high titanium concentrations a sufficiently high number of titanium aluminide particles were active in grain centres and showed a surrounding peritectic reaction [13], while the majority of boride particles were found in grain boundaries. The extension to lower titanium concentrations led to the formulation of the peritectic nucleation theory [3]. However, at low titanium concentrations off the peritectic horizontal, bulk titanium aluminide particles are not stable in contact with the melt and dissolve rapidly, while grain refinement is still observed long after dissolution of the titanium aluminide particles when carbide or boride particles are present. It is worth noting that the terms hypo- and hyper-peritectic titanium levels are often used incorrectly in the literature to describe titanium levels off the peritectic horizontal rather than respectively below and above the concentration of the peritectic point.

Borides, and to a lesser degree carbides, remain stable in aluminium melts and have therefore been suggested to be responsible for nucleation in the carbide and boride theory [3, 7]. Interestingly, boride addition without excess titanium (beyond that of TiB_2) present does not lead to significant grain refinement even if other growth restriction elements are present, suggesting involvement of titanium in the nucleation mechanism [8]. Various nucleation mechanisms have been proposed to explain the involvement of titanium with borides by:

- using bulk equilibrium phase extension arguments [14] stabilizing titanium aluminide beyond its equilibrium phase field,
- adsorption of titanium on boride particles based on local activity arguments [15],

- dissolution kinetics suggesting protective boride shells surrounding titanium aluminide in liquid aluminium [16], or
- cavities in borides preventing titanium aluminide dissolution above the liquidus temperature [17].

All of these proposed nucleation mechanisms for aluminium have been proven wrong or based on artefacts observed only in grain refiner microstructures. The key problem for conventional grain refinement tests remains that nucleation events are obscured by subsequent growth and impingement, which makes it difficult to locate nucleation centres of statistical significance [8]. Nevertheless, grain refinement is used extensively in industrial practice but is determined by empirical rules only. While differences in growth restriction between alloys can be compensated by additional excess titanium, the lack of understanding of the nucleation mechanism becomes a particular problem when alloys are difficult to grain refine, e.g. in the case where alloying elements, such as zirconium, hinder the nucleation.

The grain refinement of aerospace alloys with boride additions can be strongly hindered by zirconium present in the alloys, originally added to form strengthening zirconium aluminide particles. This effect was termed zirconium poisoning and was found empirically to depend on processing conditions where high melt temperatures, long contact times of the refiner particle with the melt and low titanium concentrations promoted zirconium poisoning. Reductions in contact times and melt temperatures, together with higher titanium concentrations, promoted improved grain refinement. The dependence on time, temperature and the concentration levels of zirconium and titanium suggested a thermally activated reaction, such as the transformation of TiB_2 to ZrB_2 assuming TiB_2 as the nucleating particle [18]. Interestingly, when carbide and titanium aluminide additions are added no significant poisoning effect can be noticed, suggesting different nucleation mechanisms between borides and carbides [19]. Carbide additions are accompanied by higher titanium concentrations and do not show the agglomeration suffered by boride particles which is unacceptable in thin section products, such as foil. The nucleation mechanism of titanium carbide is unknown but some interaction with other carbides appears likely given the typically lower casting temperature of 720 °C for carbide compared with 750 °C for boride, to avoid formation of Al_4C_3.

Hypoeutectic Al–Si casting alloys

Grain refinement of primary aluminium in hypoeutectic Al–Si alloys is hindered by lower cooling rates and lower temperature gradients in, e.g., sand casting compared with direct chill casting of wrought aluminium. However, the higher solute concentrations can facilitate better growth restriction. The dominant alloying element in aluminium casting alloys is silicon increasing the fluidity of the melt and permitting better filling. For grain refinement, silicon acts as a growth restriction element. However, empirically it was found that increased grain refiner particle addition compared with wrought alloys is necessary to facilitate grain refinement in casting alloys. Carbide additions have not been found to be as successful as boride additions because of their easy transformation to SiC. Hence boride particles remain the standard refiner addition to Al–Si alloys. The very high residual amounts of titanium >0.15 wt% in casting alloys, resulting from repeated recycling and generous specifications, do not always require further addition of titanium as a growth restriction element [20].

In casting alloys, refiner additions are in much longer contact with the melt than in wrought aluminium alloys, where refiner addition is continuous—allowing only limited time to achieve the ultimate grain size. Refiner addition for casting alloys is often in a holding furnace, where the refiner particles remain in contact with the melt for long times and fading of the refiner efficacy is observed over time. However, it has been shown that sufficient stirring can halt fading, suggesting that nucleating particles can be stopped from settling out [9]. However, after stirring the full grain refiner performance is not recovered, though refiner additions with longer contact times than those in wrought alloys can be used. Sufficient holding in the melt of a refiner addition which does not show efficient refinement after short times can lead to recuperation of the refiner efficacy [21]. This suggests that some activation of particles appears to occur in the melt if not already activated during refiner production. This can explain the use of stoichiometric TiB_2 refiner additions in casting alloys. Excess titanium in conjunction with long holding times appears to reactivate stoichiometric borides and results in successful grain refinement by nucleation and growth restriction by titanium and silicon. However, some difficulty can arise when the silicon concentration exceeds 3 wt% and a significant increase in grain size is observed, termed silicon poisoning [22].

Complications can arise during grain refinement if secondary phases are formed which deplete the melt of solute and hence decrease the solute growth restriction. Interestingly, such a phase can be found in the ternary equilibrium diagram of Al–Si–Ti in figure 12.1. The $Al_5Ti_7Si_{12}$ liquid phase field is stable at very low titanium concentrations [23] and may readily affect conventional Al–Si castings. The effect of secondary phases is not only on growth restriction, as secondary phases may also be occupying nucleation sites originally destined for aluminium. While there is no experimental evidence to suggest that the $Al_5Ti_7Si_{12}$ phase is formed on active particles, its presence is consistent with casting practice. In Al–Si alloys below 3 wt% large amounts of titanium are added to stabilize $Al_3(Ti,Si)$ thus enhancing a peritectic nucleation mechanism, whereas at silicon concentrations above 3 wt% often stoichiometric boride additions are favoured to deplete the melt of titanium and thus favour the formation of titanium aluminide rather than other silicides [20].

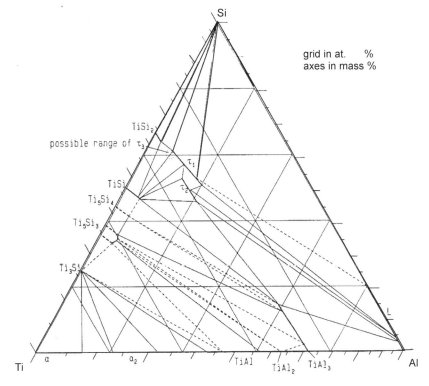

Figure 12.1. Ternary equilibrium phase diagram of Al–Si–Ti at 700 °C.

Overall, the grain refinement of wrought and cast aluminium is being held back by the lack of knowledge of the nucleation mechanisms of borides and carbides in conjunction with titanium aluminide particles.

Metallic glass technique

Recent studies have shown that amorphous aluminium alloys such as Al–Ni–Zr [24], Al–Ni–Si [25], Al–Fe–Si [26], Al–Ni–Ce [27] and Al–Y–Ni–Co [8] can be used to study nucleation processes analogous to conventional casting. The above described commercial grain refiner additions, containing titanium aluminide and boride particles, can be added to glass-forming aluminium alloys, which on rapid cooling are frozen into an amorphous matrix. Figure 12.2 illustrates schematically the metallic glass technique. For such investigations the glass-forming alloy used must conform to the following conditions.

1. There must be a driving force for nucleation of aluminium above the glass transition temperature T_g, i.e. the metastable extension of the aluminium

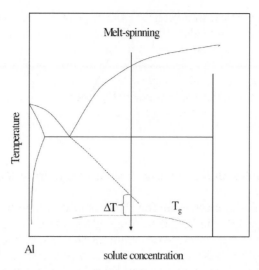

Figure 12.2. Schematic diagram of the methodology behind the metallic glass technique which investigates nucleation events frozen into a metallic glass matrix.

liquidus line must lie above the glass transition temperature for a given composition, shown by ΔT in figure 12.2.
2. Nucleation events on heterogeneous sites must be clearly distinguishable from those within the matrix, achieved by having transient nucleation and low number densities (best to be zero) of devitrification crystals.
3. Added particles must remain stable within the matrix, i.e. they must not react with the melt at the high metallic glass processing temperatures ($\sim 900\,°C$).

On quenching, nucleation of aluminium can occur below the metastable extension of the aluminium liquidus but above the glass transition temperature, below which growth of nucleation sites is halted. It is interesting to note that the glass transition is not fixed but dependent on quenching conditions. Experimentally a balance has to be found to facilitate a sufficiently wide temperature interval between the metastable extension of the aluminium liquidus (fixed) and the glass transition temperature (depressed by high cooling rates) whilst maintaining a sufficiently low density of devitrified crystals favoured by high cooling rates. At the glass transition temperature the atomic mobility has been estimated to be a factor of 10^{16} slower than in a conventional melt [8]. By this route the action of commercial grain refiner particles can be frozen into a metallic glass matrix. Subsequent electron microscopy studies readily allow identification of nucleation centres and measurement of potential orientation relationships between heterogeneous sites and aluminium crystals [8]. Higher undercoolings than in conventional casting and suppressed recalescence favour nucleation on all active particles, giving a high statistical significance to the observed nucleation mechanisms. Artefacts of the metallic glass technique occur when particles react with the melt, or when phases other than aluminium are heterogeneously nucleated because of the large solute contents in glass-forming alloys. However, the metallic glass technique gives a unique frozen-in picture of the nucleation mechanism in the undercooled melt. It is thereby possible to investigate incomplete peritectic reactions, which in conventional solidification would not be observable.

Nucleation mechanism for aluminium in undercooled melts

The same nucleation mechanism for aluminium has been observed on added grain refining particles independent of the glassy matrix used in the metallic glass experiments [15, 27, 8]. No separate titanium aluminide particles could be found within the glassy matrix, consistent with complete dissolution of titanium aluminide particles as in conventional casting at low titanium concentrations. This highlights that a peritectic reaction on isolated titanium aluminide particles is not possible in Al–Ti alloys with concentrations below the peritectic horizontal. Moreover, rapid quenching did not preserve any dissolving titanium aluminide particles and their dissolution occurred rapidly at the higher processing temperatures during melt-spinning. Nucleation of aluminium occurred on hexagonal platelets of boride particles. Such a particle is shown in figure 12.3 where the particle is tilted with its basal plane parallel to the electron beam in transmission electron microscopy. The faceted nature of the boride can be seen in the dark-field imaging mode with the objective aperture on the $\{10.0\}$ boride spot (figure 12.3(b)), which shows ledges on the boride platelet. The nucleated aluminium can be observed only on basal faces of the boride particle in bright-field (figure 12.3(a)) and in dark-field imaging with the objective aperture on the $\{111\}$ aluminium spot (figure 12.3(c)). Within the inserted diffraction pattern, which is orientated relative to the boride particle in bright field, a streak is visible in the $\langle 00.1 \rangle$ direction. Dark-field microscopy with the objective aperture on the streak (figure 12.3(d)) reveals a thin layer covering the boride on the basal planes. The layer can also be seen in bright-field microscopy on the prism faces of the boride. The thin layer has been identified as titanium aluminide from selected area electron diffraction pattern and energy dispersive X-ray analysis [28, 29]. Furthermore, in the inset selected area diffraction pattern in figure 12.3 the diffraction spots for aluminium, boride and the streak are all aligned in the $\langle 00.1 \rangle$ direction, which is indicative of an epitaxial relationship between the boride, titanium aluminide and aluminium. Closer analysis of the electron diffraction pattern reveals that the close-packed planes and directions are parallel for the boride, titanium aluminide and aluminium. With this epitaxial relationship the bulk mesh parameters on the basal plane of the boride particle of aluminium (0.286 nm) and of

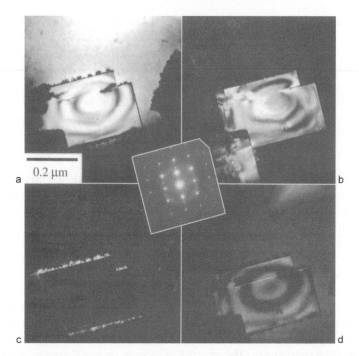

Figure 12.3. A TiB$_2$ particle tilted with its [00.1] zone axis parallel to the electron beam. (a) Bright-field TEM showing copious nucleation of aluminium on basal faces of the boride, and dark-field TEM showing (b) the faceted boride, (c) the aluminium and (d) a thin layer of Al$_3$Ti.

titanium aluminide (0.275 nm) are stretched to that of the boride (0.303 nm). Only thin epitaxial layers can accommodate such large stresses of ∼10%. However, the nucleating aluminium only grows in small patches on the stretched titanium aluminide layer and cannot accommodate such large stresses as experienced in the titanium aluminide layer. The large stress within the titanium aluminide layer is expected to reduce its thermodynamic stability; its observation suggests that chemical attraction to the underlying boride substrate must compensate for the effect of stress within the layer. Nevertheless, the low index orientation relationship underlines the high potency of the titanium aluminide covered borides to nucleate aluminium [11].

The nucleation of aluminium observed in figure 12.3 does not occur in ledges to minimize surface energy as proposed by the classical nucleation theory [17], but only on basal faces covered with the titanium aluminide layer. This suggests that the prism faces, despite being covered with titanium aluminide, are not energetically favourable nucleation sites for aluminium; for successful nucleation of aluminium, good lattice matching, as shown by the low index orientation relationship, and chemical attraction, such as by adsorption, are necessary. Moreover, the classical nucleation theory appears to break down for highly potent heterogeneous nuclei, such as titanium aluminide on the boride particle, and the nucleation process is better described as an adsorption process [30]. Elemental titanium, predicted by Jones [15], is not detected in the vicinity of the boride and any detectable titanium appears to be consumed in the titanium aluminide layer. Although the nucleated aluminium crystals are smaller (5 nm) because of larger undercoolings than those expected in conventional aluminium casts (1 µm), the key feature of the nucleation mechanism is still the thin titanium aluminide layer, which is beyond the resolution of previous microstructural investigations. It remains an unanswered question as to whether the layer observed in metallic glass experiments is already present on borides in the grain refiner before addition or whether such layers form after addition to the melt. Nevertheless, the observed nucleation mechanism can be directly related to the observed phenomena in commercial casting.

Excess titanium

The important role of excess titanium in grain refinement has been established by Jones and Pearson [18]. However, in their experiments it was not clear if excess titanium affects growth only or if a nucleation effect also exists. In metallic glass experiments the effect of excess titanium can be readily demonstrated by increasing or decreasing the amount of titanium added to the glass containing the refiner particles. At low titanium concentrations, below that required to form stoichiometric borides, no nucleation of aluminium can be observed, as shown in figure 12.4, which is consistent with the lack of a titanium aluminide layer. However, by increasing the titanium level, first a thin layer of titanium aluminide can be observed, as in figure 12.3, and then thickening of the layer and formation of equilibrium {001} facets, not parallel to the boride basal planes, can be observed, as in figure 12.5. Interestingly, the {112} of the titanium aluminide faces, parallel to the basal plane, show copious nucleation of aluminium despite being a non-equilibrium facet, while the {001} of the titanium aluminide faces seems not to nucleate aluminium [28]. This suggests that the boride particles stabilize on their basal faces a particularly effective nucleation facet of titanium aluminide. At low titanium concentrations it appears that a peritectic reaction is preserved by a strong epitaxial orientation relationship and by chemical interaction leading to adsorption of titanium aluminide on boride particles beyond the stable two-phase field of titanium aluminide and liquid.

At higher titanium concentrations boride particles readily act as nucleation centres for titanium aluminide particles and these may grow if the particle is held in the stable two-phase field. This is particularly apparent in the presence of tantalum, which forms a mixed $Al_3(Ti,Ta)$ aluminide [31]. Thus titanium is initially required to form a nucleating layer preserved on

Figure 12.4. Stoichiometric boride particle showing no nucleation of aluminium.

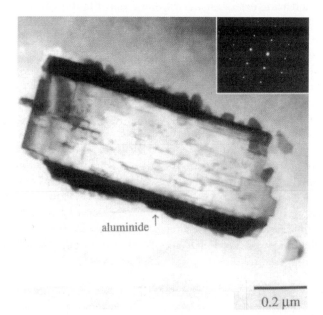

Figure 12.5. Hyperperitectic addition of titanium resulting in growth of titanium aluminide on the boride particle.

boride particles, but further additions are in equilibrium with the liquid and are available for growth restriction in conventional casting. At higher titanium concentrations titanium aluminide particles can grow in the two-phase field of liquid and titanium aluminide.

Agglomeration and fading

Agglomeration of boride particles can occur during the manufacture of the refiner where particle clustering is controlled by excessive mechanical deformation, and during holding of the grain refiner in the melt for long times, after which a loss of grain refiner efficiency is observed, termed fading [9]. Grain refiner additions held for long times (up to 45 min) in metallic glasses do not show any sign of fade relating to the previous described nucleation mechanism [8] and boride particles exhibit titanium aluminide layers similar to those observed in figure 12.3 [29]. Since borides are the effective carrier for the peritectic reaction of the titanium aluminide layer, any settling of the borides will deplete the melt of potential nucleation sites. This has been shown to affect the refiner efficacy but can only partly be restored by stirring of the settled boride particles in conventional casting experiments [9]. During sedimentation, boride particles will make physical contact with each other, whereby the titanium aluminide layers covering the borides are brought together and may adhere. It is not expected that shear forces during stirring would be sufficiently large to break boride particles joined in such a way by the layers. Only excessive superheat would result in dissolution of the titanium aluminide layer between particles [30]. Only small boride clusters can be observed by transition electron microscopy in metallic glass experiments due to thin foil preparation techniques. In figure 12.6 a high-resolution TEM micrograph of two adjoining boride particles close to their $\langle 11.0 \rangle$ zone axes are shown embedded in a metallic glass matrix. The two particles are effectively held together by a titanium aluminide layer clearly visible at the boride–boride interface and strained in between the boride particles due to lattice mismatch [29]. Agglomeration of boride particles, acting as carriers for the titanium aluminide layer, changes the effective size and number distribution of potential nucleation sites from the empirically obtained optimum and results in an overall decreased grain refiner efficiency [4], thus explaining the observed partial recovery in conventional casting experiments [9] after stirring.

Zirconium poisoning

7xxx series aluminium alloys can contain zirconium concentrations within the peritectic horizontal of the liquid + zirconium aluminide → Al peritectic

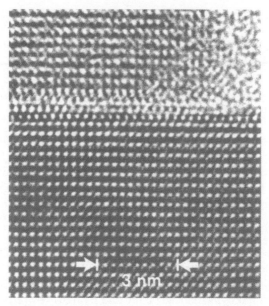

Figure 12.6. High-resolution transmission electron microscopy of two boride particles tilted close to the ⟨11.0⟩ direction, showing a titanium aluminide layer on their opposing basal planes, acting as an 'adhesive'.

system. In metallic glasses containing hyperperitecic zirconium concentrations to facilitate glass formation [24, 32] two effects have been observed on the nucleation mechanism on borides. At high temperatures and prolonged exposure times of TiB_2 to zirconium, TiB_2 particles can be completely transformed into ZrB_2, thereby affecting the delicate epitaxial orientation relationship between the boride, the titanium aluminide layer and aluminium. The lattice parameter of zirconium boride is larger than that of titanium boride and it is not possible for a titanium aluminide layer to be stable on the transformed boride. Consequently, no nucleation of aluminium is observed on the bare borides [24, 32].

At low temperatures and short exposure times a more subtle effect can be observed. Titanium boride particles appear not to have transformed completely into zirconium boride and the remainder of the layer structure can still be observed in figure 12.7. The boride particle tilted to its ⟨11.0⟩ zone axis exhibits {00.1} faces in bright field and in dark field with the objective aperture on the (10.0) spot. In contrast to previous studies [8], the titanium aluminide layer is not continuous over the boride basal planes and only weak streaking is apparent in the selected area diffraction pattern. However, some dark-field contrast can be obtained with the objective aperture placed between the (000) and (001) spots marked in figure 12.7(d) with an arrow. The small dimensions of the layer, less than 1 nm, prohibit

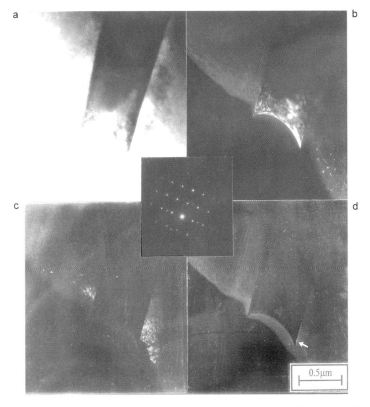

Figure 12.7. Boride particle viewed in the transmission electron microscope. (a) Bright field, with its basal plane edge-on, and in dark field with the objective aperture on (b) the (10.0) boride spot highlighting the boride, (c) the (111) aluminium spot showing some aluminium crystals close to the foil edge, and (d) the expected location for a streak in the inserted selected area diffraction pattern between the (000) and (001) spots showing a faint interface layer.

unambiguous identification of the layer by selected area diffraction or energy dispersive X-ray analysis. However, zirconium has limited solid solubility in titanium aluminide and will form zirconium aluminide at higher zirconium concentrations [33]. In the $Al_{87}Ni_{10}Zr_3$ (at%) glass-forming alloy sufficient zirconium is present to form separate zirconium aluminide particles. Such a zirconium aluminide particle is shown in bright field in figure 12.8(a) where the dendritic zirconium aluminide particle has nucleated copious amounts of aluminium. The zirconium aluminide consists of four petal-like dendrites visible in dark-field transmission electron microscopy in figure 12.8(b) resulting in four overlapping diffraction patterns of $\langle 001 \rangle$ zone axes in figure 12.8(c). Aluminium can readily nucleate on the zirconium aluminide petals, as shown in dark-field transmission electron microscopy in figure

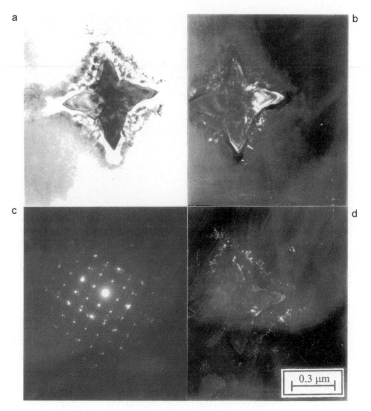

Figure 12.8. Four slightly misaligned petal-like Al$_3$Zr dendrites viewed in the transmission electron microscope. (a) Bright field, (b) dark field with the objective aperture on one of the {001} Al$_3$Zr spots, (c) selected area diffraction pattern showing the four ⟨001⟩ Al$_3$Zr patterns, and (d) dark field with the objective aperture on one of the {002} aluminium spots highlighting aluminium dendrites.

12.8(d); zirconium aluminide, similar to titanium aluminide, exhibits a peritectic reaction with liquid to form aluminium. However, the peritectic temperature for zirconium aluminide with liquid is ~4.5 °C lower than the 665 °C for titanium aluminide and liquid. This is a small difference for undercoolings in the metallic glass experiments and nucleation of aluminium on zirconium aluminide is readily observed. However, in conventional casting the columnar to equiaxed transition will be significantly influenced by even such a small difference in undercooling necessary for nucleation. Furthermore, the formation of titanium aluminide and zirconium aluminide layers results in a bimodal distribution of the number of active particles versus undercooling. Only particles with the lowest activation undercooling for heterogeneous nucleation will be active and, on their growth, recalescence will result in an overall reduced number of active nucleation sites [21, 24].

Overall, zirconium poisoning is a thermally activated process ultimately transforming titanium boride to zirconium boride. However, at intermediate transformation stages, i.e. lower temperatures or shorter contact times, zirconium will have to diffuse through aluminide layers and may initially form zirconium aluminide layers before transforming titanium boride to zirconium boride. In casting practice, at higher titanium to zirconium concentrations, this will give a sufficiently long time interval where at reduced temperatures grain refinement can be achieved with boride additions, thus explaining the observation in conventional casting of an empirical critical titanium to zirconium ratio for a given processing procedure and its time and temperature dependence.

Silicon poisoning

The grain refinement and processing of hypoeutectic casting alloys differs from that of wrought alloys. Refiner addition is often in the holding furnace, where the addition remains in contact with the melt for long times in contrast to rod addition in the launder prior to casting of wrought alloys. Apart from the above described fading and agglomeration difficulties, which are in practice overcome by continuous stirring, there is sufficient time for the melt to equilibrate so that poisoning reactions may proceed. This is in contrast to casting of zirconium-containing alloys where zirconium poisoning can be overcome by utilizing an incomplete poisoning reaction. Similar to zirconium poisoning, other elements such as tantalum, scandium and importantly silicon have solid solubility in the layer of titanium aluminide. From the metallic glass experiments simulating the nucleation mechanism in wrought alloys it appears that the strong epitaxial orientation relationship of the titanium aluminide layer with the boride, and local chemical interaction thermodynamically stabilize the layer in contact with the melt. In the case of aluminium casting alloys titanium concentrations close to or within the peritectic horizontal in binary Al–Ti are common practice and the titanium aluminide layer is expected to be further stabilized.

The phases formed on boride particles in contact with liquid aluminium can only be predicted from ternary equilibrium diagrams of aluminium, titanium aluminide and silicon without detailed knowledge of the dilute but complex boride–Al–Ti–Si multi-component systems. Furthermore, shifts in phase fields should be expected if the phases are in metastable equilibrium. In the case of low silicon concentrations up to approximately 3 wt% this will result in the formation of $Al_3(Ti,Si)$ as indicated in figure 12.1, while at higher concentrations ternary phases such as $Al_5Ti_7Si_{12}$ may be formed. Any phase isomorphous with Al_3Ti (DO_{22}) showing a peritectic reaction with the liquid will enhance the nucleation effect of the layer, while other phases can have a twofold effect. If non-peritectic phases are nucleated on basal faces of the

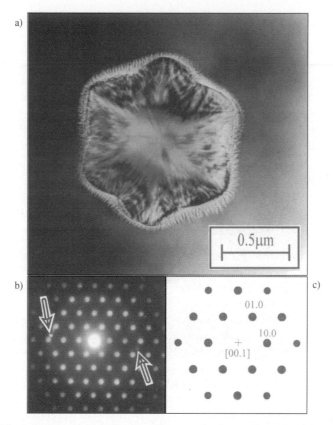

Figure 12.9. A hexagonal devitrification phase nucleating copious aluminium dendrites viewed in the transmission electron microscope. (a) Bright-field and (b) the corresponding selected area diffraction pattern of the hexagonal phase with the aluminium spots, marked, and (c) indexing of the pattern.

borides they are in direct competition with titanium aluminide. Furthermore, any ternary phase promoted by boride particles will reduce the effect of growth restriction in conventional casting, particularly when the strong growth restricter titanium is bound into a new phase.

For the simulation of nucleation in aluminium casting alloys containing silicon a metallic glass of composition $Al_{70}Ni_{13}Si_{17}$ (at%) was investigated [25] with varying amounts of excess titanium and holding times prior to quenching. $Al_{70}Ni_{13}Si_{17}$ is a marginal glass former showing no glass transition in calorimetric investigations, and only small amounts of aluminium can be detected in devitrified samples. Nevertheless, in figure 12.9 transmission electron microscopy of the as-quenched samples shows that dendritic aluminium has nucleated copiously on a hexagonal phase previously detected [34] as a devitrification product. This suggests that a driving force for the nucleation of aluminium exists.

Silicon poisoning 195

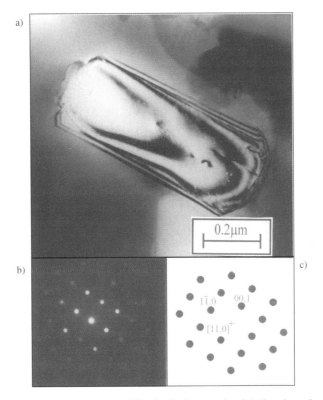

Figure 12.10. TiB$_2$ grain refiner particle tilted close to the ⟨11.0⟩ axis embedded in an Al–Ni–Si glass viewed in the transmission electron microscope. (a) Bright field showing no aluminium particles on the basal planes but small spherical caps of an unidentified intermetallic compound, (b) selected area diffraction pattern and (c) the indexed pattern.

In contrast, no nucleation of aluminium was observed on boride particles. In figure 12.10 a boride particle viewed with its basal faces close to the edge-on position exhibits no nucleation of aluminium on the basal planes. However, nucleation of an unidentified intermetallic phase occurred on the prism planes. In contrast to zirconium poisoning, energy dispersive X-ray spectroscopy analysis revealed an unchanged chemistry of the titanium boride particle itself, suggesting a change of chemistry of the nucleating layer previously observed. It is not possible to determine unambiguously the existence of a layer by energy dispersive X-ray spectroscopy. If a layer exists, it is 1 nm or less in thickness, and hence difficult to analyse. It is interesting to note that longer holding times prior to melt-spinning result in more pronounced nucleation of TiSi$_2$ on prism faces of boride particles in a metallic glass containing only 2 at% silicon [35]. This suggests that nucleation may have occurred during holding in the liquid state. The TiSi$_2$ phase is not expected to form from the equilibrium diagram in figure 12.1 (isothermal section at 700 °C) and may be the result of higher silicon and titanium concentrations or higher holding temperatures in the metallic glass experiments (950–1200 °C) compared with conventional casting practice (~750 °C).

Nevertheless, the metallic glass experiment establishes that the titanium-rich intermetallic phase can be nucleated during holding, which will be in direct competition to $Al_3(Ti,Si)$ on basal faces [25]. In conventional casting practice the formation of these phases may be avoided by decreasing the holding time of the refiner addition or, if suitable, for the casting process by a direct launder addition. The equilibrium phase diagram suggests that the aluminium nucleating phase $Al_3(TiSi)$ can be stabilized by increasing the titanium content in low silicon containing alloys, while it remains open if titanium addition beyond silicon concentrations higher than 3 wt% may be furthering the formation of titanium silicides. Interestingly, the addition of stoichiometric refiner [20] may help to decrease the amount of titanium, but the work presented here would suggest that the stoichiometric refiner particles would need to be become covered with a titanium aluminide layer to become active.

Modelling of nucleation

For simple one-dimensional temperature gradients, predictive modelling of grain sizes by grain refinement requires detailed information on growth restriction and heterogeneous nucleation undercooling. While array growth of dendrites has been successfully modelled and experimentally verified, there has been little modelling of nucleation of highly potent heterogeneous sites active at low undercoolings. Recently, heterogeneous nucleation by adsorption has been successfully modelled by Kim and Cantor [30] for binary systems. However, their model is currently limited to a thermodynamic description of regular solutions only and will require the incorporation of kinetic models if a heterogeneous nucleation in ternary systems involving peritectic reactions is to be predicted. In the case of titanium aluminide on boride particles, it appears that the adsorption of titanium aluminide on the boride is not the rate-limiting step for nucleation and growth of aluminium. The aluminium crystals nucleated in the metallic glass technique are much smaller than those of critical size (approximately 1 µm) in a slightly undercooled (approximately 0.5 °C) conventional melt. However, the nucleation of aluminium is restricted to low-energy interfaces (here the basal faces of boride particles) and results in a low index orientation relationship between the boride, titanium aluminide and aluminium [11]. Nucleation or growth is not favourable on prism faces, suggesting a high interfacial energy due to the lattice mismatch.

An aluminium crystal nucleated on a boride at low undercooling, and of similar size to the boride particle will not be able to grow freely over prism faces. In conventional melts the rate-limiting step has been suggested to be when aluminium, readily nucleated on titanium aluminide covered borides, can grow freely from the borides. A free growth model of aluminium has been proposed based on the observation that aluminium can nucleate only on basal faces of the boride [4]. Free growth from a particle occurs when the equivalent radius of the basal face of a boride particle equals that of the critical nuclei at a given temperature, thus making heterogeneous nucleation dependent on the particle size distribution of added grain refiner particles. Successful modelling of grain sizes involving growth restriction and particle size distributions has been demonstrated and is reported by Greer in chapter 13.

Summary

Heterogeneous nucleation in aluminium alloys is still the weak link in the understanding and modelling of grain refinement as a nucleation and growth restriction process. Recent investigations using a metallic glass technique have successfully identified the heterogeneous nucleation mechanisms of conventional titanium boride and titanium aluminide grain refiner additions. Boride particles are covered in a thin epitaxial layer of titanium aluminide, nucleating aluminium on basal faces of the boride particles. The nucleation mechanism found in metallic glasses is consistent with casting practice. Excess titanium, necessary for successful grain refinement, forms as a thin layer of titanium aluminide, which nucleates aluminium by a peritectic reaction outside its bulk equilibrium composition range. Additional titanium, not in the form of titanium aluminide on the boride, can act as a strong growth restricter. The limited solid solubility of elements such as zirconium and silicon in the titanium aluminide layer can lead to poisoning of the nucleation mechanism or nucleation of phases other than titanium aluminide, thereby occupying basal faces of the borides and/or affecting free solute in the liquid available for growth restriction. However, this very successful nucleation mechanism for aluminium is also the cause of agglomeration and clustering of boride particles and has led to the development of non-clustering titanium carbide based refiners which do not suffer from clustering. Future work will identify suitable metallic glasses to study the nucleation mechanism on carbides.

References

[1] Birch M E J and Fisher P 1986 in *Aluminium Technology 1986* ed T Sheppard (London: Institute of Materials) p 117
[2] Hunt J D 1986 *Mater. Sci. Eng.* **A65** 75
[3] McCartney D G 1989 *Int. Mater. Rev.* **34** 247
[4] Greer A L this Seminar
[5] Johnson M 1994 *Z. Metallkunde* **85** 781

[6] Spittle J A and Sadli S B 1995 *Mater. Sci. Technol.* **11** 533
[7] Cibula A 1949 *J. Inst. Met.* **76** 321
[8] Schumacher P, Greer A L, Worth J, Evans P V, Kearns M A, Fisher P and Green A H 1998 *Mater. Sci. Technol.* **14** 394
[9] Birch M E J and Fisher P 1988 in *Solidification Processing 1987* ed J Beech and H Jones (London: The Institute of Materials) p 500
[10] Maxwell I and Hellawell A 1975 *Acta Metall.* **23** 229
[11] Turnbull D and Vonnegut B 1952 *Ind. Eng. Chem.* **44** 1292
[12] Small C M, Pragnell P B, Hayes F H and Hardman A 1998 *Proc. ICAA-6, Toyohashi, Japan* ed T Sato, S Kumai, T Kobayashi and Y Mrakami (The Japan Institute of Light Metals) p 213
[13] Arnberg L, Backerud L and Klang H 1982 in *Grain Refinement in Castings and Welds* ed G J Abbaschian and S A David (New York: TMS–AIME) p 165
[14] Kiusallas R 1986 *Chem. Comm., Stockholm* p 1
[15] Jones G P 1988 in *Solidification Processing 1987* (London: The Institute of Materials) p 496
[16] Vader M and Noordegraaf J 1990 in *Light Metals 1990* ed C M Bickert (Warrendale, PA: TMS) p 851
[17] Turnbull D 1950 *J. Chem. Phys.* **18** 198
[18] Jones G P and Pearson J 1975 *Metall. Trans.* **7B** 223
[19] Birch M E and Fisher P 1988 in *Solidification Processing 1987* ed J Beech and H Jones (London: The Institute of Metals) p 149
[20] Sigworth G K and Guzowski M M 1985 *AFS Trans.* **93** 907
[21] Bunn A M, Evans P V, Bristow D J and Greer A L 1998 in *Light Metals 1998* ed B Welch (Warrendale, PA: TMS) p 963
[22] Spittle J A, Keeble J M and Al Meshhedani M 1997 in *Solidification Processing 1997* ed J Beech and H Jones (Sheffield) p 273
[23] Perrot P 1993 in *Ternary Alloys* ed G Petzow and G Effenberg (Weinheim: VCH) vol 8/9 p 283
[24] Bunn A M, Schumacher P, Kearns M A, Boothroyd C B and Greer A L 1999 *Mater. Sci. Technol.* **15** 1115
[25] McKay B, Cizek P, Schumacher P and O'Reilly K A Q 2001 *Mater. Sci. Eng.* **A304–306** 240
[26] Cizek P and Schumacher P 2001 *Mater. Sci. Eng.* **A304–306** 215
[27] Schumacher P and Greer A L 1997 *Mater. Sci. Eng.* **A226–228** 794
[28] Schumacher P and Greer A L 1995 *Light Metals 1995* ed J Evans (Warrendale, PA: TMS) p 869
[29] Schumacher P and Greer A L 1996 in *Light Metals 1996* ed W Hale (Warrendale, PA: TMS) p 745
[30] Kim W T and Cantor B 1994 *Acta Metall. Mater.* **42** 3115
[31] Schumacher P and Greer A L 1994 *Mater. Sci. Eng.* **A178** 309
[32] Schumacher P and Cizek P *Light Metals 2000* ed R D Peterson (Warrendale, PA: TMS) p 839
[33] Tsurekawa S and Fine M E 1982 *Scr. Metall.* **16** 391
[34] Legresy J M 1987 PhD Thesis, l'Institut National Polytechnique de Grenoble
[35] McKay B J, Cizek P, Schumacher P and O'Reilly K A Q in preparation

Chapter 13

Control of grain size in solidification

Lindsay Greer

Introduction

In metallic microstructures, the sizes and shapes of the grains can be important in determining performance, not only in structural but also in functional applications. For example, finer grain size is beneficial in alloys for ambient-temperature structural use, because it simultaneously gives greater strength and greater toughness, without requiring (possibly expensive) alloying additions. On the other hand, directional solidification, to give columnar grains with boundaries parallel to the main stress axis, or a single crystal, is beneficial for creep resistance at elevated temperatures, as required for example in turbine blades [1]. Functional uses of alloys often involve electrical conduction. For example, the design and control of grain structure is important in extending the lifetime of tungsten filaments in incandescent lamps [2]. A further example is in the metallic interconnects on integrated circuits, where a bamboo microstructure (i.e., with all grain boundaries normal to the electrical current) is particularly desirable for offering resistance to electromigration [3].

In some of the above examples, and in many practical cases, the final grain structure is the result of solid-state processes—recrystallization following mechanical deformation, or grain growth. This chapter restricts its consideration to the control of grain structure in as-solidified alloys. An interesting case is that of devitrification, i.e. the crystallization of a glass. Arguably this is a solid-state process; it is nevertheless considered in this chapter as it is very closely related to solidification, the glass being but congealed liquid. It is important to note that solidification is a near-universal part of alloy processing and that, even if an alloy is subjected to post-solidification processing, the final microstructure may still be significantly affected by the prior as-cast microstructure.

In as-cast microstructures, the prime concern is often with whether the grain structure is columnar or equiaxed. In the turbine-blade example above, a columnar structure may be desirable, but in many cases an equiaxed structure is strongly preferred. In large-scale casting, for example direct-chill (DC) casting of aluminium, columnar growth is associated with centre-line cracking and with severe macrosegregation. By having a structure of fine, equiaxed grains (obtained in industrial practice by inoculation [4], which will be described later), direct chill casting speeds can be increased. There are varied additional benefits, for example improved surface quality of the ingots. In shape casting, equiaxed grain structures are again preferred, this time to improve mould filling. Interestingly, there can be an optimum degree of grain refinement, with larger or smaller grains degrading the castability. The grain structure in weld metal is also preferably equiaxed, not only for improved strength but also to reduce elastic anisotropy thereby facilitating ultrasonic inspection of weld integrity [5].

The columnar-to-equiaxed transition (CET), determining the conditions for one or the other type of grain growth, has been very widely studied. The formation of an equiaxed grain structure requires effective nucleation of new grains in the liquid ahead of the main directional growth front (figure 13.1).

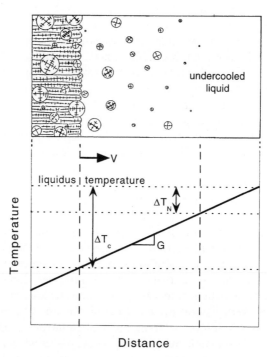

Figure 13.1. Directional solidification from left to right. A schematic view of the competition between columnar and equiaxed growth (after reference [6]). The columnar front moves with velocity V and the temperature gradient throughout the solidifying region is G. ΔT_N is the undercooling for heterogeneous nucleation on inoculant particles in the liquid, while ΔT_C is the undercooling for growth of the columnar front.

Introduction

Hunt [6] has analysed the conditions under which the new grains can successfully block the advance of the main front; this leads to an analytical estimate of the temperature gradient G under which fully equiaxed growth is ensured:

$$G < 0.617 N_0^{1/3} \left\{ 1 - \frac{(\Delta T_N)^3}{(\Delta T_C)^3} \right\} \Delta T_C. \tag{13.1}$$

In this expression, N_0 is the number of nucleant particles per unit volume, ΔT_N is the undercooling for heterogeneous nucleation on the particles, and ΔT_C is the undercooling for growth of the columnar front. The latter undercooling is related to the solidification velocity V by:

$$\Delta T_C = 2 \left\{ \frac{2m(k-1)C_0 V \sigma_{sl}}{D \Delta S_V} \right\}^{1/2}, \tag{13.2}$$

where m is the liquidus slope of the alloy, k is the solute partition coefficient, C_0 is the content of alloying element, σ_{sl} is the solid–liquid interfacial energy, D is the diffusivity of the solute in the liquid, and ΔS_V is the entropy of melting per unit volume. Hunt's treatment has been extended to include more recent dendrite growth models [7]; these can modify the constant of proportionality in equation (13.2). The CET has also been studied by stochastic microstructure modelling, which has shown that the transition is gradual, with intermediate structures of elongated grains [8].

Whether a grain structure is columnar or equiaxed is so important that most processing has been optimized to give the desired structure, aided by the good understanding of the CET. This chapter focuses instead on a less well understood aspect of grain-structure control in solidification, namely control of grain size in the usually preferred equiaxed regime. Nevertheless, it is appropriate to consider the main factors involved in the CET, as identified in equations (13.1) and (13.2). It can be seen from figure 13.1 that the nucleation of new grains occurs only in the constitutionally undercooled region ahead of the main front. Decreasing the temperature gradient increases the extent of this region and favours equiaxed growth. Increasing the solidification velocity increases the undercooling at the main growth front, also increasing the extent of the undercooled region and favouring equiaxed growth. Nucleation of new grains is facilitated if there are more nucleation sites and if these sites operate at smaller undercooling. These same factors, solidification velocity and temperature gradient, together with the alloy parameters liquidus slope, solute position coefficient and content of alloying element, site density and critical undercooling, also control the grain size in the equiaxed regime. Clearly, nucleation is an important aspect of the development of equiaxed grain structures, and this is considered next.

Difficulties of nucleation analysis

In recent years excellent progress has been made in the analysis of growth kinetics in solidification. This progress has also led to advances in the predictive modelling of as-solidified metallic microstructures [9]. The results can often be conveniently represented on a microstructure-selection map. Such a map typically has alloy composition (for a binary system) on one axis and a key processing variable on the other. The areas on the map corresponding to different microstructures can be determined experimentally and compared with modelling predictions. This approach has been widely used for surface treatments such as scanned electron beam [10] or laser [11] melting, in which the most appropriate processing parameter is scan velocity. In these cases the modelling is of growth kinetics and the microstructure selection is purely by growth competition. Nucleation does not play any significant role, because of the unmelted substrate from which growth can readily occur without a nucleation barrier for any of the main microstructures (dendritic, eutectic, etc).

However, there are many cases in which nucleation must play a key role in phase and microstructure selection. In droplet solidification (e.g. atomization), the fully liquid droplets start to solidify at an undercooling determined by the first nucleation event; this undercooling determines the microstructure and it may vary widely from droplet to droplet. In bulk processing, competitive nucleation may be involved in the selection of secondary phases [9]. Phase selection is considered in more detail in chapter 15. Unfortunately, the progress in quantitative modelling of growth kinetics has not been matched by similar progress for nucleation kinetics. The difficulties in quantitative modelling of nucleation are considered next.

Classical nucleation theory has been extensively reviewed elsewhere [12–14]. It considers that a critical nucleus arises by stochastic addition of molecules to initially sub-critical embryos. The simple classical theory uses macroscopic thermodynamics to describe the work of formation of nuclei and this is subject to criticism. Nevertheless, the basic kinetic analysis offered by the classical theory appears to be correct [15]. Many of the more recent theories, for example the density-functional theory [16] or the diffuse-interface theory [17], can be considered to be extensions of classical theory, or at least can be reduced to the classical theory in appropriate limits. Accordingly, the classical theory is retained as the basis for the present discussion.

We consider a case in which the crystalline phase nucleating in the liquid has the same composition as the liquid, in effect a one-component system. We examine first *homogeneous* nucleation in which there are no heterogeneities in the system apart from the thermodynamically expected fluctuations in the liquid itself. Standard analyses show that the homogeneous nucleation frequency I_{homo} (number of critical nuclei appearing in unit volume per second) is given by an expression of the form:

$$I_{\text{homo}} \ (\text{m}^{-3} \ \text{s}^{-1}) = \frac{A}{\eta} \exp\left(-\frac{W^*_{\text{homo}}}{k_B T}\right), \tag{13.3}$$

where η is the viscosity of the liquid at the melting point, A is a constant (proportional to the number of atoms per unit volume) which for liquid metals and alloys has the value $\sim 10^{35} \ \text{Pa m}^{-3}$ [18], W^*_{homo} is the work of formation of a critical nucleus, k_B is the Boltzmann constant and T is the temperature. The work of formation is given for a spherical nucleus by

$$W^*_{\text{homo}} = \frac{16\pi\sigma_{\text{sl}}^3}{3\Delta G_V^2}, \tag{13.4}$$

where σ_{sl} is the solid–liquid interfacial energy, and ΔG_V is the free energy change on solidification per unit volume. The key parameters needed for prediction of a nucleation frequency are thus the viscosity of the liquid, the free energy change on solidification per unit volume and the solid–liquid interfacial energy. Of these, the viscosity of the liquid and the free energy change on solidification can be measured reasonably accurately in independent experiments; the free energy change on solidification increases approximately linearly with undercooling. The main problem lies with the solid–liquid interfacial energy which is not amenable to direct determination. It can be predicted from models for the solid–liquid interface, for example the negentropy model [19]. It can also be estimated from various experiments, but these are sometimes themselves nucleation experiments. The estimates of solid–liquid interfacial energy are not very accurate, perhaps within 10%. Unfortunately, as seen in equations (13.3) and (13.4), the nucleation frequency is extremely sensitive to the value of the interfacial energy as solid–liquid interfacial energy is cubed in the argument of the exponential. With typical parameter values, even a 1% change in solid–liquid interfacial energy changes the homogeneous nucleation frequency by an order of magnitude. Such large uncertainties are likely to outweigh the corrections possible with improvements in the theory. Clearly, accurate prediction of homogeneous nucleation frequency is likely to be elusive.

The treatment so far has been for homogeneous nucleation, i.e. that occurring without any catalytic substrates. The undercooling required to initiate homogeneous nucleation is substantial, typically >20% of the absolute melting temperature [14], and much greater than the undercoolings typical of metal processing. Thus it is accepted that nucleation in most, if not all, cases of industrial importance is heterogeneous. Within the classical theory [12, 13], heterogeneous nucleation is treated as in figure 13.2. The radius of curvature of the solid–liquid interface corresponding to the critical nucleus is exactly the same as for homogenous nucleation. However, the nucleus itself is in the form of a spherical cap; its volume is less than that

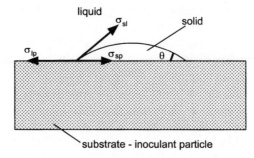

Figure 13.2. Schematic diagram of classical heterogeneous nucleation on a substrate. The nucleus is in the form of a spherical cap. The contact angle θ is determined by the relative interfacial energies, σ_{lp}, σ_{sp} and σ_{sl}.

of the full sphere relevant for homogeneous nucleation, and correspondingly its work of formation is reduced. The spherical-cap shape is stabilized by the balance of surface tensions in the plane of the surface of the nucleant substrate. The surface tensions are taken to be equivalent to the surface energies which are σ_{sl}, σ_{lp} and σ_{sp} for the solid–liquid, liquid–particle and solid–particle interfaces respectively. The contact angle θ is given by:

$$\sigma_{lp} = \sigma_{sp} + \sigma_{sl} \cos \theta. \quad (13.5)$$

The work of formation of the critical heterogeneous nucleus W^*_{hetero} is given by

$$W^*_{\text{hetero}} = \frac{(2 + \cos \theta)(1 - \cos \theta)^2}{4} W^*_{\text{homo}}. \quad (13.6)$$

As the contact angle tends to 180°, W^*_{hetero}, the work of formation of a critical heterogeneous nucleus tends to that of a critical homogeneous nucleus, but for lower values of the contact angle, it is less. As the contact angle tends to zero, the nucleation barrier also tends to zero. In the special case where the newly formed solid and the nucleant particle are the same phase, the solid–particle interfacial energy is zero, and there is no nucleation, only growth on what can be considered as a seed crystal. The heterogeneous nucleation frequency depends not only on the factors important for homogeneous nucleation, but also on the total area of nucleant substrate in contact with the liquid and on the contact angle. The initial nucleation frequency is given by

$$I_{\text{hetero}} \, (\text{m}^{-2} \, \text{s}^{-1}) = \frac{B}{\eta} \exp \left(-\frac{W^*_{\text{hetero}}}{k_B T} \right), \quad (13.7)$$

where B is a constant (proportional to the number of atoms in contact with unit area of nucleant surface) which for metals and alloys has the value

$\sim 10^{25}$ Pa m^{-2}. As the nucleation proceeds, the area of active substrate still in contact with liquid decreases and the heterogeneous nucleation frequency correspondingly decreases, ultimately to zero. In practice, the nucleating substrate is usually distributed in the melt as particles. The size of a particle determines the probability of a nucleation event occurring on it. Under most conditions nucleation is followed by rapid growth, so that each particle can support only one nucleation event. It follows that the maximum number of nucleation events is the number of nucleant particles. Site saturation describes the case where nucleation has already occurred on all the particles; further transformation is then by growth only, without further nucleation.

The problems in quantitative modelling of heterogeneous nucleation are manifold. In many practical cases the nucleant particles are unidentified; they may often be unintentional additions such as oxide from the melt surface or fragments of crucible. Even if the relevant particle is identified it is unlikely that the contact angle is known. The potency of the nucleant particle (i.e. the degree to which it promotes heterogeneous nucleation) is critically dependent on the contact angle. The uncertainties in estimating interface energies, alluded to above, mostly preclude the use of equation (13.5) to predict the contact angle, and direct measurement is difficult. Rarely, if ever, is the contact angle known with sufficient accuracy to make a quantitative prediction of the heterogeneous nucleation frequency. The potency of a particle may not be related only to the contact angle; it is possible that the surface topography of a particle (for example the existence of favourable ledges, or unfavourable surface curvature) could influence the potency [20].

Additional problems are that the number and size of nucleant particles are rarely known; also, the particles may have a distribution of size or of potency. Indeed, in most cases there are likely to be coexistent populations of quite different types of nucleant particle.

And there are yet more problems. The heterogeneities in a melt which are likely to be of most interest for nucleation are those with the greatest potency. For these, the spherical-cap model itself breaks down. Quantitative analysis of heterogeneous nucleation kinetics has been undertaken in entrained droplet experiments, in which the solidification under study is that of liquid droplets dispersed in a solid matrix. This technique and its uses have been reviewed by Cantor [21]. The samples are typically obtained by rapid solidification. They are heated to a temperature between the solidus and liquidus to obtain the dispersion of liquid droplets; these are typically 20 nm in diameter with a population of 10^{15} droplets per unit volume. Such a fine dispersion makes the droplet solidification very sensitive to nucleation. Fortunately, the droplet size distribution is often rather narrow, facilitating calorimetric analysis of the nucleation kinetics [22]. For some systems, for example solidification of tin droplets in an aluminium matrix [23], large undercoolings (>50 K) are required for significant heterogeneous nucleation. According to the classical theory, in these cases the contact angle is comparatively large, >40°. For such systems it has been found that the classical spherical-cap model (admittedly with adjustable parameters) can provide a good fit to the measured nucleation kinetics [22, 23]. On the other hand, there are cases, such as the solidification of lead droplets in an aluminium matrix [24], for which the onset undercooling for heterogeneous nucleation is much smaller (sometimes <1 K). In these cases, attempts to fit the nucleation kinetics with the classical model result in unphysical values of the contact angle and of the number of nucleant sites. It appears that the classical model breaks down for low contact angles [24, 25]. For a small critical nucleus with a low contact angle, the spherical cap cannot be a good description of the nucleus shape, and the classical model accordingly does not give reliable estimates of the work of formation of the critical heterogeneous nucleus or the heterogeneous nucleation frequency. For the regime of potent nucleation, an alternative model has been proposed, based on adsorption on to the surface of the nucleant particle [21, 26]. In this model there is a critical undercooling beyond which it is thermodynamically favourable to adsorb a layer of the new solid on the particle surface. This adsorbed layer is then the basis for further growth. The effect is that of growth being nucleated at a critical undercooling, even though nucleation (at least in the classical sense) is not involved. The critical undercoolings predicted by the adsorption model appear to be consistent with the measurements in entrained-droplet experiments. Unfortunately, the adsorption model is not yet capable of predicting effective nucleation frequencies. Thus, in the practically important case of inoculation with potent nucleant particles, there appears to be no available model for quantifying the heterogeneous nucleation frequency. How inoculation can be analysed is considered later in this chapter.

Despite all the problems with nucleation analysis, existing models have been useful in interpreting experimental results. The uncertainties in the nucleation frequency are not so severe when viewed in terms of the work of formation of the critical nucleus. Also, it is important to note that the nucleation frequency is very dependent on undercooling. In effect, there is a critical undercooling for the onset of nucleation, i.e. that beyond which the nucleation frequency exceeds the detection limit. As the undercooling is increased, the rise in the nucleation frequency defining the onset is so sharp that the onset undercooling itself can be predicted with some accuracy even with large uncertainties in the nucleation frequency. With typical parameter values for homogeneous nucleation, a 1% variation in the solid–liquid interfacial energy (which as noted above would cause an uncertainty of an order of magnitude in the nucleation frequency) would cause an uncertainty of 3% in the work of formation of the critical nucleus and of only 1.5% in the onset undercooling. Often it is the onset undercooling, defining the temperature at which solidification starts, which controls the type of microstructure which forms. Two examples follow to illustrate the usefulness of existing analyses.

Nucleation is an intrinsically stochastic process, but in a bulk sample averaging gives behaviour with negligible variability. In small droplets, however, when solidification can be triggered by a single event unaffected by the nucleation in other droplets, the behaviour may be highly variable. In particular, identical droplets under identical conditions may show a significant variation in onset temperature. It may then be expected that microstructure-selection maps of the kind developed for surface treatments [11] would be inapplicable. However, it has been shown that droplet processing can be usefully described in terms of microstructure-predominance maps [27]. The axes are composition (for binary alloys) and droplet diameter. The diameter is the key process variable as it determines droplet cooling rate as well as the volume and surface area available for nucleation. Notwithstanding the variability in onset undercooling at a given droplet diameter, there are strong trends in behaviour as the diameter is changed. In this way the maps can indicate the predominant microstructure at a given diameter. This type of mapping has been applied to simple systems, for example copper–nickel alloys which show only a cubic close packed solid solution phase [27]. The microstructural transition of interest is whether the grain structure is columnar or equiaxed, the latter case being spontaneous grain refinement. The analysis of spontaneous grain refinement is considered in the next section. As will be shown, predicted and experimental maps are in reasonable agreement.

A second example showing systematic behaviour in a situation dominated by nucleation is the phase selection in the iron–nickel system. In this case the solid solution can form with the body centred cubic (α phase) or cubic close packed (β phase) structures. The phase selection has been extensively studied in solidification of levitated droplets; these are typically 8 mm in diameter and their onset undercooling can be measured and correlated with the subsequent solidification sequence [28]. These experiments show that the cubic close packed phase is favoured by higher nickel content and by smaller undercooling. The results are in very good agreement with predictions based on comparison of the calculated values of the work of formation of the critical homogeneous nucleus for the two phases [29]; the phase with the smaller work of formation is of course favoured. Thus useful trends can be identified even in the absence of a full nucleation analysis (which in this case, for example, would have to take heterogeneous nucleation into account).

Although, as shown by these two examples, existing nucleation analyses can be used to interpret trends in experimental results, quantitative prediction of nucleation frequencies is still lacking. This appears to be a problem for the understanding of the control of grain size in equiaxed castings. The following sections select examples in which clear progress has been made in quantification of grain size.

Spontaneous grain refinement

When a clean liquid alloy without added inoculants is solidified, the expected grain structure is columnar. There are no potent substrates to enable equiaxed grains to nucleate ahead of a columnar front. Columnar structures are indeed predominant in the observed microstructures. The grains grow from the initial nucleation site or sites, and clearly show the direction of solidification. However, it is also commonly found that alloys with solidification triggered in particular ranges of undercooling show fine equiaxed grains. Figure 13.3 shows the marked refinement obtained at small and large undercoolings in copper–nickel alloys. This section is concerned with this spontaneous grain refinement; it is important to understand its mechanism before going on to consider (in the next section) grain refinement intentionally induced by added inoculants.

When castings are poured at a critical, small superheat, it is found that fine equiaxed grain structures are obtained [30]. This has been termed 'bigbang' nucleation, but the phenomenon appears not to involve nucleation at all. Rather it is attributed to the break-up of initially formed dendrites during pouring. The dendrite fragments then act as seeds for the grains. The significance of the small superheat of the melt is that the fragments would be remelted, and would be ineffective in giving an equiaxed structure, if the superheat were higher.

There have also been many studies of static liquid samples solidified at a variety of undercoolings. In such studies fine equiaxed grain structures often result (and figure 13.3 shows an example) when solidification is triggered at small undercooling. It seems that the refinement in these cases also arises from dendrite fragmentation. Such fragmentation has been directly observed

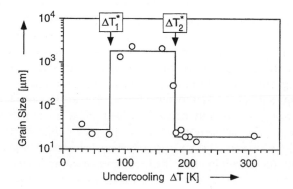

Figure 13.3. The grain size in $Cu_{30}Ni_{70}$ (at%) levitated droplets, with solidification triggered at selected undercoolings. There are clear transitions in grain structure, from refined equiaxed to coarse columnar at ΔT_1^* and back to refined equiaxed at ΔT_2^*. (From reference [33])

at the surfaces of solidifying samples [31]. Dendrites have a high surface area, and coarsening to reduce the area can naturally lead to some remelting and fragmentation [32, 33]. Even without the agitation of a poured melt, there is flow induced by the solidification shrinkage. Also, many of the experiments (as for figure 13.3) have been conducted on electromagnetically levitated droplets in which the liquid is stirred by the levitation forces. Thus liquid flow may play a role in the fragmentation. It is notable that the refinement at small undercooling is observed only for alloys and not for pure melts. However, the required solute content can be very low. For example, more than 0.01 at% oxygen in nickel [34] or more than 0.018 at% sulphur in copper [35] is sufficient to give some grain refinement at small undercooling. The solute may contribute to the effect by promoting the formation of fine dendrites and by facilitating subsequent local remelting.

As shown in figure 13.3, grain refinement is also typically found at large undercooling. Unlike the effect at small undercooling, this is found for pure metals as well as alloys. It was first reported by Walker [36], who found that when nickel melts are undercooled more than 140 to 150 K they show a fine equiaxed grain structure rather than a coarse columnar structure. The grain diameter in the equiaxed regime is <10% of the diameter of the columnar grains. Similar phenomena have now been demonstrated in other pure metals [37, 38] and in several alloys [39–41]. Observation of the temperature distribution in solidifying levitated drops shows that a thermal front, associated with the solidification front, crosses the drop from the point at which nucleation was triggered [28]. Despite this directionality of solidification, the grain structure formed behind the front is equiaxed.

Walker [36] suggested that a fine equiaxed structure could arise from copious nucleation triggered by a pressure pulse. Solidification shrinkage could indeed generate the conditions for cavitation ahead of the solid, while collapse of a cavity would generate a significant pressure pulse [42]. Alternatively, copious nucleation might naturally result from the rapid increase in nucleation frequency as the undercooling is increased. But the grain refinement is observed even when the undercooling is held fixed—in drops equilibrated at a given temperature and then touched with an external needle to trigger solidification. In addition, even if copious nucleation did arise throughout an undercooled drop, the subsequent recalescence (reheating) would most likely cause remelting of many of the small nuclei; without rapid heat extraction, this would greatly limit the refinement achievable [43]. In any case, the progression of solidification across the drop rules out the possibility that the grain refinement is the result of copious nucleation throughout the melt at a critical temperature. It can be concluded that the grain refinement observed at large undercooling is not attributable to nucleation.

At large undercooling, solidification is rapid, and the solid which forms can be highly stressed. There is therefore a driving force for recrystallization which could give grain refinement. Whether the grain structure has recrystallized or not is often discernible from the microstructure, because the composition variations corresponding to the dendritic microsegregation within the original grains are unaffected by the subsequent recrystallization. Recent studies of copper–tin offer an example where the grain refinement at large undercooling is clearly attributable to recrystallization [44]. In metallographic sections, the boundaries of the fine equiaxed grains are superposed on, but show no correspondence with, the segregation pattern of dendrites which are on a larger scale. On the other hand, the copper–oxygen system shows a fine equiaxed grain structure exactly corresponding to the segregation pattern; each grain shows coring with a spheroidal rather than dendritic shape [45]. In this case the grain refinement occurred during the solidification itself. In many cases, and the copper–nickel system illustrated in figure 13.3 is an example, grain refinement results from both solidification and subsequent recrystallization. It has been shown that in such a case, recrystallization can be prevented by rapid cooling in the solid state, but that grain refinement still occurs, at a critical undercooling set by the solidification process [46]. Thus recrystallization may have a role in the grain refinement observed at large undercooling, but there is nevertheless a clear refinement phenomenon in the solidification itself.

Just as for the refinement seen at small undercooling, the effect is attributable to dendrite break-up rather than to copious independent nucleation. There is some evidence for break-up from the preferred crystallographic orientation of grains in equiaxed structures which have been rapidly cooled to limit the time for reorientation of fragments [47].

Karma [33] has presented a simple model to account for the grain refinement both at small and at large undercooling. This model has been widely used to interpret the microstructures obtained in electromagnetically levitated drops by triggering solidification at selected undercoolings [48]. When solidification is triggered in such a case, by touching the drop with an external needle, there is rapid growth of dendrites across the drop, accompanied by recalescence. At the end of this first stage, the drop is partly solidified, with a dendritic network throughout and a temperature set by the equilibrium between that network and the interdendritic liquid. This gives a thermal plateau which has a duration Δt_{pt} depending mainly on the rate of heat extraction. During the thermal plateau the dendrites are subject to coarsening and in a time Δt_{bu} may break up into fragments; this time period necessary for dendrites to break up is determined largely by the thickness of the dendrite trunks and side-arms, related to the initial undercooling which set the growth velocity. In Karma's approach, the geometry of coarsening is simplified and is modelled as a Rayleigh instability of cylindrical dendrite trunks. It is found that for intermediate undercoolings the time period necessary for dendrites to break up is greater than the thermal plateau duration, so that break-up does not occur and columnar

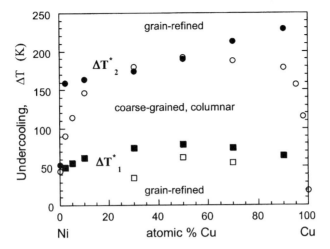

Figure 13.4. Microstructure-selection map for undercooled Cu–Ni melts, showing the critical undercoolings for transitions between coarse-grained, dendritic structures and spontaneously grain-refined, equiaxed structures. Data from droplet levitation experiments (■, ● for ΔT_1^*, ΔT_2^*) are compared with theoretical predictions (□, ○) based on the Karma model [33] for dendrite break-up.

grain structures are found. At small and large undercoolings, however, the thermal plateau duration is greater than the time necessary for dendrites to break up, so that break-up does occur. In this way, as the undercooling is increased, two microstructural transitions are predicted: from grain refined to columnar at ΔT_1^* and from columnar to grain refined at ΔT_2^*.

In figure 13.4 the composition dependence of these predicted critical undercoolings is compared with the observed transitions for the copper–nickel system. It is seen that, at least in the middle of the composition range, there is reasonable agreement between the model and experiment. In the modelling there is some adjustability of the parameters; nevertheless it is noteworthy that the model gives the correct trends with composition and gives reasonable ratios of the undercooling required for the grain-refined to columnar transition to that for the columnar to grain-refined transition. However, the agreement is not so good for low solute contents, especially for pure metals. For example, the model predicts that the undercooling for the columnar to grain-refined transition for pure nickel is 52 K [48], while most measurements (e.g. reference [49]) indicate that it is greater than 140 K.

The levitated drop experiments used to determine the undercooling at which these microstructural transitions occur can also be used to measure the velocity of the solidification front. At a given undercooling the velocity is constant across the drop. The dependence of this velocity V on undercooling, ΔT, is of interest for studies of the growth mechanism. For an undercooling less than that required for the columnar to grain-refined transition, the velocity is as expected from dendrite growth theory and increases with undercooling approximately according to $V \propto \Delta T^{2.5}$. For an undercooling greater than that required for the columnar to grain-refined transition, however, velocity varies linearly with undercooling [50]. Thus the microstructural transition at the undercooling for the grain-refined to columnar transition is not reflected in the velocity versus undercooling curve, while that at the undercooling for the columnar to grain-refined transition is.

There are other differences between the two transitions. Measurements of crystallographic texture [51] show that the grain-refined to columnar transition is diffuse, while the columnar to grain-refined transition is sharp. Also the morphology of the solidification front changes from dendritic and angular for an undercooling less than that required for the columnar to grain-refined transition to rounded for an undercooling greater than that required for the columnar to grain-refined transition [52]. There are also differences between the coring patterns typically found within the equiaxed grains formed at small and large undercooling [53]. At small undercooling, the coring normally shows branched dendritic fragments. At large undercooling, the coring is spheroidal and the segregation is often inverted (e.g. solute-rich cores are found in systems with solute partition coefficient less than one), reflecting the solute trapping expected at high growth velocities [54].

As noted above, the Karma model is certainly not fully quantitative in predicting the undercoolings at which the grain-refined to columnar and columnar to grain-refined transitions occur. Indeed, doubts have been expressed about whether dendrite break-up could occur as suggested in the model; continuing solidification during the thermal plateau would complicate the analysis of the Rayleigh instability and would inhibit break-up [55]. Also, the Karma model of break-up following dendrite growth offers no explanation of why the columnar to grain-refined transition should be accompanied by a break in the velocity versus undercooling curve.

Mullis and Cochrane [56–58] have shown that there may be a tip-splitting instability in dendrites at high growth velocities, and have associated this with the coincident transitions in velocity and microstructure at the undercooling for the columnar to grain-refined transition. The onset of instability in the morphology of the dendrite tip could account for the sharpness of the observed transition. Thus the simple Karma model is far from a complete description of the mechanisms of spontaneous grain refinement. Nevertheless, the picture it offers appears essentially valid—that recalescence followed by comparatively slow solidification can give break-up of the initial dendrites. Growth transitions of the kind identified by Mullis and Cochrane may define the exact undercooling for microstructural transition, but the Karma model correctly identifies in broad terms that the conditions are most favourable for break-up at small and at large intercooling, but not at intermediate undercooling.

In considering the grain size in as-cast equiaxed structures, the studies of spontaneous grain refinement illustrate a number of important points. The first is that not all grain refinement results from enhanced *nucleation*. It may be more correct to refer to grain *initiation*, in this case on dendrite fragments. Secondly, spontaneous grain refinement shows the importance of melt recalescence in inducing dendrite break-up, thereby leading to clear microstructural transitions at critical undercoolings. Analysis of the break-up permits a semi-quantitative prediction of the composition dependence of these undercoolings. However, it has not led to any predictions of grain size in equiaxed structures.

Conventional grain refinement by inoculation

Refinement of aluminium alloys

Inoculation is the use of additives to ensure that an alloy solidifies to a fine, equiaxed grain structure. While it is applied to a wide variety of alloy types from copper [59] to intermetallics [60], the predominant use, by far, is in aluminium casting [4]. The great majority of aluminium alloys are cast with added inoculant, i.e. grain refiner. Grain refinement is used in the direct chill casting of slabs and billets of wrought alloys, and in the modern continuous process of strip casting. In these cases, the refining reduces macrosegregation, avoids centreline cracking and permits higher casting speeds. Grain refinement is also important in shaped casting in the foundry industry. Here it can give improved feeding, reduced porosity and reduced hot-tearing. There can also be many other benefits of added grain refiner, ranging from enhanced response to subsequent heat treatment to better surface quality and improved mechanical properties. The economic benefits of inoculation in general outweigh the cost of the added refiner. Wrought alloys are, of course, subjected to heavy mechanical reduction after casting. This destroys the as-cast grain structure, but the effects of refinement are important not only in facilitating the casting itself and subsequent processing, but also in influencing aspects of the final microstructure, notably the selection of second-phase intermetallics [61, 62]. In shaped casting, there is use of eutectic and hypereutectic Al-Si alloys in which there are particular roles for melt additives to modify the eutectic microstructure [63] or to act as a nucleation catalysts for silicon as the primary solidification phase [64]. In the latter case, aluminium phosphide is the most potent nucleant phase, formed in the melt itself or added directly in an inoculant. The present coverage, however, will be of grain refinement in alloys of lower solute content in which α-aluminium is the primary and predominant solidification phase. This is relevant particularly to direct chill (DC) cast wrought alloys, including commercial-purity aluminium.

Effective grain refinement is considered to have been achieved when the grain size (i.e. mean lineal intercept) is <200 µm. The addition level of refiner required to obtain a consistent equiaxed grain structure on this scale depends on many factors. The casting method and its parameters such as cooling rate are clearly relevant, but so are alloy composition and other factors such as the fraction of recycled material in the melt. Addition levels are quoted in kg tonne^{-1}, i.e. ppt (parts per thousand). Typical addition levels are in the range 0.25 to 3.0 ppt. In direct chill casting 1 ppt addition is common, while for the more extreme conditions in strip casting levels up to 3 ppt are used. Overall, then, there is a world-wide requirement for major tonnages of refiner.

There are several commercially available inoculants for refining the grain structure of α-aluminium. Most are based on the Al–Ti–B system. The particles which are the basis for the nucleation of the aluminium grains are of the phase TiB_2. The details of the nucleation mechanism are discussed in chapter 12. The TiB_2 particles can be produced *in situ* in the melt by reaction of added salts, but are now mostly added already formed in an Al–Ti–B alloy in the form of rod, waffle or nugget. About 70% of the refiner market is taken by the composition Al–5 wt% Ti–1 wt% B (henceforth Al–5Ti–1B), sold mostly in rod form, which is found to be effective at low addition levels. Modern commercial refiners give consistent performance, but there remain significant problems, motivating much study of refinement mechanisms in recent years. These problems include the following.

- The refinement is markedly inefficient. At best only 1% of the added particles successfully nucleate aluminium grains. This inefficiency is undesirable not only for its immediate cost implications, but also because refiner particles may themselves be detrimental in the final microstructure.
- The refiner particles can agglomerate into larger clusters, giving particular quality problems in products such as thin foil and lithographic sheet.
- In the presence of certain solutes, notably zirconium [65], chromium or silicon, the refining action of an inoculant can be poisoned. This effect is shown in figure 13.5; the grain size obtained on casting rises with holding time of refiner in the melt. This poisoning action typically occurs above a threshold solute content and is faster at higher temperatures; in some cases it is so severe as to preclude effective refinement.

Recently, refiners based on the Al–Ti–C system have attracted much attention. These have TiC particles as the nucleants and have gained market share primarily because of their resistance to agglomeration. Unexpectedly, they have also been found to be resistant to poisoning by some solutes, notably by zirconium. In the following the coverage is mostly of the predominant Al–Ti–B refiners, but Al–Ti–C refiners are also considered briefly.

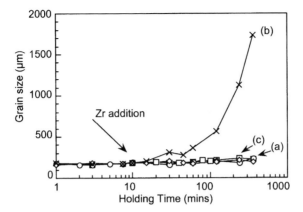

Figure 13.5. The grain size of commercial purity aluminium refined with 1 ppt Al–5Ti–1B and held at 800 °C, measured in TP-1 tests. Curve (a), two data sets, shows the lack of fade with holding time. Curve (b) shows the poisoning of the refiner action after addition of 0.05 wt% Zr. Curve (c), open squares, shows the resistance of a chemically modified refiner to the same Zr addition. (Data from reference [67].)

Basic phenomena and mechanisms

Full-scale casting is clearly expensive and inconvenient as a way of studying the grain refinement process. Most studies have been based on grain-refining tests using small melt volumes. For valid comparisons, standard test procedures are essential. Several standard tests are available. An example is the Alcan or TP-1 test (details in references [4, 66]) in which a standard ladle (of capacity $\sim 100\,\text{cm}^3$) is dipped into the melt, withdrawn and lowered into a water quench tank. The solidified melt is sectioned at a height of 38 mm from the base. At this position in the sample the cooling rate is $\sim 3.5\,\text{K s}^{-1}$, which roughly matches that in direct chill casting. The results quoted in this section, for example those in figure 13.5, are all from TP-1 tests.

When a grain refiner is added to a melt, a *contact time* is required before it becomes fully effective. For a modern Al–Ti–B refiner, the contact time is <1 min. Under typical conditions of holding the melt with added refiner, after 20 to 30 min the grain size obtained on subsequent casting starts to increase. As shown in figure 13.5, this *fade* is largely avoided if the melt is stirred just before casting [67], implying that settling of the relatively dense refiner particles is the main cause. However, there may also be a contribution to fade from particle agglomeration.

In Al–Ti–B refiners, including the common Al–5Ti–1B, all the boron is combined in TiB_2 particles. The refiner consists of an α-aluminium matrix with embedded particles of TiB_2 and Al_3Ti. On addition to the melt, the α-aluminium matrix melts and dissolves and the TiB_2 particles survive. The dilution of overall titanium content is such that Al_3Ti particles also dissolve; this appears to be so fast as to preclude any direct influence of Al_3Ti particles in the grain refinement. Thus it appears that the TiB_2 particles must be responsible for the grain refinement. However, many studies, for example those of Mohanty and co-workers [68, 69], have shown that TiB_2 particles are not always good nucleants for α-aluminium. They appear to be fully effective only if there is some excess titanium in the melt (i.e., titanium beyond that in the TiB_2 phase). On the other hand, if Al_3Ti particles are stable in the melt (at high enough titanium content) they are found to be extremely effective nucleants via the peritectic reaction. As reviewed by Schumacher et al [70] and as discussed in chapter 12, it then seems that effective grain refinement by Al–Ti–B refiners involves some type of combined action of Al_3Ti (which is a good nucleant but not stable in the melt) and TiB_2 (particles of which are stable in the melt, but not intrinsically good nucleants). The nature of this combined action has been a matter of speculation. It has been suggested that the TiB_2 could act to preserve the Al_3Ti locally, through:

- a surrounding shell of borides [71–73],
- survival in cavities [74], or
- a layer adsorbed on the borides [75–77].

It would clearly be desirable to have microscopical studies of the nucleation of α-aluminium, not only to distinguish between these suggestions, but also to examine why commercial refiners are so inefficient. There have been some useful studies, by transmission electron microscopy, of the microstructure of grain-refiner rod and of refined alloys. But key observations have been made by incorporating TiB_2 particles from added Al–Ti–B refiner in an aluminium-based metallic glass. These observations are reviewed in chapter 12 and will not be described in detail here. The main merit of the glassy-matrix technique is that growth from the nucleation events is stifled, permitting ready identification of the nucleation sites. The main conclusion is that the excess titanium facilitating nucleation is stabilized as an adsorbed Al_3Ti layer on the TiB_2 particles. The α-aluminium then nucleates directly on this layer. The TiB_2 particles are hexagonal platelets and the nucleation occurs only on the Al_3Ti coating of the $\{0001\}$ faces of the borides. The phases are found to have straightforward orientation relationships—the close-packed planes and close-packed directions in each phase are parallel, giving moderately good lattice matching.

It should be noted that the excess titanium contributing to the formation of the Al_3Ti layer also contributes to the overall solute content in the melt [78]. As discussed in the next section, solutes, and in particular titanium, inhibit the growth of nucleated grains and thereby facilitate grain refinement. Also, there is not yet a quantitative analysis of the adsorption energies which could stabilize the Al_3Ti phase in melts with low titanium contents. Nevertheless, the observations of the Al_3Ti layer (not only in the glassy-matrix studies [70], but also in the refiner itself [79]) facilitate the interpretation of several effects. The development of the layer on holding in the melt may explain the improvement in performance during the *contact time*. It may also explain the larger changes seen when badly made refiners can dramatically improve their performance on longer holding in the melt [67]. Destruction of the layer in the presence of some solutes is a likely mechanism for at least some of the observed *poisoning*. Transmission electron microscopy studies suggest that in the presence of zirconium, the Al_3Ti layer suffers from a progressive dissolution without being replaced by an Al_3Zr layer [65]. Further evidence that poisoning may act through the destruction of the Al_3Ti layer is provided by the measurements in figure 13.5 which show that a refiner chemically modified to have a more stable layer can show significantly improved resistance to poisoning [67]. And finally, the layer may also play a detrimental role as a glue aiding the agglomeration of particles [80].

The microscopical studies of the nucleation on TiB_2 particles, surveyed in reference [70] and in chapter 12, have been useful in understanding whether there is effective refinement or not. However, they have not been useful in assisting prediction of grain size under refinement. In particular, the transmission electron microscope observations offer no evidence of significant differences between the TiB_2 particles which could explain why <1% are active nucleants. At the large effective undercooling values in the glassy-matrix studies, copious nucleation of α-aluminium is found on all the boride particles. Understanding the low efficiency of refinement requires a different approach.

For a grain to be successfully initiated on a boride particle there must be not only nucleation of α-aluminium, but also growth. The nucleated α-aluminium has an interface with the liquid and this interface must break free from the inoculant particle. In effective grain refinement, nucleation on the added particles dominates over nucleation elsewhere, and so must occur at small undercoolings. The α-aluminium nuclei must therefore be rather flat on the {0001} faces of the borides. They can be considered as classical spherical cap nuclei (figure 13.2) with small contact angle θ, or (as discussed earlier in this chapter) perhaps more correctly as adsorbed layers. Initially any nucleus can grow laterally, but when it completely covers the face of the boride it can grow further only by reducing the radius of curvature of its interface with the melt (figure 13.6). This is most easily analysed by approximating the hexagonal face of the boride as a circle of diameter d. The α-aluminium crystal cannot grow beyond the point at which the radius would be less than the critical value r^* for nucleation (which depends on the temperature at that instant). If the diameter of the particle is less than twice the critical nucleus radius, then *free growth* of the crystal from the particle is not possible. It becomes possible when the undercooling is increased, thus reducing the critical nucleus radius. The critical condition for free growth of the crystal through the minimum-radius hemispherical shape is when the diameter of the boride particle is

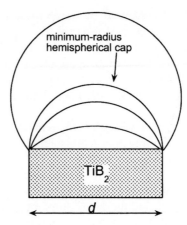

Figure 13.6. Crystal growth following nucleation on one {0001} face of a boride particle. Thickening of the crystal reduces the radius of curvature of its interface with the liquid. As this radius cannot go below the critical nucleus radius r^*, there is a barrier to free growth if $d < 2r^*$. Further growth past the critical hemispherical condition (for which the solid–liquid interface has minimum radius of curvature) is then possible only by increasing the undercooling to reduce the critical nucleus radius. (From reference [81].)

equal to twice the critical nucleus radius. The undercooling for free growth ΔT_{fg} and the nucleant particle diameter d are simply related by

$$\Delta T_{\text{fg}} = \frac{4\sigma_{\text{sl}}}{\Delta S_V d}, \qquad (13.8)$$

where σ_{sl} is the solid–liquid interfacial energy and ΔS_V is the entropy of fusion per unit volume. The variation of the free-growth undercooling with particle diameter is shown in figure 13.7 for aluminium. Actual particle diameters in typical refiners vary over a wide range (see figure 13.9 below), but many are of the order of 1 μm [81]. This diameter means that the undercooling necessary for free growth from the particles is of the order of 0.5 K. This undercooling appears small, but measured undercoolings are also very small, certainly <1 K, and typically ∼0.2 K. Thus it is clear that the free-growth barrier is significant. Even if there is perfect nucleation, the effective growth of a grain from a nucleant particle is not assured. The *nature* of the inoculant particle (including whether it is suitably coated, with Al_3Ti for example) determines how effective the nucleation is. The *size* of the particle determines the ease of subsequent growth.

The above discussion has been for the action of Al–Ti–B inoculants. In most respects the grain-refining action of Al–Ti–C inoculants is expected to be similar in mechanism. It has been verified by direct microstructural observations that the cube-shaped TiC particles in these refiners do act as nucleation centres for grains [82]. It has even been noted that the larger

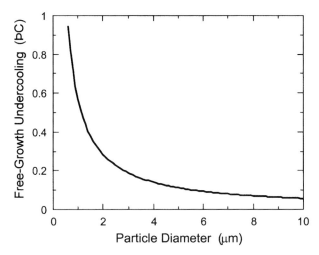

Figure 13.7. The undercooling necessary to initiate free growth from a planar circular particle face of diameter d. Such a face approximates the planar, hexagonal faces found on TiB_2 particles in Al–Ti–B refiners. This undercooling ΔT_{fg} arises from the Gibbs–Thomson shift of the solid–liquid equilibrium and is given by equation (13.8). (From reference [99])

particles are the ones found at grain centres, consistent with the free-growth model prediction that the larger particles would become active at smaller undercooling [82]. The action of the TiC particles appears to be different from that of the TiB_2 particles in that there is no evidence of the existence of, or of the need for, an intervening Al_3Ti layer to facilitate nucleation. The absence of a layer may help to explain some of the resistance to poisoning. The most significant difference between the TiB_2-based refiners and the TiC-based refiners is in the stability of these phases themselves. TiB_2 appears to be very stable, even in melts with substantial superheat. On the other hand TiC is unstable, transforming progressively to Al_4C_3 on holding in the melt [83]. This transformation is sufficiently sluggish to pose no threat to grain refinement under normal conditions, but it does speed up markedly at higher temperature. The progressive dissolution of the TiC particles, and the lack of a layer on them, may account for the alleviation of agglomeration problems with Al–Ti–C refiners. These refiners do, however, still suffer from poisoning by silicon in the melt [84].

Thermal modelling

Modelling of grain refinement, specifically addressing the issue of why it is so inefficient, was first attempted by Maxwell and Hellawell [85]. They considered the cooling of an aluminium melt containing a population of

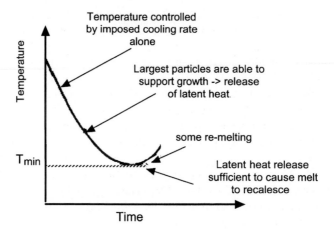

Figure 13.8. A schematic cooling curve showing the basis of the Maxwell–Hellawell model [85]. The latent heat released from growing crystals is eventually sufficient to outweigh the external heat extraction, giving recalescence. After the minimum temperature T_{min} has been passed, there is no further nucleation of new grains.

refining particles, and noted that crystal growth on an increasing number of particles would give an accelerating rate of heat release, eventually surpassing the rate of external heat extraction and giving recalescence (figure 13.8). The recalescence limits the number of nucleation events and therefore restricts the degree of grain refinement which can be achieved. This could explain why refinement efficiencies are typically so low. Being stochastic, nucleation events are naturally spread in time, and the crystal growth from the nuclei formed early stifles later events. In principle, the stifling might occur by (i) crystallization reducing the available volume of inoculated melt, (ii) a changing composition of the residual melt, i.e. soft impingement of the solute diffusion fields, or (iii) soft impingement of the thermal diffusion fields. Importantly, Maxwell and Hellawell [85] noted that the dominant mechanism had to be (iii), since the volume fraction transformed at recalescence is very small (typically 10^{-4}) and the thermal diffusivity is much greater than typically 10^4 times the solutal diffusivity. The Maxwell–Hellawell analysis is therefore essentially thermal. A complete description of the temperature distribution during casting would be complex, taking into account the nature of the external heat extraction and the localized recalescence around each growing crystal. However, Maxwell and Hellawell recognized that, to a good approximation, the melt can be treated as spatially isothermal. This is because the thermal diffusion length greatly exceeds (typically by 10^2 to 10^3 times) the separation between the nucleant particles. Maxwell and Hellawell made a further simplification of the problem by taking the crystals to be spherical rather than dendritic. As shown in reference [81], this is likely to be valid because at the onset of recalescence the grains are still so small that they have not reached the onset of the morphological instability into dendritic growth.

The classical spherical-cap model was used to calculate a heterogeneous nucleation frequency on particles in the melt. For ease of computation, the particles were taken all to have the same size. Maxwell and Hellawell computed the shape of the cooling curve (figure 13.8) and the consequent number of grain nucleation events under various conditions, but did not make any comparison with experiment. They did note, however, that the important factors controlling the grain size are the number and potency of nucleant particles, the cooling rate and the solute content in the melt. The solute content is important because it restricts the rate of growth of the crystals. Maxwell and Hellawell used a diffusion model based on the invariant-size approximation. They showed that the dominant effect of solute content can be described by the quantity Q, defined as:

$$Q = m(k-1)C_0, \tag{13.9}$$

where m is the liquidus slope, k is the equilibrium partition coefficient and C_0 is the solute content in the alloy melt. Here, we term Q the *growth-restriction parameter*, and it can be seen that it features also in equation (13.2). In the Maxwell–Hellawell case of assumed spherical crystals, the growth velocity of the crystal at a given radius and a given undercooling is approximately proportional to $1/Q$. Values of the growth-restriction parameters are additive for all the solutes in the melt, provided they do not interact to form complexes or precipitates which would effectively reduce the overall solute content. Values of the liquidus slope, the equilibrium partition coefficient and the growth restriction parameters are given in table 13.1 (using data from reference [86]), from which it can be seen that titanium is

Table 13.1. The values of liquidus slope and equilibrium partition coefficient, calculated from parameters in reference [86], for common solutes in aluminium. The solutes are compared in terms of the growth-restriction factor Q (from equation (13.9)) corresponding to a solute content of 1.0 wt%.

Solute	m (K wt%$^{-1}$)	k	Q (K)
Cr	2.6	1.75	1.9
Cu	−2.5	0.145	2.1
Fe	−2.925	0.03	2.8
Mg	−5.84	0.48	3.0
Mn	−1.2	0.62	0.5
Ni	−3.5	0.004	3.5
Si	−6.62	0.12	5.8
Ti	25.63	7.0	153.8
Zn	−1.65	0.43	0.9

the most growth-restricting solute. Addition of Al–Ti–B or Al–Ti–C refiners at typical levels (1 to 3 ppt) increases the titanium content dissolved in the melt. This extra titanium must be taken into account in analysing the growth restriction in the inoculated aluminium.

The Maxwell–Hellawell use of the spherical-cap model of nucleation, and their assumption of a single particle size, have been superseded in more recent work [81, 87, 88] on the free-growth model. It is assumed that nucleation (possibly by adsorption) occurs very readily on each inoculant particle, and that the nucleation itself is not the significant barrier to grain initiation. That barrier is the free-growth condition discussed in connection with figure 13.6. This condition is not stochastic, but dictates that a grain will be initiated on a given particle when that particle reaches the free-growth undercooling equation (13.8) set by its diameter. At no temperature is there any time dependence of the grain initiation. The model would fail if each particle had the same size, as grain initiation would occur simultaneously on all the particles. The modelling is conducted for additions of a given commercial refiner, Al–5Ti–1B (wt%) for which the particle size distribution has been measured (figure 13.9). This shows a wide spread in

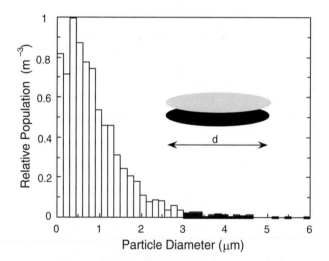

Figure 13.9. The measured diameter distribution of TiB_2 particles in a commercial Al–5Ti–1B (wt%) refiner. The particles are hexagonal platelets approximated as discs (inset). The relative populations (shown) are estimated from intersections in scanning electron micrographs of polished surfaces, corrected for sectioning effects. Absolute populations are then estimated from the known volume fraction of TiB_2 phase. The refiner is very inefficient as under normal conditions only the largest particles (shaded area) initiate grains. In the regime of active particles, the size distribution can be accurately fitted with an exponential, i.e. population proportional to $\exp(-d/d_0)$, where for this refiner the characteristic particle size $d_0 = 0.72\,\mu m$. (From reference [81].)

particle size, ensuring that grain initiation events occur progressively during cooling, starting on the largest particles.

In the calculations, the thermal history of the melt is treated as a series of short isothermal steps of duration dt. Before solidification starts, the temperature varies according to the imposed cooling rate R:

$$T_{n-1} = T_n - R \, dt \tag{13.10}$$

where T_n is the temperature of the (spatially isothermal) melt in the nth time interval. The inoculant particles are classified by diameter d, the number of particles in the range d to $d + \delta d$ being $N(d)\delta d$. For each set of particles with a given diameter range, crystal growth is started when the melt undercooling (measured relative to the liquidus) reaches or exceeds the corresponding free-growth undercooling set by equation (13.8). In all subsequent time intervals, the radius of the grains growing on this set of particles is changed according to

$$r_{n+1} = r_n + V \, dt \tag{13.11}$$

where the growth velocity V is calculated for each time increment according to the laws for diffusion-controlled growth of spherical crystals. As described in references [81, 85], the key alloy parameter is the growth-restriction factor Q (equation (13.9)). As the crystal growth rate is controlled by solute diffusion, the solute diffusivity is clearly important. It is assumed that the diffusivity is essentially the same for all solutes. Hodaj and Durand [89] have proposed a growth-restriction factor accounting for different diffusivities of different solutes; however, the diffusivities are not known with sufficient reliability to make their approach immediately useful.

For the set of crystals growing on the particles of diameter d to $d + \delta d$, there is a heat input $q(d)\delta d$ into the melt in the nth time increment:

$$(q(d)\delta d)_n = N(d)\delta d \, 4\pi r_{n-1}^2 (r_n - r_{n-1}) \Delta H_V, \tag{13.12}$$

where ΔH_V is the latent heat of solidification per unit volume. In every time increment, the heat inputs from each set of growing crystals are summed to obtain the total q_{total}, and the melt temperature in the next interval is then given by

$$T_{n+1} = T_n - R \, dt + \frac{q_{\text{total}}}{C_{pV}}, \tag{13.13}$$

where C_{pV} is the specific heat of the melt per unit volume.

As reported in reference [81], the free-growth model has been applied mostly to analyse the results of TP-1 tests on the grain refinement of commercial-purity aluminium using Al–5Ti–1B refiner. The input parameters for the calculations are given in table 13.2 (from reference [81]). The particle size distribution shown in figure 13.9 was used. All these inputs come from the literature or from independent measurements; while there can be some

Table 13.2. Parameters used in the thermal modelling of grain refinement of aluminium alloys. The material parameters are mostly for pure aluminium. (From reference [81].)

Quantity	Symbol	Units	Value
Solid–liquid interfacial energy	σ_{sl}	mJ m^{-2}	158
Entropy of fusion per unit volume	ΔS_V	J K^{-1} m^{-3}	1.112×10^6
Enthalpy of fusion per unit volume	ΔH_V	J m^{-3}	9.5×10^8
Heat capacity of melt per unit volume	C_{pV}	J K^{-1} m^{-3}	2.58×10^6
Diffusivity in melt (Ti in Al)	D	m^2 s^{-1}	2.52×10^{-9}
Cooling rate in TP-1 test	R	K s^{-1}	3.5

selectivity in the values chosen, the input parameters are essentially not adjustable.

The measured data in figure 13.10 show that the inoculation becomes less efficient as the refiner addition level is raised. This behaviour is matched well by the model predictions. The good agreement, both in the trend of the data and the absolute magnitude of the grain diameter, strongly supports the validity of the model, in particular the concept that the grain-refining efficiency is limited by recalescence. The *efficiency* can be defined by:

$$\text{efficiency} = \frac{\text{number of grains per unit volume}}{\text{number of inoculant particles per unit volume}}. \qquad (13.14)$$

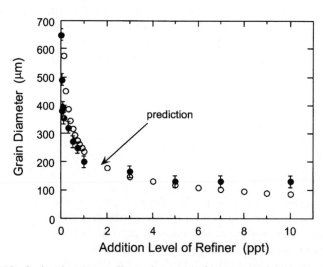

Figure 13.10. Grain size (mean linear intercept) for commercial purity aluminium inoculated with Al–5Ti–1B at various levels. The grain diameters measured in TP-1 tests (●) are compared with the model predictions (○) assuming a cooling rate of 3.5 K s^{-1}. Good agreement is found, even though the model has no adjustable parameters. (From reference [81].)

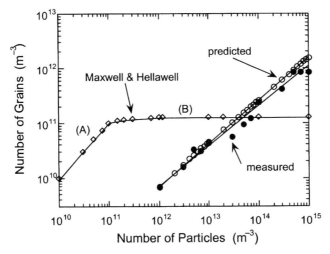

Figure 13.11. The number of grains per unit volume as a function of the number of refiner particles per unit volume. The predictions of the Maxwell–Hellawell model [85] show two regimes: (A) in which there is 100% efficiency (one grain per particle), and (B) in which the number of grains saturates. Also shown are data (●) calculated from the grain diameters measured in TP-1 tests and shown in figure 13.10. These are compared with predictions (○) of the free-growth model [81]. The free-growth predictions are qualitatively different from those of the Maxwell–Hellawell model, and are a much better fit to the data.

The data in figure 13.10 are replotted in figure 13.11; again the free-growth model predictions are in good agreement with the measured data. Figure 13.11 shows more clearly the efficiency of the refinement. Per unit volume, addition of 10^{12} particles gives only 10^{10} grains, while addition of 10^{15} particles gives 10^{12} grains—i.e. the efficiency decreases from 1% to 0.1% as the addition level is increased. Also shown in figure 13.11 are the predictions of the original Maxwell–Hellawell [85] model, showing two regimes. With a low population of particles ($<10^{11}$ m^{-3}), there is time for a grain to nucleate on every particle, giving 100% efficiency. With larger populations, the number of grains nucleated does not continue to increase with the number of particles, but instead tends to saturate, giving decreasing efficiency. The measured data do not show these two regimes, and are much better matched by the free-growth model. The prediction of two distinct regimes arises in the Maxwell–Hellawell model because of the assumption of nucleation on particles of a uniform size. In this way, if a given temperature were maintained long enough, nucleation would eventually occur on all the particles and 100% efficiency would be achieved; this is regime A, favoured by low particle populations. With the free-growth model, on the other hand, the distribution of particle diameter means that at any undercooling only a fraction of the particles could ever initiate grains. The exponential nature of the measured diameter distribution (figure 13.9)

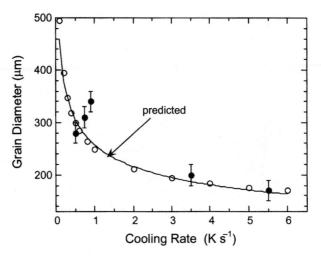

Figure 13.12. The grain size of commercial purity aluminium with 5 ppt addition of Al–5Ti–1B as a function of cooling rate, measured (●) in modified TP-1 tests, and compared predictions (○) of the free-growth model. (From reference [81].)

shows that a large fraction of the particles may never reach the undercooling at which they would become active; an efficiency of 100% is then impossible.

In the TP-1 test, the cooling rate in the standard sectioning plane is fixed at $\sim 3.5\,\mathrm{K\,s^{-1}}$, roughly matching that in direct chill casting. However, the TP-1 test can be modified to vary the cooling rate [81]; the results of such experiments are shown in figure 13.12, together with the model predictions. Despite the limited number of measurements, it can be concluded that they are consistent with the predictions. This provides evidence in support of the thermal basis of the modelling, derived from the original approach of Maxwell and Hellawell. The modelling suggests that there is a strong variation of grain size at low cooling rate, tending to saturate at high cooling rate. Typical grain-refining tests are in a regime where the grain size could be significantly affected by the cooling rate.

Figure 13.13 shows the effects of solute content on grain size. The data are taken from the work of Spittle and Sadli [90]. For each of the compositions (indicated in wt%), equation (13.9) has been used with appropriate parameters to calculate the growth-restriction factor for the added solute and has been added to the growth-restriction factor for the extra titanium arising from the 2 ppt refiner addition used in the experiments. There is fair agreement between the measurements and the predictions. This agreement, and the clear trend to smaller grain size at larger values of the growth-restriction factor, provides direct evidence for the relevance of growth restriction as included in the modelling. Although there is considerable scatter of the data around the trend line, this is less than when

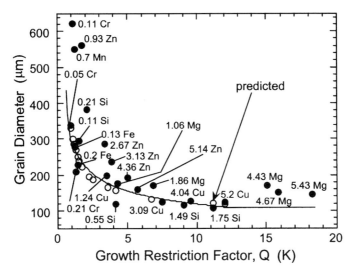

Figure 13.13. Grain size as a function of growth-restriction parameter Q (equation (8)) for a standard TP-1 test with 2 ppt addition of Al–5Ti–1B refiner. The measured data (●) from Spittle and Sadli [90] are compared with predictions (○) from the free-growth model, taking parameters appropriate for each chosen solute addition. Solute elements and addition levels in wt% are indicated for each data point. (From reference [81].)

the data are plotted as a function of the growth-restriction factor divided by the equilibrium partition coefficient (the steady state undercooling of a planar solidification front) as in reference [90]. At low growth-restriction values, the measured grain sizes can considerably exceed the predicted values. However, examination of TP-1 test samples with stated grain sizes greater than ∼400 μm suggests that these grain structures are columnar, and therefore not treated by the modelling at all.

Overall, the modelling appears very successful in predicting the grain size obtained in small-scale melts (in particular in the TP-1 test) in which there is known to be recalescence. But the failure to treat columnar structures arises naturally from the underlying assumption of an isothermal melt, and suggests that the existing model may have problems in the presence of significant temperature gradients. Even in TP-1 tests, the measured cooling curves are not fully matched by the free-growth model predictions. The measured and predicted maximum undercoolings are in good agreement, but the measured temperature changes are significantly slower than predicted; as discussed in reference [81], the measurements are likely to show a broadening of the recalescence because of the temperature gradients and progressive solidification through the melt.

The above discussion has been based on Al–Ti–B refiners, in particular Al–5Ti–1B. In Al–Ti–C refiners, the TiC particles are cubes rather than hexagonal platelets and tend to be smaller than TiB_2 particles. Nevertheless, the free-growth model applied for Al–Ti–B refiners [81] seems to work also for Al–Ti–C refiners [88]. Indeed the smaller particle size correlates with the greater nucleation undercooling which has been suggested for Al–Ti–C refiners [91]. General comparisons of Al–Ti–B and Al–Ti–C refiners can be found in references [91, 92].

Microstructural modelling

To treat directional solidification, and especially to treat the CET, it is appropriate to undertake microstructural modelling. Approaches to such modelling have been reviewed in references [93, 94]. For realistic, non-regular grain structures to be obtained, a stochastic or probabilistic model is required, and this has mostly been based on a cellular-automaton (CA) approach to grain growth. The basics of this type of modelling are described in references [8, 95], with further developments outlined in reference [96]. Mostly the modelling has been in two dimensions, though no fundamental difficulties are involved in the extension to three dimensions. The CA modelling is available as a commercial software package, calcoMOS^{TM} [97], and that is used in the present work.

A CA grid is defined in which solidification proceeds from cell to cell. The CA approach takes account of randomly dispersed heterogeneous nuclei, random grain orientation, and growth kinetics. In the case of dendritic growth, preferred growth directions ($\langle 100 \rangle$) are included, and the growth law for the dendrite tips is calculated with the model of Kurz *et al* [98], and approximated as a power-law dependence of velocity on local undercooling. Nucleation is assumed to be on sites which become instantly active at critical undercoolings. In this way, the basic nucleation behaviour matches that in the free-growth modelling. However, in the CA modelling, the shape of the distribution of nucleation undercoolings is arbitrarily taken to be Gaussian. The parameters of the Gaussian, including the total number of nucleation sites, can be set arbitrarily. Both nucleation and growth are simulated in the same CA grid, within which the cell size must be somewhat less than the spacing between the dendrite tips for reliable modelling of dendritic growth.

As already noted by Maxwell and Hellawell [85], the length scale appropriate for thermal modelling is much greater than the grain size. An enthalpy-based finite-element (FE) algorithm is used to calculate temperature distributions and heat flow. The FE mesh is much coarser than the CA grid, but the thermal computation is fully coupled with the latent heat released in the CA cells and with changes in specific heat.

The CET has been studied using this CA–FE modelling [8]. In this work, Gandin and Rappaz considered the solidification of an Al–7 wt% Si ingot set on a chill plate, a geometry giving a continuously decreasing

Conventional grain refinement by inoculation

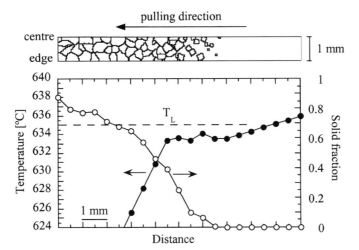

Figure 13.14. A grain structure (in a 2 mm × 10 mm area, half shown) developing in the two-dimensional CA–FE model during directional solidification of Al–4.15 wt% Mg with a solidification velocity of 0.1 mm s^{-1} and $N_{max} = 10^{11}$ m^{-3}. The corresponding profiles of temperature and solid fraction are also shown. (From reference [99].)

temperature gradient and forcing a CET. The results of CA–FE modelling were qualitatively compared with experimental observations. The modelling showed that the CET is broad. As the temperature gradient decreases, grains can nucleate ahead of the main columnar front. These grains progressively dominate the microstructure. But when such grains are first formed, they are elongated in the heat flow direction. Only when the temperature gradient falls still further are fully equiaxed grain structures obtained. The diffuse nature of the CET appears to be in agreement with the observed microstructures.

Gandin and Rappaz [8] did not focus on the grain refinement process. More recent modelling of a similar kind [99] has examined the effects of solidification velocity V, and temperature gradient G, on the effectiveness of inoculation. This has included a quantitative comparison between measurements and predictions of the equiaxed grain size as a function of refiner addition level. Figure 13.14, from this work, shows two-dimensional CA–FE modelling of Bridgman solidification of an Al–4.15 wt% Mg alloy inoculated with an Al–3Ti–0.15C–1Fe (wt%) refiner. The parameters of the Gaussian distribution of nucleation undercoolings were set to give the first nucleation events at an undercooling of ~1.5 K, roughly as expected from measurements of the particle size distribution in the refiner [82]. The figure shows one example of a microstructure, together with the corresponding profiles of temperature and solid fraction along the length of the sample. The plotted profiles are averages of the profiles along three lines spaced across the sample, but still show some noise from the specific features of the microstructure. In the Bridgman experiment which is being modelled, the temperature gradient is $10\,\text{K}\,\text{mm}^{-1}$. For this case, the CA–FE modelling can provide semi-quantitative fits to the measured grain sizes. It is now of interest to examine how the grain refinement in this case relates to the free-growth modelling described earlier.

The development of the equiaxed microstructure can be clearly seen in the modelling in figure 13.14. Grains nucleate independently, grow and impinge. At a solid fraction, in this particular alloy of ~25%, the dendritic envelopes have completely impinged, so that the remaining solidification is slower and between the dendrite side-arms. Of particular interest is the corresponding temperature profile, which shows a quasi-isothermal zone, undercooled by 1 to 2 K. This zone interrupts the linear temperature profile which would be observed if there were no release of latent heat within the sample in the Bridgman furnace. Evidently, dendritic growth is sufficiently rapid to counteract the external heat extraction in this zone.

As shown above, the free-growth model can make quantitative predictions of the grain size despite assuming an isothermal melt. Figure 13.14 shows how the basic concepts of the model can still apply even in the presence of a significant temperature gradient. In the case of an isothermal melt, grain initiation is halted by recalescence. In directional solidification, there cannot be a similar reheating in the temperature profile, but the grain initiation is similarly halted by the quasi-isothermal zone. On the cold side of this zone, cooling resumes, but the residual liquid is now (in a eutectic system such as Al–Mg) enriched in solute; the depressed liquidus ensures that no further nucleation or grain initiation takes place. Previous treatments of directional solidification and of the CET have not taken into account the quasi-isothermal zone, assuming instead a linear temperature profile [6, 7].

Prediction of grain size

It appears that the simple, free-growth model described earlier in the chapter is capable of giving fully quantitative predictions of grain size in inoculated melts, at least when the melt volume is small and overall recalescence can be observed. (It should be noted that the model has not yet been quantitatively tested on large, industrial-scale melts such as direct chill casting.) Furthermore, the predictions are based on parameters which have been determined independently and are not adjustable. In view of the difficulties with nucleation analysis, already identified, the origins of this success need to be considered.

The first point is that for inoculated aluminium alloys, grain initiation is totally dominated by the added grain refiner. When equiaxed grain structures are formed, there is only one initiation mechanism to consider, on particles which can in principle be well characterized. Secondly, as already pointed out, commercial grain refiners are so effective that grain initiation is limited by the free-growth condition and not by nucleation itself. This initiation is therefore not stochastic. Importantly, the key parameter determining the critical undercooling for grain initiation is the particle size, which in principle and practice is measurable. In contrast, for heterogeneous nucleation the undercooling depends on the contact angle, which is not normally measurable nor predictable with sufficient accuracy. Furthermore, for the most potent nucleants, such as those used in aluminium alloys, the classical nucleation theory may not be applicable in any case. If the nucleation on inoculant particles is made more difficult, for example by poisoning [65], then the nucleation stage itself could become controlling. In that case, the free-growth model would break down, and the quantitative predictability of the grain size would be lost.

The free-growth model relies on knowledge of the size distribution of the inoculant particles. In the work reported above, all the modelling has been based on the measured size distribution (figure 13.9) which can be approximated as an exponential. However, even with its limitations the model is sufficiently successful to justify its use in exploring the effects of different size distributions. In this way, design guidelines may be obtained for improving refiner performance.

In preliminary work [100], the particle diameter distribution has been taken to be Gaussian. The Gaussian is defined by an average particle diameter λ and standard deviation in diameter s, measured as a fraction of the average particle diameter. Comparison of different distributions is undertaken for a given refiner addition level, i.e. for a constant volume fraction of refiner particles. Thus when the particles are larger, their populations are correspondingly much smaller. The refiner performance can be measured in terms of the grain size obtained for a given addition level (its *effectiveness*), or in terms of its *efficiency* (equation (13.14)). Effectiveness is of practical concern, while efficiency is of fundamental interest.

For a given average particle diameter, the predicted grain size varies only weakly with distribution width s in the range investigated (10% to 50% of λ). In general, a narrower size distribution gives a finer grain size. Within the present model, 100% efficiency is expected if the particles have a single size. As discussed briefly earlier, this is a limit in which the present model is inapplicable. If all the particles were to initiate grains at one instant, not all the grains would survive. But the calculation of the fraction surviving would require analysis of thermal fluctuations of a kind which has not been attempted. In any case, the limit of mono-sized particles is not a realistic one for a manufacturable grain refiner.

The predicted grain size is much more strongly dependent on the average particle diameter than on the distribution width. Figure 13.15 shows that the grain size goes through a minimum as the average particle diameter is varied. In this particular case, the minimum occurs at an average particle diameter of approximately $2\,\mu m$ when the standard deviation in

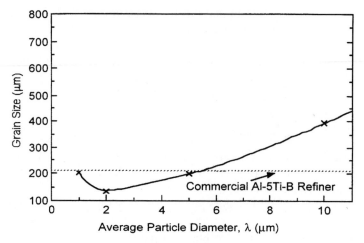

Figure 13.15. Free-growth model predictions of the grain size obtainable in commercial purity aluminium with 2 ppt addition of Al–5Ti–1B refiner. The grain size as a function of average inoculant particle diameter is shown for an assumed Gaussian distribution of particle diameter. The refiner efficiency decreases markedly as the average diameter is reduced and correspondingly the number of particles increases. The grain size predicted for a commercial refiner (with particle distribution as in figure 13.9) is shown for comparison.

diameter is 0.5 μm. This optimum value of the average particle diameter is not significantly dependent on s. It does decrease for higher refiner addition level, higher cooling rate, and higher growth restriction. The minimum grain size arises from competing trends. As the particle size is increased, the number of particles decreases, and so does the number of grains. On the other hand, as the particle size is decreased, the refiner efficiency dramatically decreases, again giving fewer grains when the particle size is small enough. The efficiency is nearly 100% for average particle diameters of greater than 5 μm and tends to zero for average particle diameters less than 1 μm. There are two effects underlying the trend in efficiency. The first is that, under otherwise identical conditions, efficiency is lowered if there are more particles; this is evident from the behaviour shown in figures 13.10 and 13.11. The other is that, for constant population, smaller particles are less efficient; their undercooling for grain initiation on them is greater, leading to faster growth, a hastened onset of recalescence and impaired refinement.

At the optimum particle size shown in figure 13.15, the efficiency of the refiner is ~20%, roughly 10 times better than the efficiency of 0.2% predicted for the particle size distribution found in a commercial refiner. Yet the grain size is only decreased by a factor of two. In adopting a Gaussian size distribution, the large populations of small particles found in the exponential distribution in a commercial refiner are eliminated. These small particles are not active in nucleating grains, so efficiency is greatly improved with

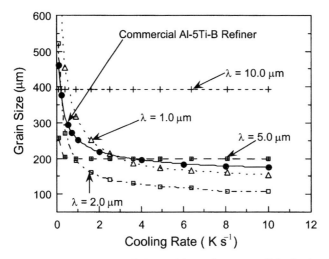

Figure 13.16. Free-growth model predictions of the performance of idealized grain refiners in which the particle size distribution is Gaussian (with, in this case, a standard deviation of $s = 0.5\,\mu m$). The dependence of grain size of commercial purity aluminium on cooling rate for idealized refiners with different average particle diameters, and for a commercial refiner (with particle distribution as in figure 13.9).

their removal. But the small particles do not make a dominant contribution to the overall volume fraction of TiB_2, so for a constant volume fraction the number of active particles is not greatly increased when the small particles are removed. Thus a commercial refiner is predicted to be almost as effective as the 'optimized' refiner, despite having a very different size distribution and a much lower efficiency. This suggests that efficiency, at least as defined in equation (13.14), is not a very useful parameter, and may even be misleading.

As noted earlier, the reason for inoculating aluminium alloys is rarely, if ever, to obtain the finest possible grains. Rather, an important aim may be uniformity of grain size. A key processing parameter which varies through an industrial-scale casting is cooling rate, and the sensitivity of grain refinement to this parameter can readily be tested using the free-growth modelling. As shown in figure 13.16, a refiner with a larger average particle size would give a grain size which, though coarser, would be much less sensitive to cooling rate. The insensitivity arises because the efficiency is 100% for large particle size.

The preliminary results shown in figures 13.15 and 13.16 suggest that there is some scope for improving refiner performance by tailoring of the particle size distribution. The calculations have been based on the commercial refiner composition Al–5Ti–1B (wt%). It remains to be seen whether such tailoring can be achieved, and if so whether it would be commercially worthwhile. An additional reason for interest in these calculations is the choice of refiner type. Commercial refiners of composition Al–3Ti–0.15C (wt%) have active TiC inoculant particles with average diameters ∼23% of those of active TiB_2 particles. Thus it is expected that the Al–Ti–C and Al–Ti–B refiners should show different performance characteristics. Some of these comparisons have been made in references [91, 92]. It is suggested from experiment that Al–Ti–C refiners are preferable to Al–Ti–B at higher addition levels, higher growth restriction, and lower cooling rate or temperature gradient. Some of these trends appear consistent with the modelling results in figures 13.15 and 13.16.

Finally, it can be noted that the degree of grain refinement which can be achieved by inoculation is limited by recalescence. Thus, just as for spontaneous grain refinement, recalescence is crucial in controlling grain size. The next section explores what can be achieved if recalescence can be avoided.

Grain refinement by rapid quenching and devitrification

Rapid solidification can be achieved by prior undercooling of the bulk melt, or by rapid heat extraction. The spontaneous grain refinement arising from dendrite break-up on growth into highly undercooled melts has already been discussed. In this section we consider the effects of rapid quenching. Even in very early work on splat-quenched metals, it was clear that grain sizes as fine as 10 nm could be reached [101]. More recently attention has turned to the production of nanophase composites in which an nm-scale dispersion of crystallites in a glassy matrix is achieved by controlled rapid quenching or by partial devitrification of a fully glassy precursor. The production of these ultra-fine microstructures will be considered later.

Single-phase polycrystals

In the laboratory, the most common technique for rapid quenching of metals and alloys is *melt spinning* in which a jet of the melt is directed at the rim of a rapidly rotating wheel (typically of copper), giving continuous production of a ribbon, 10 to 100 μm thick, which has been cooled at approximately 10^5 to 10^6 K s^{-1} [102]. Industrially, wider sheet is produced by the closely related process of *planar flow casting* [103]. When a non-glass-forming alloy is subjected to such processing, a predominantly single-phase polycrystalline product results. There is a fine equiaxed chill zone at the surface next to the quenching wheel and columnar growth from this [104]. Measurements of grain size are complicated by the natural coarsening of the columnar structure which occurs as growth continues through the ribbon thickness, less favourably oriented grains being suppressed by their neighbours. There have been few quantitative studies of the dependence of grain size on cooling rate, and in particular direct measurements of cooling rate are rare. Reviews

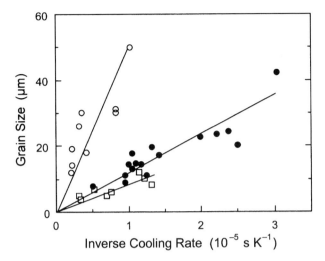

Figure 13.17. Grain size in melt-spun ribbons plotted against inverse cooling rate to test the relationship suggested in reference [43]. The data are collected in reference [105]: Fe–6.5 wt% Si (●); Ni–5 wt% Al (○); 316L stainless steel (□). The lines are regression fits to the data.

of the available data [105] do, however, show a clear pattern in which the grain size is inversely proportional to cooling rate (figure 13.17, from reference [43]). The origin of this dependence can be examined by approximate analysis of nucleation and growth rates.

Direct measurements of temperature in melt spinning and the analysis of the microstructures obtained [106–108] suggest that the liquid is significantly undercooled (by up to a few hundred degrees) in contact with the quenching surface before solidification starts. The cooling of melt-spun ribbons is approximately Newtonian, at a rate R proportional to the peripheral wheel speed U. The initial grain diameter l at the wheel surface is related to the crystal growth rate V (in m s^{-1}) and the heterogeneous nucleation rate I_{hetero} (in m^{-2} s^{-1}) by [43]:

$$l = \left\{ \frac{8V}{\pi I_{\text{hetero}}} \right\}^{1/3}. \tag{13.15}$$

A key assumption here is that stifling of nucleation by recalescence does not arise because of the ease of heat extraction. Measurements and analysis [43] show that the undercooling ΔT in contact with the wheel is not strongly dependent on the wheel speed. Furthermore, at high undercoolings the nucleation rate is much more dependent on undercooling than is the growth rate. Thus in equation (13.15), the variation of grain diameter at the wheel surface depends mainly on the heterogeneous nucleation rate.

Taking the local solidification time to be proportional to $1/U$, it follows that

$$\frac{l}{V} \propto \frac{1}{U}. \tag{13.16}$$

Since, in comparison with the other parameters in equation (13.16), the growth rate is constant, we arrive at

$$l \propto \frac{1}{UR}, \tag{13.17}$$

as observed in figure 13.17. More detailed analyses do not seem justified given the scatter in the available data. The key point is that if recalescence can be suppressed, the grain size can continue to be refined by faster cooling. However, these single-phase grain structures are columnar, and 'grain size' is the diameter of the columns. The nucleation itself is heterogeneous, on the quenching surface. To obtain similarly small grain sizes in three dimensions would require copious nucleation throughout the melt. The required population density of nucleation centres would be so high that homogeneous nucleation would be a more plausible way of achieving it than heterogeneous nucleation. The prospects for successful exploitation of homogeneous nucleation are considered next.

Nanophase composites

Greer [43] has analysed crystal nucleation and growth in undercooled liquids to assess the prospects for microstructural refinement by rapid quenching. For simplicity only polymorphic solidification (i.e. without solute partitioning) was analysed. The atomic rearrangement processes at the solid–liquid interface may be *diffusion limited* or *collision limited*. The former case refers to the need for diffusive-type jumps as the interface moves and does not imply any need for long-range partitioning. In this diffusion-limited case, the maximum interface speed V is the diffusive speed D/Λ, where D is the liquid diffusivity and Λ is the jump distance or atomic diameter. This case applies to, for example, the growth of intermetallic compounds and is consistent with glass formation. In the collision-limited case, on the other hand, the interfacial processes are not diffusive. The maximum speed of the solid–liquid interface is then approximately the speed of sound, precluding glass formation. Collision-limited kinetics were analysed first for pure metals [18], but apply also for random solid solutions and permit solute trapping [109]. These two cases thus serve to illustrate a range of glass-forming ability even though only polymorphic solidification is considered.

Figure 13.18 (from reference [43]) shows homogeneous nucleation rates and growth rates for the diffusion-limited and collision-limited cases, calculated with parameters typical for metallic (e.g. nickel-based) alloys. The calculations are for isothermal transformations, even though (as will

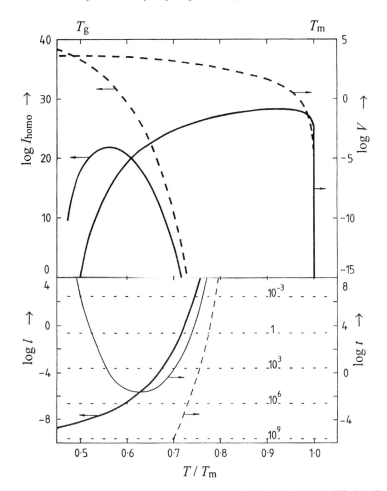

Figure 13.18. Calculated solidification kinetics for typical alloys showing diffusion-limited (solid lines) and collision-limited (dashed lines) interfacial kinetics. Shown are the $\log_{10}$ of: I_{homo} (homogeneous nucleation rate, m^{-3} s^{-1}), V (crystal growth rate, m s^{-1}), l (grain size, $(V/I)^{1/4}$, m), and t (solidification time, $l/2V$, s). The grain size is the same for both types of kinetics. The horizontal lines are labelled with cooling rates in K s^{-1}. If the solidification time falls below these lines, recalescence will preclude grain refinement and the calculated value of grain size will not apply. (From reference [43].)

be emphasized below) such conditions cannot always be maintained during solidification. For the diffusion-limited case, a glass-forming alloy is considered with a glass transition temperature $T_g = 0.5 T_m$ (the equilibrium melting point), and a Vogel–Fulcher–Tammann temperature dependence of viscosity from which the interfacial kinetics are derived [110]. The crystal growth rate is assumed to be limited by the interfacial kinetics. In the collision-limited case a similar assumption would be unrealistic, as the growth is limited by the thermal and curvature undercoolings. Since planar growth under these conditions is not stable, a simple dendrite growth law $V \propto \Delta T^3$ (found for pure nickel [111]) is used, modified at high undercoolings to limit the maximum interfacial speed to the speed of sound. In each case, classical homogeneous nucleation theory was used [110].

Figure 13.18 shows that substantial undercooling ΔT is required for the onset of homogeneous nucleation. At lower undercooling, heterogeneous nucleation will dominate. If latent heat release during solidification is ignored, isothermal transformation is possible, and the resulting grain size l is related to the growth rate V and the bulk nucleation rate I_{homo} by [112]

$$l = \left\{ \frac{V}{I_{\text{homo}}} \right\}^{1/4}. \tag{13.18}$$

The grain size calculated from equation (13.18) is shown in figure 13.18 and is the same for both diffusion-limited and collision-limited kinetics. The curve for the resultant grain size demonstrates considerable refinement at large undercooling, into the nm regime. However, this refinement can be observed only if recalescence during growth does not stifle further nucleation.

The latent heat release rate has already been considered in an earlier section. Here a simple analytical approach is adopted. As the crystals grow, even at a constant rate V, the rate of release of latent heat accelerates as the surface area of the crystals increases. As previously discussed, the rate of latent heat release can become greater than the external heat extraction rate, characterized by the cooling rate R which would be found in the absence of the heat release. Taking the volume associated with one growing crystal to be $(4\pi/3)(l/2)^3$, and adapting an analysis of Boettinger and Perepezko [113], the time t^* at which the heat release and extraction rates are equal is

$$t^* = \left\{ \frac{l^3 C_{\text{pV}} R}{24 V^3 \Delta H_{\text{V}}} \right\}, \tag{13.19}$$

where C_{pV} and ΔH_{V} are the liquid heat capacity and latent heat per unit volume. We can conclude that the effects of latent heat release are not important if the solidification time $(l/2V)$ meets the condition

$$\frac{l}{2V} \geq \frac{3 \Delta H_{\text{V}}}{C_{\text{pV}} R}. \tag{13.20}$$

Contours of $(3\Delta H_{\text{V}}/C_{\text{pV}} R)$ are plotted on figure 13.18 for comparison with the solidification time for the diffusion-limited and collision-limited cases. At the large undercooling where grain refinement due to copious homogeneous nucleation might occur, the nucleation will certainly be stifled in the collision-limited case, even for imposed cooling rates greatly in excess of the maximum ($\sim 10^6$ K s^{-1}) achievable in melt spinning. In the diffusion-limited case, however, cooling rates greater than (10^6 K s^{-1}) are sufficient to render the latent heat release unimportant. In that case, grain refinement due to copious nucleation should become observable just as glass formation becomes possible. Figure 13.18 is based on isothermal kinetics. Although this may be useful in considering whether or not recalescence is likely to be important, and in demonstrating the refinement achievable at large undercooling, it cannot be used to obtain quantitative estimates of grain size.

The prediction arising from figure 13.18, that ultra-fine microstructures can be attained under conditions on the margin of glass formation, has been amply verified by experiment. When glass-forming alloy compositions are melt-spun, they can show a transition from a glassy to a fine-grained structure through the thickness, which appears to arise from copious homogeneous nucleation in the bulk [114]. The combination of slow growth rate and rapid external heat extraction ensures that a high undercooling (and therefore rapid nucleation) can be maintained even as solidification progresses. Recently the interest in such systems has focused on aluminium–transition-metal–lanthanide alloys which can give nanophase composites in which α-aluminium crystallites are dispersed in an Al-based glassy matrix [115, 116]. These microstructures can be obtained directly by rapid quenching on the margin of full glass formation, or by subsequent annealing and partial devitrification of fully glassy precursors obtained by more rapid quenching. The crystallites are 3 to 10 nm in diameter and occupy up to 25% volume fraction. The partially devitrified alloys are up to 50% harder than the fully glassy alloys, while avoiding the embrittlement which often accompanies annealing of metallic glasses. The similarity of structure and properties of material made by different routes confirms that the microstructural refinement achieved by devitrification of a glassy alloy is very closely related to that achieved in melts undergoing rapid quenching.

Apart from the aluminium-based nanophase composites of interest for their mechanical properties, similar nanophase composite microstructures in iron-based systems are of interest for their magnetic properties. Commercially important microcrystalline hard-magnetic materials based on Fe–Nd–B alloys can be made (apart from a powder metallurgical route) equally by controlled quenching directly to the microcrystalline structure or by devitrification of faster-quenched glassy material [117]. Also of interest are soft-magnetic materials based on the formation of α-iron nanocrystallites in compositions based on the Fe–Si–B system [118]. In this case, a typical composition, $Fe_{73.5}Cu_1Nb_3Si_{13.5}B_9$, has added copper to stimulate crystallite nucleation [119] and niobium to inhibit growth of the α-iron crystallites. These additions, of course, are similar in intent to those involved in conventional grain refinement. While the copper is evidently successful in giving a fine dispersion of nucleant particles, such additions are not always necessary; nm-scale dispersions of α-iron crystallites giving good soft-magnetic properties can also be obtained by partial devitrification of amorphous alloys such as $Fe_{91}Zr_7B_2$ [120].

The aluminium-based and iron-based glasses just quoted do not show polymorphic crystallization. The solute partitioning accompanying the growth of the α-aluminium or α-iron crystallites plays an important role in the microstructural refinement by impeding growth. The compositions in these systems must be chosen to avoid polymorphic crystallization to make glass formation possible in the first place. Nevertheless, the basic forms of behaviour shown by the curves for the diffusion-limited case in figure 13.18 still apply.

The slow crystal growth in devitrification greatly facilitates quantitative analysis of solidification kinetics. There is good evidence in a number of glasses that the crystal nucleation during crystallization is homogeneous [121]. In others, growth occurs on nuclei which appear to have been formed homogeneously during the quench; the population of these nuclei is strongly dependent on quench rate [121]. In either case there is the prospect, as indicated in figure 13.18, of obtaining very fine grain sizes. The heat release which could stifle such an effect is not significant for the glass-forming system until near the minimum in the solidification time curve shown in figure 13.18. At lower temperatures a metallic glass can remain closely isothermal as crystallization proceeds. However, near the minimum, external heat extraction may be necessary to prevent the heat of crystallization leading to accelerating nucleation and growth rates and a runaway, or *explosive*, reaction. In explosive crystallization much of the grain refinement effect would be lost.

Grain size in thin films

In the processing of integrated circuits and other devices, there is some interest in transient melting to obtain useful microstructures. By using short heat pulses, melting can be achieved in the top layer of a device structure without raising the temperature to deleterious levels in other parts of the structure. This has been demonstrated in *sequential lateral solidification* of silicon thin films, used for the fabrication of single-crystal silicon thin-film transistors [122, 123]. The microstructural design in the thin films is vital for this application as the good performance of the transistors is dependent on the absence of high-angle grain boundaries within their active region. Transient heating has also been used to process aluminium lines on devices. Melting and resolidification can give single-crystal or 'bamboo'-structure lines which are of interest for increased resistance to electromigration damage [124, 125]. The capability for precise specification of nucleation sites has also recently been demonstrated in transient melting of thin films, as discussed next.

In the experiment reported by Jakkaraju *et al* [126], a copper film 1.6 μm thick was deposited over an oxidized silicon substrate with a square array of holes formed in it. These holes in the silica are of diameter 0.3 to 1.0 μm, and their centre-to-centre distance is 1.85 to 3.2 μm. When the copper is deposited on top, it bridges over these holes. The entire structure is put under pressure (70 MPa) and then pulse-heated with a neodymium-YAG laser giving 200 nm pulses. The heating melts the copper, partially or completely, and the metal flows down into the holes. The extent of this flow was the focus of the original experiment, designed to test procedures for filling via contact holes in the dielectric layers of modern integrated circuits. The resolidification of the copper starts in the metal which has flowed down the holes; this metal is much colder than that on the top of the structure. The growth from each hole progresses up to the film on top and spreads laterally, impingement giving the regular array of square grains seen in figure 13.19. The image in

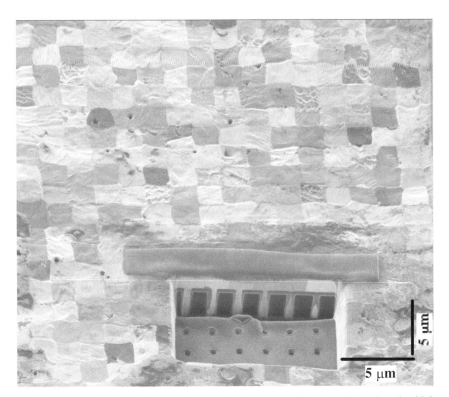

Figure 13.19. Focused-ion-beam micrograph of a copper film pulse-heated under high pressure giving an evident chequer-board grain pattern on resolidification. A trench has been ion-milled into the sample. At the top of this is a strip of platinum, deposited to aid in making a clean section. The base of the trench shows the square array of holes etched in the SiO_2 substrate. The side of the trench sections the holes, showing that these have been filled with copper as a result of the transient melting. (From reference [126].)

this figure has been obtained in a focused-ion-beam workstation. This has been used to mill away the trench which reveals the pattern of holes in the underlying substrate. The vertical side wall of the trench shows the penetration of the copper into the holes that takes place when melted under pressure.

The regularity of this 'chequer-board' pattern demonstrates the degree of control which has been achieved. Subsequent work has demonstrated other patterns, including hexagonal grains [127]. The steep temperature gradient induced by the sharp heating pulse to the top of the device is crucial in achieving the microstructural control. It ensures that the nucleation in the holes takes place essentially simultaneously. This prevents growth from one nucleation site overgrowing other sites. Also, the limited range for growth defined by the thickness of the metal film prevents subsequent growth competition in a columnar zone which would lead to grains of larger diameter. The very limited extent of the liquid (a thin layer on top of a relatively massive substrate) gives a very high cooling rate ($\sim 10^9$ K s^{-1}) and precludes any influence of recalescence.

As described in reference [126], the degree of microstructural control which can be exercised in the thin-film processing goes beyond control of grain shape. By adjusting whether the copper is partially or fully melted, one of two different crystallographic textures can be selected.

Summary

The work illustrated in figure 13.19 demonstrates that considerable microstructural control can be achieved with enough constraints on the solidifying system. However, in most cases there are not such tight constraints. An overall theme in the various situations examined in this chapter is the role of recalescence. This is often the dominant factor in controlling the final grain size of a solidified alloy, whether the solidification takes place at small undercooling as in conventional processing or at large undercooling when spontaneous grain refinement is observed.

Figure 13.20 compares the grain sizes obtainable by different processing techniques in terms of the number of grains per unit volume. It shows that much finer grain sizes can be attained if recalescence is suppressed by rapid heat extraction or by slowing the crystallization by forcing it to occur in the glassy state. The figure also shows how near the attainable microstructures are to the final limit of grain refinement, which is one atom per grain. The nanophase composites have moved remarkably far towards this limit compared with conventionally grain-refined alloys. The limit itself could be interpreted as reaching a glassy structure, to which of course the nanophase composites are closely related.

The examples chosen, and in particular the conventional grain refinement of aluminium alloys, show that some degree of predictability is possible

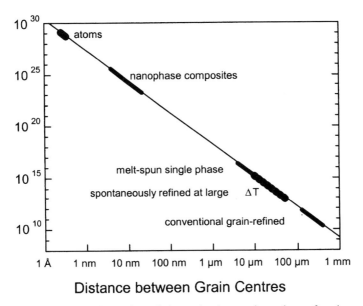

Figure 13.20. Schematic illustration of the grain sizes and numbers of grains per unit volume for as-cast alloy microstructures. The degrees of grain refinement which can be achieved vary widely with processing technique.

for the grain sizes achieved in equiaxed solidification. The mechanisms determining the grain size are surprisingly varied, however. It is particularly important to note that a dispersion of classical nucleation events is only very rarely responsible for initiating grains. Nucleation may apply in rapid quenching and in devitrification, but in conventional grain refinement by inoculation, grain initiation appears to be determined by a free-growth criterion; and in spontaneous grain refinement at large undercooling, the grains arise from dendrite break-up.

References

[1] Versnyder F L and Shank M E 1970 *Mater. Sci. Eng.* **6** 213
[2] Briant C L 1993 *Metall. Trans.* **24A** 1073
[3] Cho J and Thompson C V 1989 *Appl. Phys. Lett.* **54** 2577
[4] McCartney D G 1989 *Int. Mater. Rev.* **34** 247
[5] Tyas N H 2000 PhD Thesis, University of Cambridge
[6] Hunt J D 1984 *Mater. Sci. Eng.* **65** 75
[7] Gäumann M, Trivedi R and Kurz W 1997 *Mater. Sci. Eng.* **A226–228** 763
[8] Gandin C-A and Rappaz M 1994 *Acta Metall. Mater.* **42** 2233
[9] Boettinger W J, Coriell S R, Greer A L, Karma A, Kurz W, Rappaz M and Trivedi R 2000 *Acta Mater.* **48** 43

[10] Boettinger W J, Shechtman D, Schaefer R F and Biancaniello F S 1984 *Metall. Trans.* **15A** 55
[11] Gill S C and Kurz W 1995 *Acta Metall. Mater.* **43** 139
[12] Christian J W 1975 *The Theory of Transformations in Metals and Alloys* 2nd edition (Oxford: Pergamon) chapter 10
[13] Porter D A and Easterling K E 1992 *Phase Transformations in Metals and Alloys* 2nd edition (London: Chapman & Hall) section 4.1
[14] Kelton K F 1991 *Solid State Phys.* **45** 75
[15] Greer A L and Kelton K F 1991 *J. Am. Ceram. Soc.* **74** 1015
[16] Oxtoby D W 1992 *J. Phys.: Condens. Matter* **4** 7627
[17] Granasy L 1993 *J. Non-Cryst. Solids* **162** 301
[18] Turnbull D 1969 *Contemp. Phys.* **10** 473
[19] Spaepen F and Meyer R B 1976 *Scr. Metall.* **10** 257
[20] Perepezko J H and Uttormark M J 1996 *Metall. Mater. Trans.* **27A** 533
[21] Cantor B 1994 *Mater. Sci. Eng.* **A178** 225
[22] Kim W T, Zhang D L and Cantor B 1991 *Metall. Trans.* **22A** 2487
[23] Kim W T and Cantor B 1991 *J. Mater. Sci.* **26** 2868
[24] Kim W T and Cantor B 1992 *Acta Metall.* **40** 3339
[25] Kim W T and Cantor B 1994 *Acta Metall.* **42** 3045
[26] Kim W T and Cantor B 1994 *Acta Metall.* **42** 3115
[27] Norman A F, Eckler K, Zambon A, Gärtner F, Moir S A, Ramous E, Herlach D M and Greer A L 1998 *Acta Mater.* **46** 3355
[28] Herlach D M, Cochrane R F, Egry I, Fecht H J and Greer A L 1993 *Int. Mater. Rev.* **38** 273
[29] Eckler K, Gärtner F, Assadi H, Norman A F, Greer A L and Herlach D M 1997 *Mater. Sci. Eng.* **A226–228** 410
[30] Chalmers B 1964 *Principles of Solidification* (New York: Wiley) p 265
[31] Schaefer R J and Glicksman M E 1967 *Trans. AIME* **239** 257
[32] Jackson K A, Hunt J D, Uhlmann D R and Seward T P 1966 *Trans. AIME* **236** 149
[33] Karma A 1998 *Int. J. Non-Equil. Proc.* **11** 201
[34] Jones B L and Weston G M 1970 *J. Aust. Inst. Metal.* **15** 189
[35] Southin R T and Weston G M 1973 *J. Aust. Inst. Metal.* **18** 74
[36] Walker J L 1959 in *The Physical Chemistry of Process Metallurgy* part 2, ed G R St Pierre (New York: Interscience) p 845
[37] Powell G L F 1965 *J. Aust. Inst. Metal.* **10** 223
[38] Costa Agra Mello M and Kiminami S 1989 *J. Mater. Sci. Lett.* **8** 1416
[39] Kattamis T Z and Flemings M C 1966 *Trans. AIME* **236** 1523
[40] Tarshis L A, Walker J L and Rutter J W 1971 *Metall. Trans.* **2** 2589
[41] Kobayashi K F and Shingu P H 1988 *J. Mater. Sci.* **23** 2157
[42] Horvay G 1965 *Int. J. Heat Mass Transf.* **8** 192
[43] Greer A L 1991 *Mater. Sci. Eng.* **A133** 16
[44] Cochrane R F, Battersby S E and Mullis A M 1998 in *Solidification 1998* ed S P Marsh, J A Dantzig, R Trivedi, W Hofmeister, M G Chu, E J Lavernia and J H Chun (Warrendale, PA: TMS) p 245
[45] Cochrane R F, Battersby S E and Mullis A M in *Solidification 1998* ed S P Marsh, J A Dantzig, R Trivedi, W Hofmeister, M G Chu, E J Lavernia and J H Chun (Warrendale, PA: TMS) p 223

[46] Cochrane R F and Herlach D M 1989 in *Proc. 7th Eur. Symp. Materials and Fluid Sciences under Microgravity* (Paris: ESA SP-295) p 147
[47] Cochrane R F, Herlach D M and Feuerbacher B 1991 *Mater. Sci. Eng.* **A133** 706
[48] Schwarz M, Karma A, Eckler K and Herlach D M 1994 *Phys. Rev. Lett.* **73** 1380
[49] Bassler B T, Hofmeister W H, Carro G and Bayuzick R J 1994 *Metall. Mater. Trans.* **25A** 1301
[50] Willnecker R, Herlach D M and Feuerbacher B 1989 *Phys. Rev. Lett.* **62** 2707
[51] Gärtner F, Norman A F, Greer A L, Zambon A, Ramous E, Eckler K and Herlach D M 1997 *Acta Mater.* **45** 51
[52] Matson D 1998 in *Solidification 1998* ed S P Marsh, J A Dantzig, R Trivedi, W Hofmeister, M G Chu, E J Lavernia and J H Chun (Warrendale, PA: TMS) p 233
[53] Kattamis T Z and Flemings M C 1967 *Mod. Casting* **52** 97
[54] Munitz A and Abbaschian G J 1986 in *Undercooled Alloy Phases* ed E W Collings and C C Koch (Warrendale, PA: TMS) p 23
[55] Mullis A M and Cochrane R F 2000 *Int. J. Non-Equil. Proc.* **11** 283
[56] Mullis A M 1997 *Mater. Sci. Eng.* **A226–228** 804
[57] Mullis A M and Cochrane R F 1997 *J. Appl. Phys.* **82** 3783
[58] Mullis A M and Cochrane R F 1998 *J. Appl. Phys.* **84** 4905
[59] Bustos O L and Reif W 1994 *Metall.* **48** 123
[60] Cheng T T 2000 *Intermetallics* **8** 29
[61] Meredith M W, Greer A L and Evans P V 1998 in *Light Metals 1998* ed B Welch (Warrendale, PA: TMS) p 977
[62] Allen C M, O'Reilly K A Q, Evans P V and Cantor B 1999 *Acta Mater.* **47** 4387
[63] Flood S C and Hunt J D 1981 *Met. Sci.* **15** 287
[64] Ho C R and Cantor B 1995 *Acta Metall. Mater.* **43** 3231
[65] Bunn A M, Schumacher P, Kearns M A, Boothroyd C B and Greer A L 1999 *Mater. Sci. Technol.* **15** 1115
[66] *Standard Test Procedure for Aluminum Alloy Grain Refiners: TP-1, 1987* (Washington, DC: Aluminum Association)
[67] Bunn A M, Greer A L, Green A H and Kearns M A 1997 in *Proc. 4th Dec. Int. Conf. on Solidification Processing (SP'97)* ed J Beech and H Jones (Sheffield: University of Sheffield) p 264
[68] Mohanty P S and Gruzleski J E 1995 *Acta Metall. Mater.* **43** 2001
[69] Mohanty P S, Samuel F H, Gruzleski J E and Kosto T J 1994 in *Light Metals 1994* ed U Mannweiler (Warrendale, PA: TMS) p 1039
[70] Schumacher P, Greer A L, Worth J, Evans P V, Kearns M A, Fisher P and Green A H 1998 *Mater. Sci. Technol.* **14** 394
[71] Johnson M and Bäckerud L 1992 *Z. Metallkd.* **83** 774
[72] Vader M and Noordegraaf J 1989 in *Light Metals 1989* ed P G Campbell (Warrendale, PA: TMS) p 937
[73] Vader M and Noordegraaf J 1990 in *Light Metals 1990* ed C M Bickert (Warrendale, PA: TMS) p 851
[74] Turnbull D 1950 *J. Chem Phys.* **18** 198
[75] Jones G P 1983 'New ideas on the mechanism of heterogeneous nucleation in liquid aluminium', NPL Report DMA (A) 19 (Teddington: National Physical Laboratory)

[76] Jones G P 1985 in *Proc. Int. Sem. on Refining and Alloying of Liquid Aluminium and Ferro-Alloys* ed T A Engh *et al* (Düsseldorf: Aluminium Verlag) p 212
[77] Jones G P 1988 in *Proc. Conf. on Solidification Processing 1987* ed J Beech and H Jones (London: The Institute of Metals) p 496
[78] Easton M and StJohn D 1999 *Metall. Mater. Trans.* **30A** 1613, 1625
[79] McKay B J, Cizek P, Schumacher P and O'Reilly K A Q 2000 in *Light Metals 2000* ed R D Peterson (Warrendale, PA: TMS) p 833
[80] Schumacher P and Greer A L 1996 in *Light Metals 1996* ed W Hale (Warrendale, PA: TMS) p 745
[81] Greer A L, Bunn A M, Tronche A, Evans P V and Bristow D J 2000 *Acta Mater.* **48** 2823
[82] Tronche A and Greer A L *Phil. Mag. Lett.*, in press
[83] Vandyoussefi M, Worth J and Greer A L 2000 *Mater. Sci. Technol.* **16** 1121
[84] Tronche A and Greer A L *Aluminum Trans.*, submitted
[85] Maxwell I and Hellawell A 1975 *Acta Metall.* **23** 229
[86] Massalski T B (ed) 1990 *Binary Alloy Phase Diagrams*, vol 1 2nd edition (Metals Park, OH: ASM International)
[87] Bunn A M, Evans P V, Bristow D J and Greer A L 1998 in *Light Metals 1998* ed B Welch (Warrendale, PA: TMS) p 963
[88] Greer A L and Tronche A 2000 in *Continuous Casting* ed K Ehrke and W Schneider (Weinheim: DGM and Wiley–VCH) p 149
[89] Hodaj F and Durand F 1997 *Acta Mater.* **45** 2121
[90] Spittle J A and Sadli S B 1995 *Mater. Sci. Technol.* **11** 533
[91] Schneider W, Kearns M A, McGarry M J and Whitehead A J 1998 in *Light Metals 1998*, ed B Welch (Warrendale, PA: TMS) p 953
[92] Van Wiggen P C and Belgraver J K 1999 in *Light Metals 1999* ed C E Eckert (Warrendale, PA: TMS) p 779
[93] Rappaz M and Rettenmayr M 1998 *Curr. Opin. Solid State Mater. Sci.* **3** 275
[94] Rappaz M, Gandin Ch-A, Desbiolles J-L and Thévoz Ph 1996 *Metall. Mater. Trans.* **27A** 695
[95] Rappaz M and Gandin Ch-A 1993 *Acta Metall. Mater.* **41** 345
[96] Gandin Ch-A, Charbon Ch and Rappaz M 1995 *ISIJ Int.* **35** 651
[97] Calcom SA, Lausanne, Switzerland
[98] Kurz W, Giovanola B and Trivedi R 1986 *Acta Metall.* **34** 823
[99] Greer A L, Tronche A and Vandyoussefi M 2000 *Mater. Res. Soc. Symp. Proc.* **578** 425
[100] Tronche A and Greer A L 2000 in *Light Metals 2000* ed R D Peterson (Warrendale, PA: TMS) p 827
[101] Jones H 1977 in *Vacancies '76* ed R E Smallman and J E Harris (London: Metals Society) p 175
[102] Liebermann H H 1980 *Mater. Sci. Eng.* **43** 203
[103] Narasimhan M C 1979 US Patent 4 142 571
[104] Batawi E, Morris M A and Morris D G 1988 *Mater. Sci. Eng.* **98** 161
[105] Kim W T and Cantor B 1990 *Scr. Metall. Mater.* **24** 633
[106] Hayzelden C, Rayment J J and Cantor B 1983 *Acta Metall.* **31** 379
[107] Cantor B, Kim W T, Bewlay B P and Gillen A G 1991 *J Mater. Sci.* **26** 1266
[108] Clyne T W 1984 *Metall. Trans.* **15B** 369
[109] Aziz M J 1982 *J Appl. Phys.* **53** 1158

[110] Thompson C V, Greer A L and Spaepen F 1983 *Acta Metall.* **31** 1883
[111] Schleip E, Willnecker R, Herlach D M and Görler G P 1988 *Mater. Sci. Eng.* **98** 39
[112] Johnson W F and Mehl R F 1939 *Trans AIME* **135** 416
[113] Boettinger W J and Perepezko J H 1985 in *Rapidly Solidified Crystalline Alloys* ed S K Das, B H Kear and C M Adam (Warrendale, PA: TMS) p 21
[114] Boettinger W J 1988 *Mater. Sci. Eng.* **98** 123
[115] Kim Y H, Inoue A and Masumoto T 1990 *Mater. Trans. JIM* **31** 747
[116] Chen H, He Y, Shiflet G J and Poon S J 1991 *Scr. Metall. Mater.* **25** 1421
[117] Coehoorn R and Duchateau J 1988 *Mater. Sci. Eng.* **99** 131
[118] Yoshizawa Y, Oguma S and Yamauchi K 1988 *J Appl. Phys.* **64** 6044
[119] Ohnuma M, Hono K, Linderoth S, Pedersen J S, Yoshizawa Y and Onodera H 2000 *Acta Mater.* **48** 4783
[120] Suzuki K, Kataoka N, Inoue A and Masumoto T 1991 *Mater. Trans. JIM* **32** 93
[121] Greer A L 1988 *Mater. Sci. Eng.* **97** 285
[122] Sposili R S and Im J S 1998 *Appl. Phys.* **A67** 273
[123] Im J S, Crowder M A, Sposili R S, Leonard J P, Kim H J, Yoon J H, Gupta V V, Song H J and Cho H S 1998 *Phys. Status Solidi* **A166** 603
[124] van den Homberg M J C, Alkemade P F A, Verbruggen A H, Dirks A G, Hurd J L and Radelaar S 1997 *Microelectron. Eng.* **35** 277
[125] van den Homberg M J C, Alkemade P F A, Radelaar S, Hurd J L and Dirks A G 1997 *Appl. Phys. Lett.* **70** 318
[126] Jakkaraju R, Dobson C D and Greer A L 2000 *Mater. Res. Soc. Symp. Proc.* **594** 111
[127] Jakkaraju R 2001 PhD Thesis, University of Cambridge

Chapter 14

Step casting

Masayuki Kudoh, Tatsuya Ohmi and Kiyotaka Matsuura

Introduction

Grain refinement of an ingot is required in order to improve its mechanical properties. Many techniques have been suggested. Nucleation and crystal multiplication are the fundamental concepts to make fine grains in an ingot. Inoculation and mechanical stirring are typical grain refinement techniques which utilize nucleation and crystal multiplication. There is, however, still ample scope for novel methods of grain refinement which reduce the cost of manufacturing ingots. Chapters 12 and 13 describe general aspects of nucleation and grain refinement respectively. This chapter describes two related novel methods for grain refinement and the cheap fabrication of intermetallic compounds, named 'duplex casting' and 'synthesis casting', respectively, with the term 'step casting' encompassing both techniques. The principle of the methods is the mixing of two liquids, which are sequentially poured into a mould with a particular time interval between pouring. Careful control of the relative liquidus temperature of the two alloys and cooling conditions can lead to significant grain refinement of the resulting ingot. This chapter will present a number of examples illustrating the benefits of step casting, including the refinement of primary aluminium in hypoeutectic Al–Fe alloys, the refinement of primary silicon in hypereutectic Al–Si alloys, the simultaneous use of duplex casting and inoculants to grain refine a hypereutectic Al–Si alloy, and the use of synthesis casting for the low cost manufacture of NiAl and its *in situ* joining to structural materials.

Step casting

Step casting, including both duplex casting and synthesis casting, is carried out as follows. Two liquids with different liquidus temperatures and different volumes are prepared. The liquid with the lower liquidus temperature should have the largest volume, and is poured first into the mould. The second

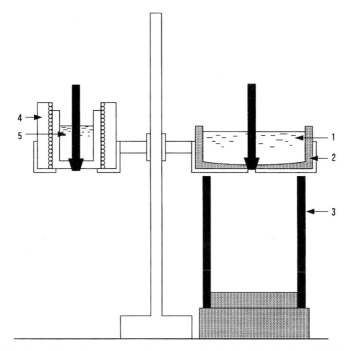

Figure 14.1. Experimental apparatus for step casting. 1: first liquid, 2: tundish, 3: mould, 4: electric furnace, 5: second liquid.

liquid, with the higher liquidus temperature and smaller volume, is poured into the mould after a predetermined time, once solidification of the first liquid has begun. In order to obtain grain refinement of the primary phase it is necessary to ensure that a solid shell of the first liquid has formed before pouring in the second liquid. That is, the second liquid must be rapidly cooled by heat loss to the first liquid and by partial remelting of the solidified shell.

Figure 14.1 shows schematically typical apparatus for step casting [1]. The first liquid is kept at its predetermined temperature in a tundish prior to pouring into the mould, whilst the second liquid is kept at its predetermined temperature in an electric furnace which is rotated into position above the mould to allow the second liquid to be poured after the required time has lapsed. For aluminium-based binary alloys, for example, the superheat of each liquid is usually 50 K above the liquidus temperature.

Refinement of primary aluminium in hypoeutectic Al–Si alloys

The grain structure of a conventionally cast Al–3%Si ingot weighing 2.6 kg is shown in figure 14.2. The primary phase is aluminium. The grain structure comprises zones of columnar and equiaxed crystals with the equiaxed

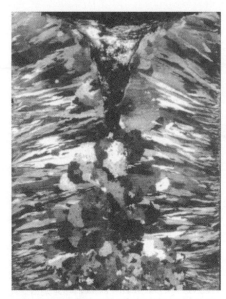

Figure 14.2. Grain structure of a conventionally cast ingot for Al–3% Si alloy.

crystals being coarse and occupying only a narrow zone. This is a typical structure for a conventionally cast ingot.

The grain structure of an Al–3%Si ingot fabricated by duplex casting is shown in figure 14.3 [1]. Two liquids both of composition Al–3%Si, weighing 2.2 and 0.4 kg, were used as the first liquid and second liquid respectively. The pouring temperature of both liquids was 973 K. The second liquid was poured 14.6 s after the pouring of the first liquid. The equiaxed crystals are significantly refined and the size of the equiaxed zone extended compared with the conventionally cast ingot, even though both liquids have the same solute content.

Figures 14.4(a) and (b) show grain structures of ingots fabricated by duplex casting [1]. The first liquid, Al–3%Si weighing 2.2 kg, and the second liquid, pure aluminium weighing 0.4 kg, were poured into a graphite mould of inner diameter 100 mm. The pouring temperatures of the two liquids were 973 and 1023 K, respectively. In figures 14.4(a) and (b), the time intervals between the pouring of the first and second liquids were 7.9 s and 14.6 s respectively. Zones of both columnar and equiaxed crystals are seen in the ingot, with the refinement of the equiaxed crystals being promoted by increasing the time interval between pourings. In these experiments, the columnar zone was not completely remelted by the second liquid. However, the grain size of the equiaxed crystals was reduced below 1 mm and the equiaxed zone was enlarged.

Comparing figures 14.3 and 14.4 shows that grain refinement of the primary aluminium phase during duplex casting is enhanced by using a

Refinement of primary aluminium in hypoeutectic Al–Si alloys

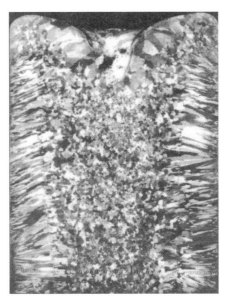

Figure 14.3. Grain structure of an ingot fabricated by duplex casting. First and second liquids are Al–3% Si alloy.

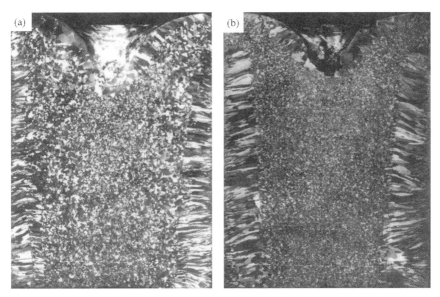

Figure 14.4. Grain structures of duplex casting. The first liquid is Al–3% Si and the second is pure aluminium. (a) $\Delta t = 7.9$ s, (b) $\Delta t = 14.6$ s.

second liquid with a higher liquidus temperature than the first liquid, and by increasing the time interval between pourings.

Refinement of primary silicon in hypereutectic Al–Si alloys

Alloy AC 9C (Japan Industry Standard) has a high tensile strength and good wear resistance when the primary silicon is satisfactory. This aluminium alloy contains 22% silicon and minor solute elements. Conventional refinement of the primary silicon is by inoculation with phosphide. However, the formation of inclusions due to this inoculant causes practical problems. Duplex casting of an Al–22%Si alloy [2] has been used to refine the primary silicon without the use of an inoculant.

Figure 14.5 shows the relationship between the primary silicon particle size and the composition of the first liquid. The composition of the second liquid was kept constant at 32% silicon and the final solute content after mixing was maintained at 22% silicon by adjusting the volume of the second liquid. The time interval between pouring the first and second liquids was 7 s. Two kinds of mould were used; mould 1 was made of plain graphite, and mould 2 was also made of graphite but with a bottom plate made of brick. Solidification in mould 1 was faster than that in mould 2. The primary silicon particle size shows a minimum value for compositions of the first liquid in the range 2–17% silicon. It is usually said that a primary silicon

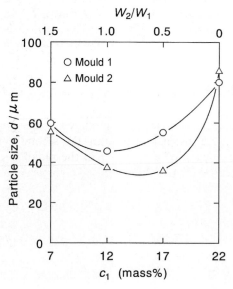

Figure 14.5. Relationship between the content of the first liquid, C_1, and the particle size of the primary Si, d, in a stepwise cast ingot.

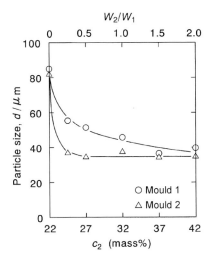

Figure 14.6. Relationship between the content of the second liquid, C_2, and the particle size of the primary Si, d, in a stepwise cast ingot.

particle size below 40 μm is useful in practical terms. In these experiments, the primary silicon was refined below this size.

Figure 14.6 shows the relationship between the primary silicon particle size and the composition of the second liquid. The final solute content after mixing is the same as shown in figure 14.5. The composition of the first liquid was 12% silicon. The time interval between pouring the first and second liquids was again 7 s, and moulds 1 and 2 were used as described above. The primary silicon refined rapidly with increasing composition of the second liquid and also with increasing cooling rate.

Effect of solidified shell

The primary silicon particle size can be related to the average temperature T_{m1} of the first liquid just before pouring the second liquid. For the first liquid, the temperature profile at time t in the radial direction of the first liquid is assumed to be $T(r)$, and its average temperature T_{mL} is expressed as

$$T_{mL} = \frac{1}{\pi R^2} \int_0^R 2\pi R T(r) \, dr, \qquad (14.1)$$

where R is the inner diameter of the mould. Considering the thermal effect corresponding to the latent heat of the solidified shell of the first liquid, an effective average temperature in the first liquid, T_{mL^*} can be expressed as:

$$T_{mL^*} = T_{mL} - f_s \frac{L}{C_p}, \qquad f_s = \frac{W_s}{W_s + W_L}, \qquad (14.2)$$

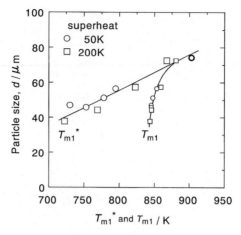

Figure 14.7. Relationship between the particle size of primary Si and T_{m1}^* for different amounts of superheat of the first liquid.

where L is the latent heat of fusion, C_p is the specific heat, f_s is the fraction solid of the first alloy, and W_s and W_L are respective masses of the solidified shell and the remaining liquid. An Al–12%Si alloy and an Al–32%Si alloy were used as the first and second liquids, respectively. Changing the time interval between the pourings changes T_{mL} and T_{mL^*}. Figure 14.7 shows the primary silicon particle size as a function of T_{mL^*} by considering the local solidification of the first liquid. Hence, the refinement of primary silicon may be caused by rapid quenching of the second liquid by mixing with the first liquid, and by the partial remelting of the solidified shell of the first alloy.

Simultaneous use of duplex casting and inoculants for a hypereutectic Al–Si alloy

In practice, for hypereutectic alloys, phosphorus and sodium are used to refine the primary silicon and to modify the eutectic silicon, respectively. However, the simultaneous use of phosphorus and sodium inhibits the refining effect of primary silicon because of their mutual interference. Figures 14.8(a) and (b) show the refinement of the primary silicon and the modification of the eutectic silicon for conventionally cast ingots and for duplex cast ingots [3]. In the conventional cast alloys, the addition of either phosphorus or sodium allows refinement of the primary silicon or modification of the eutectic silicon respectively, whilst their simultaneous use results in coarse primary silicon. On the other hand, by using duplex casting and adding sodium to the first liquid and phosphorus to the second liquid, the primary and eutectic silicon were both satisfactorily refined.

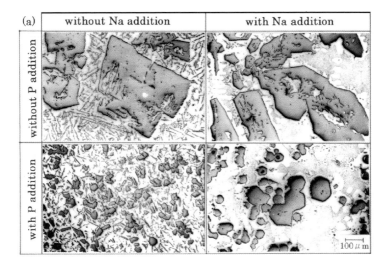

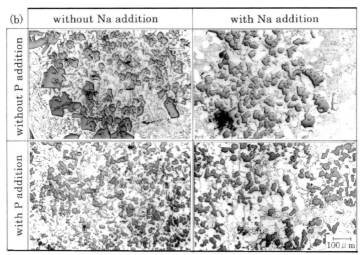

Figure 14.8. Comparison of microstructures in ingots fabricated by (a) conventional method and (b) duplex casting.

Manufacture of intermetallic compounds and *in situ* joining

The intermetallic compound NiAl has several excellent properties including low density and high strength at elevated temperatures, high melting temperature (1911 K) and good oxidation resistance [4]. It is usually fabricated by the technique of combustion synthesis from a mixture of nickel and aluminium powders. However, this technique is expensive and results in high levels of porosity due to the low density of powder compacts.

Synthesis casting has been used to manufacture the intermetallic compound NiAl [5] by pouring molten nickel into molten aluminium. The intermetallic compound was successfully fabricated even though its melting point is higher than those of either of the liquids. The temperature of the mixed liquid rose quickly above the melting temperature of the NiAl because of heat generation due to the synthesis of NiAl. One effective use of this heat generation may be to join NiAl to structural materials, usually difficult to achieve because NiAl has no ductility. The joining of NiAl to structural materials such as carbon steels, stainless steels and superalloys has been tried using synthesis casting. The standard method used is as follows. First, molten aluminium is poured on to the structural material that was previously set in a container and, subsequently, molten nickel is poured into the container. NiAl is successfully formed at the surface of the structural material. The quality of the joint between the NiAl and the structural material is good, because the surface of the material is partly remelted in the process.

Summary

Duplex casting for the refinement of the primary phase during ingot solidification and synthesis casting for the cheap fabrication of intermetallic compounds have been introduced. Together, these two techniques are collectively known as step casting. In step casting, one liquid is poured into a mould and a second liquid is subsequently poured into the mould during solidification of the first liquid. In duplex casting, refinement of the primary phase is enhanced by controlling the combination of liquidus temperatures of the two liquids and by increasing the cooling rate. Primary silicon in hypereutectic Al–Si alloys been successfully refined using a combination of duplex casting and two inoculants. Primary silicon is coarse when the same two inoculants are used simultaneously without duplex casting. Recent research has shown that synthesis casting can be used to manufacture NiAl at low cost and for *in situ* joining of NiAl to structural materials.

References

[1] Kudoh M, Ohmi T and Takahashi T 1987 *J. Japan. Inst. Met.* **51** 948–955
[2] Ohmi T, Kudoh M and Ohsasa K 1992 *J. Japan. Inst. Met.* **56** 1064–1071
[3] Ohmi T, Nakadera K and Kudoh M 1992 *J. Japan. Inst. Light Met.* **42** 132–137
[4] Darolia R 1991 *J. Met.* **43** 44–49
[5] Matsuura K, Jinmon H and Kudoh M 2000 *ISIJ Int.* **40** 167–171

Chapter 15

Intermetallic selection during solidification of aluminium alloys

Keyna O'Reilly

Introduction

Control of intermetallic type, size, morphology and distribution in wrought aluminium alloys is necessary to obtain the desired material properties uniformly throughout the final product. This chapter gives an overview of two casting processes, and describes how the processing conditions affect intermetallic selection. As the commonly used solidification processes are non-equilibrium in nature, it is necessary to understand the factors that govern intermetallic phase selection under such conditions. Hence, the roles of competitive growth and competitive nucleation of two intermetallic phases are reviewed, together with the experimental techniques available for the assessment of their relative influences. Chapters 12 and 10 discuss nucleation and growth of primary phases during solidification, but the present chapter is concerned with nucleation and growth of secondary intermetallic phases.

Wrought aluminium alloys are used in a variety of commercially important applications, including packaging, beverage cans, extrusions and aerospace and automotive components. The type, size and distribution of secondary and ternary intermetallic phases critically influence the material properties [1], including strength, toughness, formability, fatigue resistance, surface quality, corrosion resistance and anodizing response [2]. Surface quality, etching response and anodizing behaviour are particularly important in surface critical products such as extrusions, lithographic printing sheet, and sheet used in architectural and automotive applications. Solid solution content is important in controlling properties such as electrical conductivity, corrosion resistance and recrystallization characteristics.

The particle population and the solid solution content in the final product are determined by the as-cast state and by subsequent ingot homogenization and thermo-mechanical processing. The non-equilibrium nature of solidification during typical commercial casting processes commonly results in the failure of thermodynamic considerations to predict the phase content and solid solution levels of the as-cast microstructure correctly. As such, as-cast solid solution levels and secondary and ternary phase crystallography and morphologies are dependent on complex kinetic competitions for nucleation and growth. Subsequent homogenization reduces segregation, encourages the transformation of phases into equilibrium phases, and acts to equilibrate solid solution levels of soluble elements, resulting in the precipitation of dispersoids. Hence, particular microstructural features can be manipulated during downstream processing. However, improved control of intermetallic phase selection during casting leads to lower costs by reducing homogenization time, and more consistent, uniform material properties throughout the final product.

Solidification processing

Approximately 80% of worldwide aluminium production (approximately 19 million tonnes/year) is currently processed using direct chill (DC) casting. DC casting of aluminium alloys is discussed in detail in chapter 1. The DC casting process is shown schematically in figure 15.1 [3]. Most commonly ingots are cast by the vertical process, in which molten metal is fed through a distribution system into one or more water-cooled moulds having retractable bases. The first stage of solidification is therefore the formation of a solid shell at the chilled mould surface. The base of the mould is then withdrawn at a constant rate and the emerging casting sprayed directly with water, resulting in the solidification of the remaining ingot cross section. Rectangular ingots, typically $\sim 0.6 \text{ m} \times (1-2) \text{ m} \times (6-11) \text{ m}$ in dimension are cast for subsequent rolling to sheet products. Cylindrical logs, typically $(0.1-0.5) \text{ m}$ diameter $\times (3-6) \text{ m}$ in dimension are cast for subsequent extrusion.

Industry has been looking for an alternative to conventional DC casting for the production of strip and sheet products. There is considerable interest in by-passing the downstream hot rolling stages and directly casting strip to near-gauge dimensions. Continuous casting of aluminium alloys is discussed in detail in chapter 2. Twin roll casting can produce strip of thickness 3–10 mm whilst belt or block cast material is more typically 10–20 mm in thickness. Figure 15.2 illustrates the twin roll casting process. Molten metal is fed directly between two water cooled rotating rolls, where it solidifies to produce thin strip which undergoes deformation. Intimate contact between molten metal and the rolls during solidification results in high rates of heat extraction compared to DC casting.

Local growth velocities in a cast are difficult to analyse but are estimated from predicted steady-state shapes of the solidification front. Figure 15.3(a)

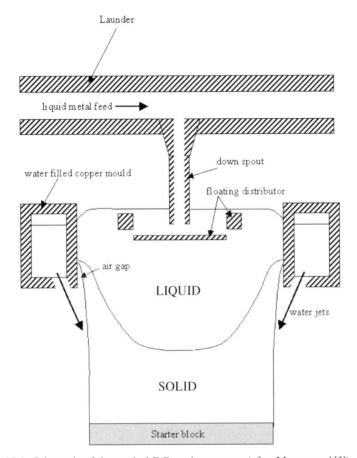

Figure 15.1. Schematic of the vertical DC casting process (after Maggs *et al* [3]).

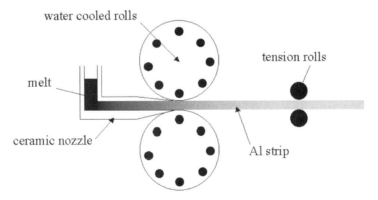

Figure 15.2. Schematic of the twin roll casting process.

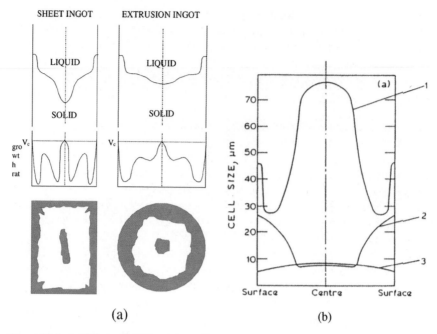

Figure 15.3. (a) Schematic DC sump profiles (top), velocity profiles (centre) and resultant phase intermetallic selection (bottom) in rectangular ingots and cylindrical billets (after Brusethaug et al [4]); and (b) variation in cell size across cast section for (1) DC block (500 mm thick), (2) DC strip (25 mm) and (3) twin roll cast strip (7 mm) (after Odolz and Gyougyos [7]).

shows schematically the shape of the solidification front and the resultant phase selection in DC cast ingots [4]. The maximum growth velocity will be less than or equal to the casting speed which is typically of the order of 1.7 mm s^{-1} for DC casting and 17 mm s^{-1} for continuous strip casting [5]. Cooling rates during casting are more commonly reported than growth velocities, because cooling rates can be estimated from the cast microstructure. The secondary dendrite arm spacing has been determined empirically to be proportional to t_f^n, where t_f is the freezing time and n usually lies between 0.3 and 0.4 [6]. The scale of the as-cast dendritic or cellular structure is demonstrated in figure 15.3(b) for DC and twin roll casting processes [7]. Note the complex growth velocity and cooling rate profiles, and that positions of maximum growth velocity in the ingot do not necessarily correspond to positions of maximum cooling rate.

The difference in cooling rates experienced in DC and twin roll casting is seen to have a significant effect on the cast microstructures produced by the two routes. The grain size, cell/dendrite scale and intermetallic particle size within a cast are all refined with the faster cooling rates, as experienced in twin roll casting. The cell size is more homogeneous in twin roll casting than in DC cast material. A refined cell structure dictates a fine distribution of intermetallic phases in the interdendritic regions as well as reducing annealing times required to eliminate microsegregation due to reduced diffusion distances. The growth velocities and cooling rates in both DC and twin roll casting lead to several non-equilibrium effects, including the extension of solute solid solubility and the formation of new metastable intermetallic phases, with these effects being more marked the more rapid the solidification.

The current restrictions to the use of twin roll casting include lack of control over the final sheet texture and the presence of defects such as centre-line segregates [8], where alloying elements have segregated to the centreline of the strip to form stringers of intermetallic phase. These effects are alloy and process dependent. The process is also limited to use with alloys of narrow freezing range so that material is fully solidified while in contact with the rolls. Further steps are required to be taken to develop alloys for twin roll casting to take full advantage of the novel microstructures and intermetallic phases that can be produced and to develop downstream processing so that the final product performance can match that produced via the conventional DC casting route.

Nucleation and growth of intermetallic phases

A wide range of equilibrium and metastable intermetallic phases have been observed in most as-cast wrought aluminium alloys (e.g. 1xxx series alloys [9]) resulting from the non-equilibrium nature of solidification during commercial processing. Varying solidification conditions can lead to variations in intermetallic phase content at different positions in the casting. Thermodynamics are not usually sufficient to predict intermetallic phase selection, and hence an understanding of the factors governing phase selection under non-equilibrium conditions is required. Allen *et al* [9] discuss the features of non-equilibrium solidification, which include: competitive growth of different phases; competitive nucleation of different phases; suppression of solidification reactions involving solid state diffusion of slowly diffusing species over large distances; non-equilibrium solute partitioning; and the role of alloy chemistry, specifically of impurities and grain refining additions. Here we will briefly review the roles of competitive growth and competitive nucleation.

Competitive growth

The temperature at which a phase grows (T_g) is a function of the velocity at which growth is being driven and the cooling rate during solidification. One phase, say A, can be kinetically displaced by another, say B, if the growth temperature of A is depressed to below that of B, i.e. if $T_{g,A} < T_{g,B}$, assuming that both phases can nucleate under the given solidification conditions. Assume that this displacement takes place at a critical growth velocity U_{crit} or at some critical cooling rate $(dT/dt)_{crit}$. For a given temperature gradient across the solid/liquid interface, G, the cooling rate dT/dt and the growth velocity U are related under one-dimensional steady-state growth conditions by

$$\frac{dT}{dt} = \frac{dT}{dx}\frac{dx}{dt} = GU,$$

so that the critical cooling rate and solidification velocity for the displacement of phase A by phase B are related by:

$$U_{crit,A}(dT/dt)_{crit,B} = U_{crit,B}(dT/dt)_{crit,A}.$$

Which of the two parameters out of the cooling rate and the critical growth velocity is the more fundamental in determining whether the transition from phase A to phase B will occur will depend upon the growth kinetics of the two phases. Adam and Hogan [10,11] studied intermetallic phase selection in unidirectionally solidified Al–(2.0–4.0)wt% Fe alloys as a function of growth velocity and temperature gradient, and later Liang and Jones [12] carried out similar studies on Al–(2.2–6.2)wt% Fe alloys. These investigations determined that with increasing growth velocity the undercooling required for growth of the equilibrium Al–Fe$_4$Al$_{13}$ eutectic increased faster than that for the metastable Al–FeAl$_6$ eutectic, such that the equilibrium eutectic was kinetically displaced by the mestable eutectic at a critical growth velocity of 0.1 mm s^{-1}. No unique cooling rate was determined for the transition. Figure 15.4 shows experimentally determined results of Liang and Jones, showing this transition as a function of growth velocity. As a result of these investigations, confirmed by later work, it is now more commonly accepted that *growth velocity* rather than cooling rate is critical in the determination of intermetallic phase selection under conditions of competitive growth.

Competitive nucleation

Most experiments to date have concentrated upon establishing a critical cooling rate or critical growth velocity for the transition from one intermetallic phase to another. In the previous section, where it was suggested that growth velocity and not cooling rate was the more fundamental parameter in determining the transition from Al–Fe$_4$Al$_{13}$ eutectic to Al–FeAl$_6$ eutectic, it was assumed that both phases could nucleate relatively easily at low undercoolings.

Bäckerud [13] proposed a nucleation-based argument to explain the observed transition from Al–Fe$_4$Al$_{13}$ eutectic to Al–FeAl$_6$ eutectic on

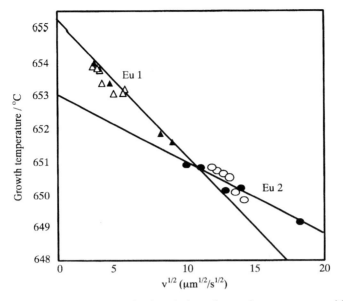

Figure 15.4. Experimentally determined variation of growth temperatures with solidification velocity. Al–Fe$_4$Al$_{13}$ eutectic (Eu1) and Al–FeAl$_6$ eutectic (Eu2) (after Liang and Jones [12]).

increasing cooling rate. One phase, say A, can be kinetically displaced by another, say B, if the nucleation temperature of A is depressed to below that of B, that is if $T_{n,A} < T_{n,B}$, assuming that both phases can grow under the given solidification conditions. He proposed that Fe$_4$Al$_{13}$ required a high undercooling to nucleate, irrespective of whether it was a primary or secondary phase, and interpreted this as evidence that Fe$_4$Al$_{13}$ was poorly nucleated by a primary aluminium solid solution, suggesting that Fe$_4$Al$_{13}$ nucleated on impurity particles in the liquid instead. Conversely, he proposed that FeAl$_6$ needed no detectable undercooling to nucleate, indicating that primary aluminium solid solution is an efficient nucleant for FeAl$_6$.

A nucleation-based argument can still be invoked to explain the kinetic displacement of the equilibrium eutectic by the mestable eutectic with increasing cooling rate, even if aluminium is a good nucleant for Fe$_4$Al$_{13}$ at low cooling rates. Figure 15.5 illustrates schematically isothermal nucleation diagrams for the equilibrium and metastable eutectics. Assume that nucleation of the equilibrium eutectic is easier at higher temperature and lower undercooling than for the metastable eutectic, contrary to Bäckerud's proposal, and that growth of equilibrium eutectic is harder at higher temperature and lower undercooling than the metastable eutectic, as proposed above. At low cooling rates the equilibrium Al–Fe$_4$Al$_{13}$ will be

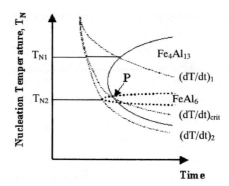

Figure 15.5. Curves for the start of nucleation of Fe_4Al_{13} and $FeAl_6$. At $dT/dt = (dT/dt)_1$, Fe_4Al_{13} nucleates first at T_{N1}. At $dT/dt = (dT/dt)_2$, $FeAl_6$ first nucleates at T_{N2}. The critical cooling path, at $dT/dt = (dT/dt)_{crit}$ passes through point P, the intersection of the two TTT curves (after Maggs et al [3]).

the preferred eutectic phase. At high cooling rates, however, the converse is true, and Al–$FeAl_6$ will displace Al–Fe_4Al_{13}.

Nucleation and growth experimental studies

It is critical to our understanding to be able to determine whether an intermetallic phase is selected over another phase due to reasons of competitive growth or competitive nucleation. Direct study of commercial casting processes such as DC or twin roll casting is typically too complex for the determination of such mechanisms. Hence, a series of laboratory scale experiments have been determined to attempt to either (i) exaggerate the role of nucleation while minimizing growth effects or (ii) exaggerate the role of growth while minimizing nucleation effects.

Nucleation studies

The entrained droplet technique

The nucleation of solidification usually takes place by a heterogeneous process. There are serious experimental difficulties in studying heterogeneous nucleation, because of the difficulty of excluding extraneous impurities which can act as uncontrolled catalysts leading to non-reproducible results. The entrained droplet technique was first devised by Wang and Smith [14] to improve the reproducibility of liquid undercooling measurements. Alloys are thermomechanically manipulated to produce a microstructure of low melting point particles embedded in a high melting point matrix, and are then heat treated to melt the particles and monitor their solidification behaviour during subsequent cooling. Cantor and O'Reilly [15] have shown that rapid solidification often produces entrained droplets, <100 nm in size, leading to temperature and undercooling measurements reproducible in some cases to better than 0.2 K. Early work using the entrained droplet technique focused on immiscible alloys [15], but more recently the technique has been used to study eutectic and peritectic systems [15], and now commercial alloys, where primarily it has been used to investigate the influence of trace alloying elements and grain refiner additions on heterogeneous nucleation of intermetallic phases [16].

In a DC as-cast microstructure, in addition to interconnected cell and grain boundary eutectic phases there are isolated pockets of aluminium–intermetallic eutectics. These pockets form in the final stages of solidification, where they get pinched off between the growing dendrites. Hence, the microstructure produced in the entrained droplet technique exaggerates the nucleation and growth of intermetallics occurring within such isolated pockets in conventional microstructures. Partly as a result of using the entrained droplet technique, it has recently been realized that particular impurities and grain refiners can play a major role in intermetallic phase selection. In particular, grain refiner additions, added only to refine the primary aluminium solid solution grain size, were not thought to have any influence on intermetallic phase formation.

Figure 15.6 shows entrained droplet differential scanning calorimeter traces taken from a range of 1xxx series aluminium alloys (Al–Fe–Si) with different purity and impurity levels. The peaks represent the melting of aluminium–intermetallic eutectics. In a low purity material, two peaks are observed, representing the melting of the equilibrium Al–Fe_4Al_{13} eutectic and the metastable Al–$FeAl_m$ eutectic. The high purity base material only contains the equilibrium Al–Fe_4Al_{13} eutectic. Additions to this base material of high levels of vanadium (approximately 500 ppm) or a combination of low levels of vanadium (approximately 100 ppm) with Al–Ti–B grain refiner, lead to the formation of the metastable Al–$FeAl_m$ eutectic [16]. This metastable eutectic is responsible for the commercially occurring fir-tree defect.

The metallic glass technique

The metallic glass technique described in more detail in chapter 12 [17] has been developed to study the nucleation of primary aluminium solid solution by commercial grain refiner additions, such as Al–Ti–B refiners containing TiB_2 particles. This technique employs a metallic glassy alloy, to which the refiner particles are added above the glass transition temperature. The alloy is then rapidly quenched to form a glass. Nuclei form during the quench, but once the temperature is below the glass transition temperature growth of these nuclei is halted. This limiting of the growth results in

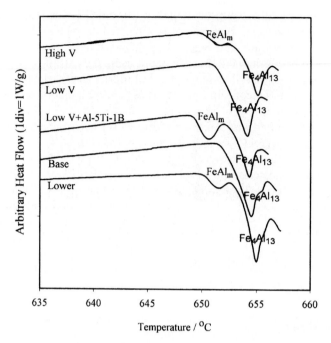

Figure 15.6. Entrained droplet melting differential scanning calorimeter traces from 1xxx series aluminium alloys with a high purity base, a low purity base, high purity base plus high and low levels of vanadium, and high purity base with a combination of low vanadium and grain refiner addition (after Allen *et al* [16]).

the formation of isolated aluminium nuclei, embedded within the glassy matrix, which can be observed using transmission electron microscopy. Hence, the technique can be readily used to determine the mechanisms of nucleation on grain refiner particles. In the past, however, the technique has been criticized, because the requirement that the matrix alloys must be good glass-formers has meant that their compositions are not good models of real commercial alloys. Hence, large additions of such elements as yttrium and cobalt may have influenced the nucleation mechanisms in operation.

Recent work [18] has modified the technique slightly, to look at the nucleation of *intermetallic phases* by grain refiner additions in aluminium alloys. New glassy matrix alloys have been determined which are more realistic models of commercial alloys. Relatively large quantities of alloying elements are still required, but these alloying elements are commonly used in commercial alloys. Two matrix systems are currently under investigation, Al–Si–Ni and Al–Fe–Si.

Figure 15.7 shows a boride grain refiner particle in an Al–Si–Ni glass [18]. In this alloy, there is little evidence of the thin layer of Al_3Ti on the basal planes of the boride particle, responsible for the nucleation of aluminium observed in

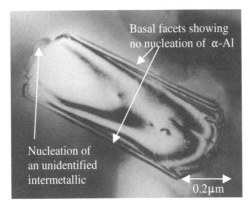

Figure 15.7. Boride grain refiner particle showing no nucleation of aluminium but nucleation of an unidentified intermetallic phase (after McKay *et al* [18]).

previous work [17]. However, an, as yet unidentified, intermetallic phase has nucleated on the non-basal planes, as shown.

Growth studies

Bridgman growth

Bridgman growth is a well established technique designed to achieve unidirectional solidification under steady-state growth conditions. The range of growth velocities it is possible to study (5–120 mm min^{-1}) spans typical DC casting rates, though the temperature gradients involved (5–15 K mm^{-1}) are somewhat higher [19]. A vertical furnace is constructed so that there is a constant temperature gradient through the furnace, with the hottest region at the top. The specimen consists of a thin rod, typically 2 mm in diameter and 150 mm in length, encased in an alumina crucible. The specimen is either fully melted or melted such that the bottom 20–30 mm of the specimen remains solid to act as a seed for subsequent growth. The specimen is then drawn down through the furnace at constant speed [19]. Intermetallic phase selection as a function of growth velocity can then be investigated. The aluminium matrix is dissolved to concentrate the intermetallic phases. X-ray diffraction analysis can then be performed to determine the majority intermetallic phases selected at different growth velocities.

Unidirectional growth occurs locally in solidification processes such as DC casting. Hence, Bridgman growth studies are useful for simulating intermetallic phase selection in the interconnected cell and grain boundary eutectics in as-cast materials. In this interconnected material, growth dominates, and there is little if any requirement for subsequent nucleation of new phases.

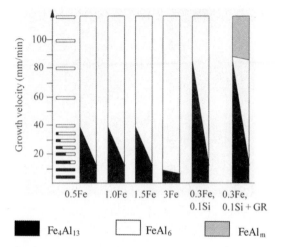

Figure 15.8. Qualitative diagram of intermetallic phase selection in model 1xxx series aluminium alloys (after Evans *et al* [19]).

Figure 15.8 shows the intermetallic phase selection in a range of model 1xxx series aluminium alloys as a function of growth velocity. In addition to the equilibrium Al–Fe$_4$Al$_{13}$ eutectic, the metastable Al–FeAl$_6$ and Al–FeAl$_m$ eutectics are also observed, dependent on alloy composition and growth velocity [19].

Electron beam surface melting

Whilst Bridgman growth studies are a useful experimental simulation of DC casting growth conditions, limitations on heat extraction preclude steady-state growth velocities much in excess of 2 mm s^{-1} (120 mm min^{-1}). Hence Bridgman techniques cannot be used to simulate the growth velocities experienced in some of the more modern solidification processes such as twin roll casting. Electron beam [20] or laser surface melting, in which an energetic beam is focused on the surface of a sample to generate a molten pool, and then traversed at a constant velocity, is an experimental technique capable of generating steady-state growth velocities in the continuous strip casting regime. The unmelted bulk alloy encourages epitaxial growth so that nucleation conditions are known, and acts as a perfectly efficient heat sink which facilitates the faster growth speeds. For aluminium alloys, an electron beam power source is preferred to a laser due to the high reflectivity of laser light by an aluminium surface (up to 99%) which reduces the energy coupling efficiency.

Electron beam surface melted microstructures consist of aluminium dendrites with interdendritic aluminium–intermetallic eutectic phases, as in Bridgman grown microstructures. However, specimens demonstrate a

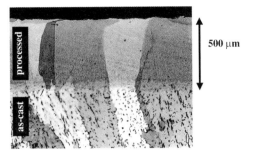

Figure 15.9. Epitaxial grain growth from the as-cast substrate into the processed zone in a model 3xxx series aluminium alloy (after Carroll *et al* [20]).

greatly refined sub-grain structure. The orientation of the aluminium–intermetallic eutectic phases varies systematically through the processed zone, and can be used to determine the local growth velocity as a function of depth through the zone [20]. Intermetallic phase content can be determined for the entire processed zone or for specific depth and hence growth velocity ranges through the zone. As in Bridgman studies, intermetallic phase identification is facilitated by dissolving away the aluminium matrix to concentrate the intermetallic phases, and then carrying out X-ray diffraction analysis. Alternatively electron microscopy techniques can be employed on the extracted intermetallic particles, or on deep etched samples.

Figure 15.9 shows an electron beam surface melted model 3xxx series alloy, etched to reveal the grain structure rather than the sub-grain structure [20]. It can clearly be seen that growth has occurred epitaxially from the as-cast substrate, and is predominantly columnar in nature. Hence, solidification is growth dominated and nucleation effects are minimized.

Conclusions

Control of intermetallic phase selection during solidification of aluminium alloys is important for control of material properties. The non-equilibrium nature of commercial solidification processes causes variations in intermetallic phase content at different positions in the ingot, as a result of different solidification conditions. In order to understand phase selection it is necessary to investigate the respective roles of nucleation and growth for particular alloy compositions and solidification conditions. Direct study of commercial casting processes such as DC or twin roll casting is typically too complex to decouple easily the roles of nucleation and growth. Laboratory scale experiments, designed to exaggerate either the role of nucleation or of growth, can provide valuable insight into intermetallic phase selection during solidification of aluminium alloys. Nucleation may play a significant role when the aluminium–intermetallic eutectics solidify in isolated liquid pockets. Such divided liquid can occur in commercially cast materials, especially in dilute alloys. Nucleation of intermetallic phases can also be promoted by grain refiner additions. Growth plays a significant role when the interdendritic liquid is highly interconnected. This condition commonly occurs in more highly alloyed materials and when solidification is strongly directional.

References

[1] Mondolfo L F 1976 *Al Alloys: Structure and Properties* (London: Butterworths)
[2] Kosuge H 1984 *Keikinzoku* **34** 4
[3] Maggs S J, Cochane R F, Flood S C and Evans P V 1995 *Symposium on Shape Casting Technology VI: Vertical Casting and Microstructural Development 2*, TMS Annual Meeting, Las Vegas
[4] Brusethaug S, Porter D and Vorren O 1987 Hydro aluminium, Suundal Verk. In: ILMT 477
[5] Emley E F June 1976 *Int. Met. Rev.* 75
[6] Campell J 1991 *Castings* (London: Butterworth-Heinemann) p 146
[7] Odok A N and Gyongyos I 1975 *Continuous Casting Seminar Papers*, Aluminium Association, Paper 14
[8] Lockyer S A, Yun M, Hunt J D and Edmonds D V 1996 *Mater. Characterisation* **37** 301
[9] Allen C M, O'Reilly K A Q, Cantor B and Evans P V 1998 *Prog. Mater. Sci.* **43** 89
[10] Adam C McL and Hogan L M 1972 *J. Austr. Inst. Met.* **17** 81
[11] Adam C McL and Hogan L M 1975 *Acta Metall.* **23** 345
[12] Liang D and Jones H 1992 *Z. Metallkde.* **82** 224
[13] Bäckerud L 1968 *Jernkont* **152** 109
[14] Wang C C and Smith C S 1950 *Trans. Metall. Soc. AIME* **188** 136
[15] Cantor B and O'Reilly K A Q 1997 *Curr. Opin. Solid State Mater. Sci.* **2** 318
[16] Allen C M, O'Reilly K A Q, Evans P V and Cantor B *Acta Mater.* in press
[17] Schumacher P and Greer A L 1998 *Mater. Sci. Technol.* **14** 394
[18] McKay B J, Cizek P, Schumacher P and O'Reilly K A Q submitted to *Mater. Sci. Eng.*
[19] Evans P V, Worth J, Bosland A and Flood S C 1997 *Proceedings of the 4th International Conference on Solidification Processes* p 531
[20] Carroll L, O'Reilly K A Q, Cantor B and Evans P V 1997 *Proceedings of the 4th International Conference on Solidification Processes* p 546

Chapter 16

Cooling rate and structure of commercial aluminium alloys

Jim Kellie

Introduction

This chapter compares the structures of aluminium alloy powders produced by different atomization techniques with the structures found in cast ingots, and outlines some implications of the changes for the eventual use of the powder. The most important variable is the interaction between changes in cooling rate and the alloy phase diagram. As with much industrial research, the emphasis is on the production of the required final properties in an economic manner rather than on rigorous proof of the proposed mechanisms by which the structures are achieved. The two atomizing techniques are:

- centrifugal granulation by pouring molten metal onto a spinning disc,
- inert gas atomization via a concentric nozzle.

Centrifugal atomization is the simpler of the two techniques but it tends to produce an elongated and rather coarse particle compared with gas atomization. Both techniques produce very much higher cooling rates than ingot casting although they are slower than those used to produce amorphous metals or in splat casting. The effects of varying the cooling rate depend strongly on the phase diagram of the alloy. This chapter illustrates some of the features that are observed in the following alloy systems:

- aluminium metal matrix composites containing insoluble TiB_2 particles,
- a superplastic Al–Mg–Zr alloy containing Al_3Zr intermetallics precipitated from the melt,
- Ni–Al alloys in which Al_3Ni_2 and Al_3Ni intermetallics are precipitated by peritectic reactions.

Before studying the effects of different casting systems on these alloys, it is worth outlining the features of the two atomization techniques and the cooling rates they achieve.

Aluminium granulator

Figure 16.1 shows a schematic diagram of the equipment used in this work. Molten aluminium alloy is fed at a controlled rate down a launder on to a disc rotating at 8000 rpm. The metal stream is broken up by shear as it strikes the spinner and forms droplets as the metal leaves the spinner (a simple carbon disc of 200 mm diameter); the droplets solidify as they are thrown out towards the collector. Changes in the metal feed rate and spinner geometry change the cooling rate by varying the weight of metal on the spinner and its residence time; this in turn controls the particle size of the product. The most important factor is the ratio of the tip speed of the disc to the metal feed rate. The rate of heat extraction as the metal spreads over the spinner is high but it needs to be less than the latent heat of the

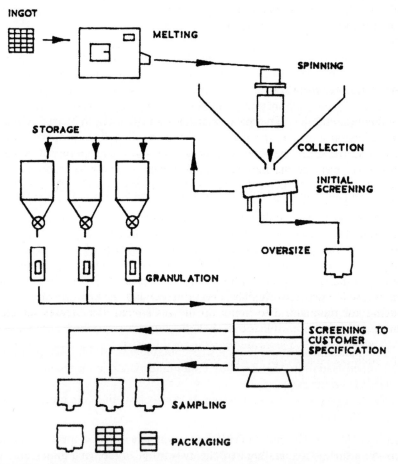

Figure 16.1. Schematic diagram of the equipment used for centrifugal atomization.

metal stream (120 kW at 1 tonne per hour). It is essential that the particles then solidify before hitting the walls of the vessel (a distance of 2 m, traversed in 0.25–0.75 s). In air, this implies a similar rate of heat extraction to that on the spinner, limiting the upper particle size to a few mm in diameter. By using different spinner designs and more conductive gases such as helium or a water spray, the heat extraction can be increased and although these arrangements are rarely commercially viable, it is worth noting that Cox *et al* [1] produced nearly spherical particles as fine as 10 μm diameter on similar equipment using a turbo-spinner and helium as a cooling gas.

Under production operating conditions, relatively coarse elongated particles are produced (typically 500 μm × 2500 μm). Because of the elongated shape, the as-cast product is often referred to as a needle. These needles can be milled to an isotropic shape known as granules but the diameter is generally similar to that of the original needle because of the high ductility of most aluminium alloys. Granules are coarser than gas-atomized powders but they are cheaper to produce and have applications where fine size is not required.

Inert gas atomization

A schematic diagram of a typical pilot gas atomizer is shown in figure 16.2(a). Molten metal is sucked upwards out of the furnace by the venturi effect of the gas jet surrounding the atomizing nozzle, as shown schematically in figure 16.2(b). In practice the flow is more complex and, although this has been modelled by US workers, the results are confidential and details of nozzle designs are proprietary.

For a given nozzle design, the as-atomized particle size is largely determined by the ratio of gas flow to metal flow (typically 5–10 m^3/kg metal for 50 μm particles). The choice of inert gas has an important but complex effect. Argon is generally used as the atomizing gas with 2% oxygen added to provide a passivating oxide film on the powder particles. Helium is sometimes used despite its higher cost. It produces a finer powder despite having less momentum to break up the gas stream as it freezes the metal quicker. Another variable is the gas temperature as it cools due to adiabatic expansion in the jet. Gases are sometimes preheated to produce finer powders although the mechanism of this effect is unclear.

Cooling rates

Cox *et al* [1] showed that cooling rates are inversely proportional to particle size on a turbo-spinner. Figure 16.3 shows that estimated cooling rates for gas atomization are one to two orders of magnitude slower than for the

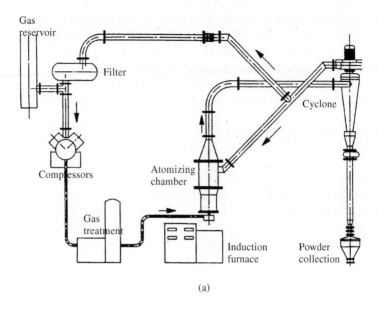

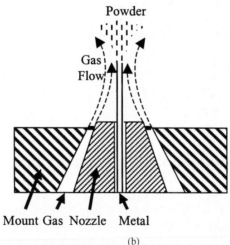

Figure 16.2. Schematic diagrams of the equipment used for inert gas atomization: (a) general arrangement and (b) atomizing nozzle.

turbo-spinner with cooling rates in the centrifugal atomizer being intermediate between these two graphs.

Cooling rates have been estimated from dendrite arm spacing (DAS) in a similar manner to that reported by Grant *et al* [2]. Figure 16.4 shows the relationship between dendrite arm spacing and cooling rate. Note that the

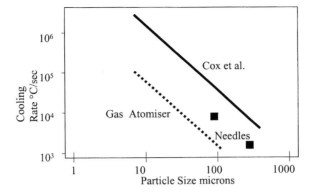

Figure 16.3. Cooling rates achieved in gas atomization and by Cox et al [1] using a turbo-spinner. Black squares indicate conditions experienced in this study using centrifugal atomization.

logarithmic scale means that cooling rates can only be estimated to an order of magnitude. Figure 16.5 shows typical gas atomized particles in cross-section. From this micrograph the dendrite arm spacing is estimated to be approximately 2 μm for the central 100 μm particle and at the resolution limit in the finer particles. This is broadly consistent with the graph in figure 16.3 and seems to be independent of alloy type for casting temperatures in the range 500–1000°C.

The micrograph also shows other interesting features. The central particle has been impacted by a smaller particle while both were in the semi-solid state. This is quite a common phenomenon when atomizing metals with a wide freezing range and is known as satelliting. Some workers

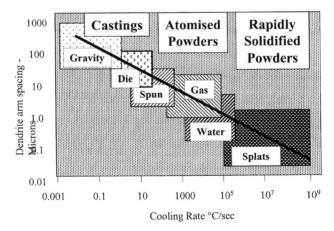

Figure 16.4. Dentrite arm spacing observed as a function of cooling rates achieved during various atomization processes.

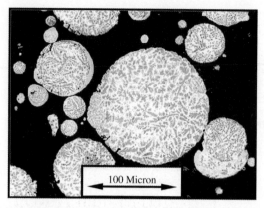

Figure 16.5. Cross-sections of gas atomized particles.

have commented that particle impacts are statistically improbable because of the low concentrations. The fact that so many can be seen in this micrograph suggests that conditions in the atomizing zone are extremely turbulent, increasing the probability of collision. The micrograph also shows that the small impacting particles have ultra-fine microstructures, indicating faster cooling than single particles of similar size. Other examples can be found where a thin, apparently amorphous, structure surrounds the main particle. It is believed that the larger particles quench liquid or semi-solid particles that impact on them, thus increasing the cooling rate in a similar manner to spray-forming.

In view of the wide variation of cooling rate about the mean, the relationship between particle size and cooling rate shown in figure 16.3 has to be regarded as indicative of trends rather than as absolute. These observations mean that liquid–liquid impact may significantly increase the particle size of droplets leaving the atomizing zone. Previously this phenomenon has been considered improbable.

It is concluded from these studies that commercial atomization produces very much faster cooling rates than ingot casting or die-casting although these are still significantly slower than those found in splats and ultra-high-speed discs. However, unlike splat casting, these techniques can be cost competitive with conventional casting so that the property improvements achieved are commercially viable in a wide range of applications.

Effect of cooling rate on the distribution of insoluble particles in aluminium

The London and Scandinavian Metallurgy Company produces TiB_2 containing grain-refiners for aluminium by reacting mixed potassium

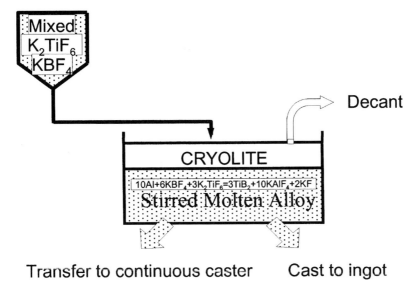

Figure 16.6. Production of TiB$_2$ particles in aluminium.

fluorotitanate and potassium fluoroborate salts with molten aluminium as shown schematically in figure 16.6. The reaction is well established and produces fine insoluble TiB$_2$ particles *in situ* in the melt and a cryolite slag which is decanted off the metal. For optimum grain refinement, the titanium addition is super-stoichiometric and is present in the alloy as partially soluble Al$_3$Ti. The same technology is used to produce metal matrix composites [3] with a stoichiometric titanium:boron ratio.

Figure 16.7 compares structures of a 5% Ti–1% B grain-refiner cast by different routes. Figure 16.7(a) shows a gravity cast alloy with a cooling rate of 1–$10\,°C\,s^{-1}$; the TiB$_2$ forms a network of 1–2 μm particles at the grain

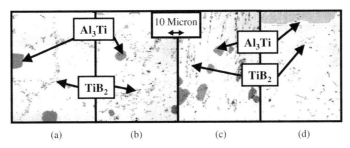

Figure 16.7. Microstructures of TiBAl grain refiners: (a) TiBAl gravity cast ingot, (b) TiBAl direct chill cast bar, transverse section, (c) TiBAl direct chill cast bar, longitudinal section and (d) TiBAl centrifugally atomized needle, longitudinal section.

boundaries. Figures 16.7(b) and (c) show the more uniform structure of a direct chill cast bar which is in part due to a faster cooling rate (approximately $50\,°C\,s^{-1}$) but shear is also a factor (note the longitudinal alignment of particles in figure 16.7(c)). Figure 16.7(d) shows that when the alloy is centrifugally atomized to form needles, the TiB_2 particles are randomly dispersed and are non-aligned. Despite the wide variation in TiB_2 dispersion, all of these samples have the same grain refining efficiency, indicating that the fundamental particle is more important than its dispersion. Metal matrix composites of this type can be inert gas atomized and this further improves the TiB_2 distribution, but a more unexpected effect is that most of the TiB_2 is present as nanoparticles, which cannot be resolved optically. The mechanism for this is unclear as TiB_2 is effectively insoluble at normal casting temperatures, so they have not been precipitated from solution. Dube at Imperial College studied this in more detail as part of a final year project [4]. The work was directed at TiB_2 added to a eutectic Al–Si alloy to improve wear resistance. Alloys with and without TiB_2 were atomized and properties evaluated on extruded billet. Optical micrographs showed the expected fine distribution of silicon in both alloys, but there was no evidence of TiB_2 particles although chemical analysis showed titanium and boron to be present and X-ray diffraction confirmed the presence of TiB_2. X-ray diffraction also showed various TiSi phases to be present, although TiB_2 does normally react with silicon at these temperatures; this may have accounted for an apparent increase in the number of silicon precipitates. Transmission electron microscopy showed that the TiB_2 was actually present as 20–40 nm particles.

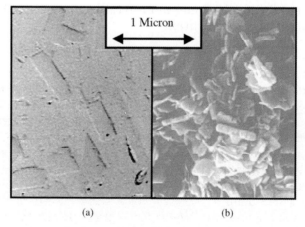

Figure 16.8. Morphology of TiB_2 particles: (a) transmission electron micrograph of TiB_2 particles in a TiBAl alloy and (b) scanning electron micrograph of TiB_2 particles extracted from the matrix by dissolution of the aluminium.

Table 1. Mechanical properties of extruded Al-12%Si alloy powder.

		Hardness (H_v)	σ 0.2 ps (MPa)	UTS (MPa)	Modulus (GPa)	Elongation (%)
Atomized Al–Si	Mean	68	130	219	65	17
	Spread	8	5	2	5	1
Atomized Al–Si + 4.6% Ti–1.65% B	Mean	80	160	250	90	17
	Spread	8	6	1	3	0

There is no clear explanation of this phenomenon but the particles may break up during atomization. Figure 16.8(a) shows a transmission electron micrograph of typical TiB_2 particles in a grain refiner and figure 16.8(b) shows a scanning electron micrograph of particles chemically extracted from the alloy. It can be seen that the 1–2 µm TiB_2 particles seen in optical micrographs are actually composed of numerous crystallites in the form of bars and platelets, and the apparent particles may actually be agglomerates. The crystallites are coarser than nanoparticles but they may break up further during atomization, or the breakage may expose reactive surfaces which lead to the TiSi phases that have been observed.

Table 16.1 shows that the TiB_2-containing alloy is superior in hardness, strength and stiffness. The nanoparticles appear to have achieved this with no loss in ductility. Indeed the alloy even showed some superplasticity due to the fine grain size. Another benefit is improved high-temperature properties (at up to 350 °C). There is considerable potential for improvement in this system.

Aluminium alloys containing intermetallics precipitated from solution by rapid solidification

Zirconium is soluble in aluminium although the slope of the liquidus is steep; even at 900°C the limit of solubility is only 3%. This means that when an alloy containing zirconium in solution is cast as an ingot, Al_3Zr precipitates as coarse platelets, often 100 µm in diameter as shown in figure 16.9(a) and these segregate severely due to the high density of the intermetallic. This limits the use of zirconium to low levels in commercial alloys. Commercial development of aluminium alloys which exhibit superplastic behaviour at high strain rates, depends on the production of a micron-scale grain size in the final alloy. This can be achieved by cold-working an aluminium alloy containing a fine uniform dispersion of a stable second phase material, albeit the strains necessary to achieve such a grain size may be impracticably large. Work at Imperial College [5] has investigated the superplasticity of atomized Al–Zr alloys. Centrifugally atomized needles achieve the required structure, although the casting conditions have to be carefully controlled.

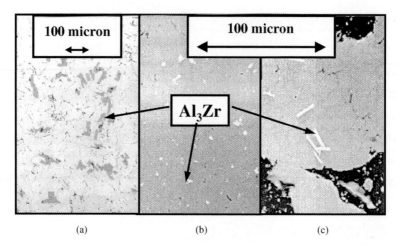

Figure 16.9. Morphology of Al_3Zr particles: (a) optical micrograph of conventionally cast Al–0.5% Zr alloy, (b) and (c) scanning electron micrographs of centrifugally atomized Al–1.0% Zr–4.0% Mg needles cast with superheats of 100 and 50 K respectively.

Figure 16.9(b) shows well dispersed μm-sized Al_3Zr particles in an Al–1% Zr–4% Mg alloy (the Al/Mg matrix was selected as being appropriate for automotive applications). Achievement of this structure requires careful control and figure 16.9(c) shows Al_3Zr platelets 10–20 μm in length can form if pouring temperatures are too low; they are believed to have been precipitated as metal flows down the launder to the centrifugal atomizer. The platelets are smaller than the 100 μm aluminides formed during ingot casting but are still big enough to affect properties adversely. Although this work has concentrated on improved zirconium containing alloys, needle atomization may also improve the properties of established superplastic and other alloys.

Ni–Al catalysts

Figure 16.10 shows the Ni–Al phase diagram. It can be seen that various intermetallics can form by peritectic reactions. This alloy system is commercially important as a means of producing activated nickel catalysts, a process which was patented by Murray Raney in the 1920s and is still in widespread use to form hydrogenation catalysts. Conventionally an aluminium alloy containing 50 wt% nickel is cast as a large block and then ground to the required size. This ground alloy is then treated with caustic soda to dissolve the aluminium matrix, leaving a microporous nickel catalyst. Alloys containing more than 50% aluminium can produce catalysts with higher

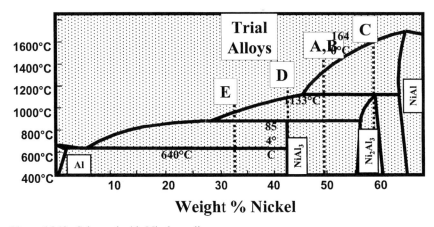

Figure 16.10. Schematic Al–Ni phase diagram.

surface areas but these catalysts are often unstable and friable [6]. Furthermore the extra cost per unit due to extracting more aluminium can outweigh any quality benefits. The result of this balance of technology and commerce is that Murray Raney's original selection of a 50% Ni–Al alloy precursor remains the most popular choice today. It forms a mixture of the two intermetallics, Ni_2Al_3 and $NiAl_3$. The precise ratio of the intermetallics in the alloy depends on the cooling rate and this is also likely to affect the optimum Ni:Al ratio. Ground alloys have the disadvantage that they produce a wide size range of angular particles which are prone to breakdown in use. In general, fine particle size is associated with high catalytic activity by providing good access for the reactants to the catalyst surface, but the downside is that they can be difficult to separate from the reactants and products; a further disadvantage of ultra-fine particles is that they are unstable. The aim has been to produce stable catalysts with high surface area in a narrow size range, with reduced friability.

Atomized intermetallics

Various Ni–Al alloys have been atomized at high cooling rates with the aim of producing improved catalysts, as has been shown in work on activated copper produced from atomized Cu–Al alloys at Osaka University [7]. Table 16.2 compares the properties of catalysts made from atomized alloys with the conventional material. Some improvements in catalytic performance can be seen but the relationship between metallurgy and catalytic behaviour is complex. Catalyst A is a typical activated nickel catalyst produced by conventional technology. Figure 16.11(a) shows a back-scatter electron image of a complex phase structure. The phase diagram shows that Ni–Al

Table 16.2. Properties of NiAl powders and activated catalyst product.

Alloy	% Ni	Casting method	Alloy properties					Catalyst	
			Cooling rate (°C s^{-1})	Particle size (μm)	Ni$_2$Al$_3$ (%)	NiAl$_3$ (%)	Al (%)	BET Area (m^2 g^{-1})	Activity
A	50	Large block	10	50	65	25	10	115	High
B	50	Small ingot	50	50	60	40	–	105	Medium
C	59	Gas atomized	2000	100	100	–	–	55	Low
D	42	Gas atomized	2000	100	10	70	20	75	Medium
E	32	Needles	500	1000	–	70	30	90	Medium

precipitates first from the melt, and subsequently reacts peritectically with the high Al content liquid at temperatures below 1133 °C to form Ni$_2$Al$_3$. At temperatures below 854 °C a second peritectic reaction takes place, forming NiAl$_3$. By this time the remaining liquid is richer in aluminium than Ni$_2$Al$_3$ so that an aluminium phase is also formed. The back-scattered electron image shows pale Ni$_2$Al$_3$ at the centre of the darker NiAl$_3$ and there is a discontinuous black Al phase. Another important feature of this microstructure is the network of fine cracks which can be seen in the Ni$_2$Al$_3$ phase. The cause of these cracks is thought to be the transformation of solid NiAl to Ni$_2$Al$_3$ which involves a 10% reduction in density. The cracks may make the alloy easier to grind but this is probably at the expense of increased generation of fine particles. Figure 16.11(b) shows an example of particle breakage that probably initiated at such a site. This is one reason why micrographs of ground particles rarely show NiAl$_3$ surrounded by Ni$_2$Al$_3$. The cracks are believed to perform an important role in opening up the structure when the catalyst is formed. The scanning electron micrograph of a catalyst

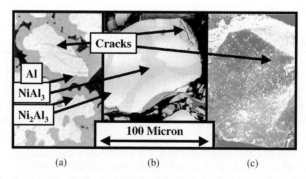

Figure 16.11. Conventionally cast Al–50% Ni (alloy A): (a) as cast alloy, (b) ground alloy and (c) final catalyst after dissolution of the aluminium.

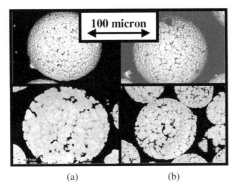

Figure 16.12. Gas atomized Al–59% Ni (alloy C): (a) atomized particle surface and cross-section and (b) final catalyst particle surface and cross-section after dissolution of the aluminium.

particle in figure 16.11(c) clearly shows numerous fine cracks although it is difficult to be sure that they originated in Ni_2Al_3.

In sample B, 50% NiAl was cooled more rapidly than in sample A; X-ray diffraction shows a simple mixture of the Ni_2Al_3 and $NiAl_3$ phases. In this case, it is believed that undercooling leads to little or no NiAl being formed even as an intermediate phase. In subsequent cooling, the liquid phase remains within the Ni_2Al_3 and $NiAl_3$ boundaries so that no free aluminium is precipitated. The microstructure confirms this, showing few examples of the cracked Ni_2Al_3 phase. This may indicate that less NiAl forms on cooling. As expected, the reduced cracking reduces the surface area and activity of the catalyst but the grindability was not assessed quantitatively.

Alloys C and D were gas-atomized with the aim of producing pure Ni_2Al_3 and pure $NiAl_3$ respectively. Figure 16.12 shows scanning electron micrographs of particles and particle cross-sections of alloy and catalyst type C. The dark phase is Ni_2Al_3 but the lighter phase is also Ni_2Al_3 of lower nickel content. Figure 16.12 shows that Ni_2Al_3 has a wide range of stoichiometry enabling this to happen (the sharp delineation between the two phases probably originates from intermetallics formed above and below the peritectic temperature). The catalyst produced from alloy C was free from microcracks although the particles are somewhat porous. Table 16.2 shows the expected lower surface area and activity.

Figure 16.13 shows similar micrographs of alloy D. This alloy is not quite single phase, as presumably there was insufficient time in the temperature range 854–640 °C to form the equilibrium structure. The activity of this material is lower than that of alloy A but higher than expected for a conventional catalyst as coarse as this. It is therefore a significant improvement over alloy B although the low surface area is surprising. A major benefit of alloy D is that the coarse and narrow particle size distribution give the catalyst

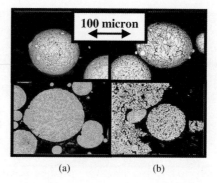

Figure 16.13. Gas atomized Al–42% Ni (alloy D): (a) atomized particle surface and cross-section and (b) final catalyst particle surface and cross-section after dissolution of the aluminium.

exceptionally good separability from reactants in use and the catalyst is more robust than conventional ones. Figure 16.14 shows similar results for alloy type E, where the alloy nickel content was reduced further to 32% and the product cast as needles. Table 16.2 shows that this forms an alloy free from Ni_2Al_3, which produces a catalyst with higher surface area than alloy D, despite an even coarser particle size. The very open structure gives a high surface area even in the absence of cracks; the dendrite arms of the original $NiAl_3$ are clearly visible and confirm a cooling rate of $500\,°C\,s^{-1}$.

The combination of a narrow size range with controlled high cooling rate yields significant improvements in quality. The objective is now to produce similar structures in alloys with a higher nickel content, as alloys D and E would be expensive to convert to catalyst because of the large amount of aluminium which has to be dissolved.

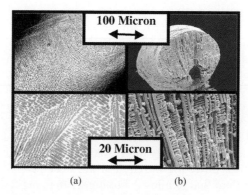

Figure 16.14. Centrifugally atomized Al–32% Ni (alloy E): (a) atomized needle surface and cross-section and (b) final catalyst needle surface and cross-section after dissolution of the aluminium.

Summary

By taking three different alloys with very different phase diagrams, it has been shown how changes in cooling rate can improve product properties in areas as diverse as metal matrix composites, superplastic alloys and catalysts. Knowledge of the alloy phase diagrams has helped explain many of the effects observed and has underpinned the experimental programmes. On the industrial side, it is worth emphasizing that although the use of production plant limits the range of conditions that can be studied, it has the advantage that once procedures are established, the products can be commercialized easily and this stimulates further research. An additional benefit of industrial participation is the accumulated operational experience which enables practical hurdles to be overcome. Unexpected side effects of full-scale operation have sometimes produced improvements as big as the planned changes. The engineering is as important as the metallurgy.

References

[1] Cox et al 1976 *3rd International Conference on Superalloys*
[2] Grant et al 1968 'Effect of cooling rate on dendrite arm spacing' *JIM* **96**
[3] Wood et al October 1993 'Properties of reactively cast aluminium TiB_2 alloys' *Mater. Sci. Technol.*
[4] Dube 1996 'Al based in-situ composites' 3rd year project, Imperial College
[5] Grimes et al June 1998 'Development of high strain rate superplastic aluminium' *Materials Congress*, Cirencester
[6] Bakker et al 1988 'Selective leaching of $NiAl_3$ and Ni_2Al_3 intermetallics to form Raney nickels' *J. Materials Sci.* **23**
[7] Ohnake et al 1992 'Effects of solidification rate of mother alloys on structure of Raney copper catalysts' *J. Japan. Inst. Metals* **56**(8)

Chapter 17

Solidification structure control by magnetic fields

Itsuo Ohnaka and Hideyuki Yasuda

Introduction

Although there has been much work on the general effect of electromagnetic fields on solidification, this is not the case for the application of high static magnetic fields to materials in the semi-solid state. This area has recently attracted new research because high static magnetic fields are now relatively easily available. This chapter reviews the application of magnetic fields to solidification processing and considers the conditions necessary to control solidification structures. Application of magnetic fields during solidification and crystal growth processing has been used practically for continuous casting and the Czochralski process. For the case of imposition of a static magnetic field, the Lorenz force is used to control fluid flow, as it operates by moving conductive fluids and reduces flow velocity. In such processes, conventional electromagnets and permanent magnets have been used and the typical magnitude of the magnetic field is less than 2 T. Recently, cryocooled superconducting magnets have been developed. They have many advantages such as easy operation, long-term continuous running, room-temperature bore and compact size. Under a high magnetic field, the magnetization force, which is caused by interaction between the magnetization of a substance and the magnetic field, can operate even on paramagnetic and diamagnetic substances. For example, water can be magnetically levitated under a high magnetic field of about 20 T [1, 2]. Therefore, it is to be expected that a high magnetic field may not be used only for measurement of physical properties of materials but also for various kinds of materials processing.

Control of macrosegregation with Lorenz force

The effect of high magnetic fields on macrosegregation in Bridgman growth has been examined by Matthesen et al [3]. They applied the magnetic field in the growth direction and found that the concentration profile in the growth direction agreed well with the diffusion-controlled profile. Becla et al [4] reported that macrosegregation in the radial direction was reduced under a magnetic field of 3 T for HgMnTe specimens. Sha et al [5] reported that the misalignment of the cartridge/sample (HgZnTe) inside the furnace can cause a pronounced radial segregation, while liquid flow was damped by a 5 T axial magnetic field. On the other hand [6], we found that a magnetic field of 8 T enhances macrosegregation in the radial direction for $(Bi,Sb)_2Te_3$. Numerical calculation of the solute profile at a solidifying front indicates that macrosegregation in the radial direction is enhanced if convective transfer of solute is reduced to the same order of magnitude as diffusive transfer [7]. Therefore, although application of a high magnetic field significantly reduces convection of the melt, it is not clear if it is useful for practical applications. However, if we can eliminate thermal convection by appropriate thermal design and the magnetic susceptibility changes significantly with concentration, it may be possible to decrease the solutal concentration with the magnetization force, resulting in a more uniform concentration profile.

Control of melt flow using a magnetization force

The magnetic energy per volume, U, under a magnetic field H for an intensity of magnetization M, and the magnetization force F_M are given by

$$U = -MH \quad \text{for ferromagnetic substances,} \tag{17.1a}$$

$$U = -\frac{\chi}{2} H^2 \quad \text{for para- or diamagnetic substances,} \tag{17.1b}$$

$$F_M = -\nabla U = M\left(\frac{\partial H}{\partial x}\right), \tag{17.2a}$$

$$F_M = \chi H\left(\frac{\partial H}{\partial x}\right). \tag{17.2b}$$

Equation (17.2) indicates that the magnetization force is proportional to the magnetic susceptibility χ. In the case of conventional magnetic fields (<1 T), the magnetization force is negligible, except for ferromagnetic substances.

Cryocooled superconducting magnets can achieve $400 \, T^2 \, m^{-1}$ at $H = 10 \, T$. In such cases, the magnetization force can be used even for paramagnetic and diamagnetic substances. Table 17.1 shows the estimated magnetization force compared with gravity. For chromium, manganese

Table 17.1. Magnetic susceptibility (SI unit) and comparison between magnetization force and gravity. F_m = magnetization force, ρg = gravity.

Materials	Susceptibility (H m^{-1})	$F_m/\rho g$
Al	2.62×10^{-11}	0.249
Al$_2$O$_3$	-2.29×10^{-11}	0.148
Bi	-2.09×10^{-10}	0.545
Cr (273 K)	3.91×10^{-10}	1.408
Cr (1713 K)	4.87×10^{-10}	1.753
Cu	-1.22×10^{-11}	0.035
Gd (350 K)	1.47×10^{-10}	0.479
Mn (α)	1.11×10^{-9}	3.918
Mn (β)	1.02×10^{-9}	3.577
Pb	-1.99×10^{-11}	0.045
Si	-7.74×10^{-12}	0.056
SiO$_2$	-1.72×10^{-11}	0.201
H$_2$O	-1.14×10^{-11}	0.293

and graphite, the magnetization force is larger than the gravitational force. This means that chromium, manganese and graphite can be magnetically levitated by using a superconducting magnet, and pseudo-microgravity can be achieved on earth. Alternatively, although the magnetization force for paramagnetic aluminium is about one-fifth that of gravity, it is still much larger than the driving force for convection, which is caused by concentration and/or temperature differences in the melt. Although high magnetic fields have the potential to control convection in melts, not only by the Lorenz force but also by the magnetization force, no studies of this effect have as yet been published.

Structure control with magnetization force

Recently the magnetization force under the high magnetic field has attracted much attention, because it can promote textured structures. For example, magnetic alignment of superconducting oxides has been achieved [8–16]: the critical current density in YBaCuO is anisotropic and therefore it is advantageous to obtain polycrystalline structures with aligned grains. YBaCuO is paramagnetic and has a large anisotropy in its paramagnetic susceptibility. Its magnetic energy becomes sufficiently large compared with its thermal energy kT when a high magnetic field, such as 4 T is imposed. During solidification, grains tend to rotate to a favoured direction that minimizes the magnetic energy. As a result, this anisotropy in magnetic susceptibility leads to an aligned structure. In general, to get aligned structures, the magnetic energy change should be sufficiently large compared with the thermal energy kT. As an example, let us consider paramagnetic substances with a hexagonal crystal structure. When the magnetic susceptibilities in the a axis and c axis directions are χ_a and χ_b respectively, the magnetic energy of a grain $E(\theta)$ is expressed as

$$E(\theta) = \tfrac{1}{2}[(\chi_c - \chi_a)\sin^2\theta]H^2 V \sin^2\theta, \qquad (17.3)$$

where θE_m is a measure of the anisotropy of the magnetic energy, and is given by

$$\theta E_m = \tfrac{1}{2}(\chi_c - \chi_a)\theta^2 V. \qquad (17.4)$$

Here, θ is the angle between the magnetic field and the c axis, and V is the volume of the grain. If the angle distribution is assumed to be a Boltzmann distribution, the probability function $F(\theta)$ of the angle θ is given by

$$F(\theta)\,d\theta = \pi \sin\theta \exp\left[-\frac{E(\theta)}{kT}\right] d\theta. \qquad (17.5)$$

Figure 17.1 shows the probability function as a function of the angle θ and $\Delta E_m/kT$. The c axis of the grains tends to align in the magnetic field as the magnetic energy ΔE_m increases. The average angle $\bar{\theta}$ can be derived by

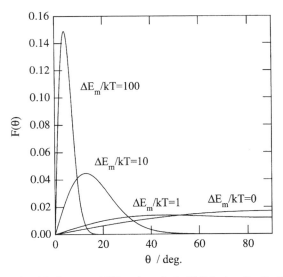

Figure 17.1. Relationship between $F(\theta)$ and angle θ. $F(\theta)$ is the distribution function of crystal grains, and θ is the angle between the c axis and the magnetic field.

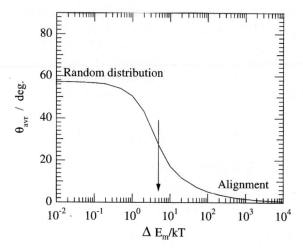

Figure 17.2. Relationship between the average angle θ or θ_{avr}, and $\Delta E_m/kT$.

integration of equation (17.5):

$$\bar{\theta} = \frac{\int_0^{\pi/2} \theta F(\theta)\, d\theta}{\int_0^{\pi/2} F(\theta)\, d\theta}. \tag{17.6}$$

Figure 17.2 shows the average angle as a function of the magnetic energy of the grain divided by the thermal energy kT. In the case of $\Delta E_m/kT > 10$, alignment is readily achieved. The average angle is 5° when $\Delta E_m/kT = 100$. From the viewpoint of magnetic energy, $\Delta E_m/kT > 10$ is a required condition in order to obtain aligned structures.

The anisotropy of magnetic energy, expressed by equations (17.5) and (17.6), is a function of the magnetic susceptibility, the magnetic field and the volume of the grain. The magnitude of the magnetic field is at most 15 T, if a conventional cryocooled superconducting magnet is used. The susceptibility is constant, for a particular material. However, the volume depends on how the specimens are prepared, so from a materials processing perspective, control of grain size should be taken into account.

Requirement of solidification processing

In the above explanation it is assumed that each grain can rotate to a favoured direction and that grain rotations can be achieved by fluctuation. In practice, we must consider the kinetics of grain rotation. Therefore, the above conditions are minimum requirements of physical properties and

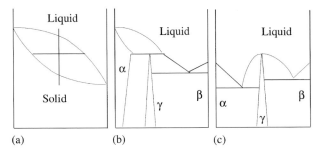

Figure 17.3. Examples of phase diagrams. (a) solid solution, (b) peritectic compound and (c) congruent compound.

grain size in order to obtain aligned structures. It is important that the materials processing allows such favourable conditions such that each grain can move and rotate to a favoured direction. Figure 17.3 shows a series of schematic phase diagrams. In the case of a solid solution such as that shown in figure 17.3(a), the coexistence of the solid and liquid phases can be easily obtained by keeping the temperature between the liquidus and solidus lines. In this case the application of a magnetic field can produce aligned structures rather easily. On the other hand, a peritectic compound can exist with liquid in equilibrium only if the average concentration is off stoichiometric, as shown in figure 17.3(b). Therefore, it is not easy to produce aligned structures consisting only of peritectic compounds. This is also the case for congruent melting compounds, as shown in figure 17.3(c). Alternative methods such as those described below must be developed otherwise magnetic alignment through a solidification route is restricted only to exceptional materials.

Nucleation and growth

There are some reports [17] claiming that high magnetic fields can affect the nucleation and growth of organic crystals. It might be interesting to examine the effect for metallic systems, though the mechanism is not clear.

An example: grain alignment in a BiMn alloy

Although the BiMn compound is a ferromagnetic substance with a large anisotropy energy [18], it is not easy to produce a magnetically anisotropic ingot with a macroscopically uniform structure through conventional solidification processing. This is because BiMn forms via a peritectic reaction [19]. Rapidly solidified particles with compositions Bi–20, 30, 40 and 50 at%

Mn have been produced by the rotating-water-atomization technique [20, 21]. Bulk specimens with dimensions 8 mm in diameter and 10 mm in length, or 6 mm in diameter and 5 mm in length, were produced from these rapidly solidified powders by cold pressing at 300 MPa. Hereafter, these bulk specimens will referred to as rapidly solidified specimens. For comparison, bismuth powder and manganese powder, of particle size 30 µm in diameter, were mixed and bulk specimens were produced, referred to as Bi + Mn specimens. Bulk specimens made from bismuth powder and BiMn powder were also produced, referred to as Bi + BiMn specimens. Bulk specimens were sealed in SiO_2 tubes under a pressure of several kPa of Ar gas and annealed at 300 °C for 2 h. Some specimens were heated to above the peritectic temperature, 480 °C, for 1 h, and then cooled at a cooling rate of 0.026 K s^{-1}. This second heat treatment corresponds to solidification from the semi-solid state. X-ray diffraction, magnetization measurements and scanning electron microscope observations were carried out on the solidified specimens.

X-ray diffraction patterns of the rapidly solidified Bi–20 at% Mn specimen annealed at 300 °C for 2 h showed that the c axis tended to be parallel to the magnetic field. Magnetization curves of the rapidly solidified Bi–50 at% Mn specimen, annealed at 300 °C for 2 h are shown in figure 17.4(a). Magnetization was easily saturated in the parallel direction (i.e. when the magnetic field was applied in the same direction as the magnetic field applied previously during solidification), while magnetization in the perpendicular direction did not tend to saturate even at 18 kOe. The specimen had a large magnetic anisotropy due to alignment of the BiMn grains. Figure 17.4(b) shows the magnetization curves of the rapidly solidified Bi–50 at% Mn specimen annealed at 480 °C for 1 h, i.e. solidified from a semi-solid state above the peritectic temperature. There is no difference between parallel and perpendicular magnetization curves, indicating that no magnetic anisotropy was induced. Annealing the rapidly solidified specimens below the peritectic temperature was also highly effective for magnetic alignment in the BiMn, although more typical solidification from the semi-solid state above the peritectic temperature did not produce alignment in the BiMn compound. Magnetic anisotropy energy, as a quantitative parameter of the magnetic alignment, was estimated from the parallel and perpendicular magnetization curves to be in the range 0–15 kOe.

Figure 17.5(a) shows the magnetic anisotropy energy for different specimens and procedures when the applied magnetic field was 0.3 T. Rapidly solidified specimens annealed at 300°C for 2 h had relatively large magnetic anisotropy energies over the range of composition, while specimens solidified from the semi-solid state above the peritectic temperature had low anisotropy energies. The Bi + Mn and Bi + BiMn specimens did not exhibit large magnetic anisotropy in comparison with the rapidly solidified specimens. Figure 17.5(b) shows the magnetic anisotropy energy for different

An example: grain alignment in a BiMn alloy 293

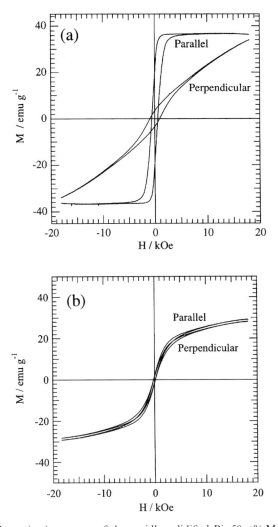

Figure 17.4. Magnetization curves of the rapidly solidified Bi–50 at% Mn specimen. (a) 300 °C for 2 h (annealed below the peritectic temperature), and (b) 480 °C for 1 h and 0.026 K s^{-1} cooling (solidification from the semi-solid state above the peritectic temperature). Parallel means when the applied magnetic field is applied parallel to the magnetic field applied previously during solidification. Perpendicular means when the applied magnetic field is applied perpendicular to the magnetic field applied previously during solidification.

specimens and procedures when the applied magnetic field was 4 T. For the rapidly solidified specimens, the magnetic anisotropy was enhanced. However, even at 4 T, other specimens did not tend to have large magnetic anisotropy in comparison with the rapidly solidified specimens.

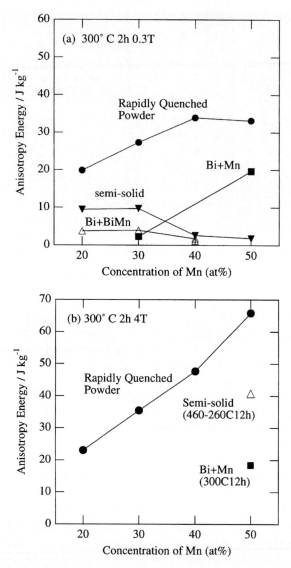

Figure 17.5. Magnetic anisotropy energy estimated from the parallel and perpendicular magnetization curves. Applied magnetic field of (a) 0.3 T and (b) 4 T during solidification. 'Rapidly solidified powder' is a bulk specimen manufactured from rapidly solidified powder and then annealed at 300 °C for 2 h. 'Semi-solid' is a bulk specimen manufactured from rapidly solidified powder and then reheated and solidified from the semi-solid state (480 °C for 1 h and 0.026 K s^{-1} cooling). 'Bi + Mn' is a bulk specimen manufactured from Bi and Mn powders (300 °C for 2 h), and 'Bi + BiMn' is a bulk specimen manufactured from Bi and BiMn powder (300 °C for 2 h).

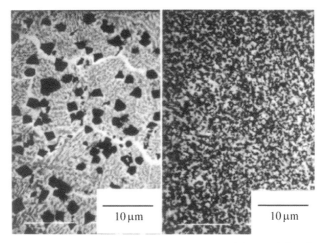

Figure 17.6. Scanning electron micrographs of a rapidly solidified particle (Bi–50 at% Mn) showing fine and coarse regions.

Figure 17.6 shows scanning electron micrographs of the rapidly solidified Bi–50 at% Mn powder. A fine Bi/BiMn eutectic structure was observed, containing primary Mn particles, several μm in length. During conventional solidification, BiMn and Mn particles are evenly distributed with the Bi/BiMn eutectic only observed between BiMn and Mn particles. By rapid solidification, the volume fraction of the Bi/BiMn eutectic was much higher than that from conventional solidification. During annealing of the specimens at 300 °C, that is above the melting point of Bi; and above the eutectic temperature of Bi/BiMn, but below the peritectic temperature, melting occurs. As a result, BiMn particles were surrounded by liquid and consequently grew into the melt. In this situation, BiMn particles could easily rotate because the particles were surrounded by liquid.

The solidification process is clearly important for magnetic alignment. The semi-solid conditions necessary for the magnetic field to rotate the grains to favoured directions depend significantly on the initial microstructure and solidification procedure.

Summary

The application of high magnetic fields to solidification processing is an interesting topic. In particular, the use of the magnetization force and anisotropy in magnetic susceptibility is a promising route for the production of aligned structures. However, to get effective results, it is necessary to select not only the correct chemical compositions but also appropriate solidification conditions. As demonstrated in the Bi–Mn example, rapid solidification and re-heating into the solid/liquid state, followed by solidification under a magnetic field is one promising process route.

References

[1] Beaugnon E and Tournier R 1991 'Levitation of organic materials' *Nature* **349** 470

[2] Tagami M, Hamai M, Mogi I, Watanabe K and Motokawa M 1999 'Solidification of levitating water in a gradient strong magnetic field' *J. Cryst. Growth* **203** 594–598

[3] Matthiesen D H, Wargo M J, Motakef S, Carlson D J, Nakos J S and Witt W F 1987 'Dopant segregation during vertical Bridgman–Stockbarger growth with melt stabilization by strong axial magnetic fields' *J. Cryst. Growth* **5** 557–560

[4] Becla P, Han J-C and Motakef S 1992 'Application of strong vertical magnetic fields to growth of II–VI pseudo-binary alloys HgMnTe' *J. Cryst. Growth* **121** 394–398

[5] Sha Y-G, Su C-H and Lehoczky S L 1997 'Growth of HgZnTe by directional solidification in a magnetic field' *J. Cryst. Growth* **173** 88–96

[6] Yasuda H, Ohnaka I, Furukubo Y, Koh H J, Tozawa S and Fukuda T 1996 'Macrosegregation of unidirectionally solidified $(Bi,Sb)_2Te_3$ in a magnetic field' *Solidification Science and Processing* (TMS) pp 311–317

[7] Yasuda H, Ohnaka I, Furukubo Y, Fukuda T, Watanabe K and Koh K 1997 'Effect of magnetic field on macrosegregation in $(Bi,Sb)_2Te_3$' *J. Japan. Inst. Met.* **61** 1288–1295

[8] Farrell D E, Chandrasekhar B S, DeGuire M R, Fang M M, Kogan V G, Clem J R and Finnemore D K 1987 'Superconducting properties of aligned crystalline grains of $Y_1Ba_2Cu_3O_{7-d}$' *Phys. Rev.* **B36** 4025–4027

[9] Lusnikov A, Miller L L, McCallum R W, Mitra S, Lee W C and Johnston D C 1989 'Mechanical and high-temperature (920 °C) magnetic field grain alignment of polycrystalline $(Ho,Y)Ba_2Cu_3O_{7-d}$' *J. Appl. Phys.* **65** 3136–3141

[10] Tkaczyk J E and Lay K W 1990 'Effect of grain alignment and processing temperature on critical currents in $YBa_2Cu_3O_{7-d}$ sintered compacts' *J. Mater. Res.* **5** 1368–1379

[11] de Rango P, Lees M, Lejay P, Sulpice A, Tournier R, Ingold M, Germi P and Pernet M 1991 'Texturing of magnetic materials at high temperature by solidification in a magnetic field' *Nature* **349** 770–772

[12] Sarkar P and Nicholson P S 1992 'Magnetically enhanced reaction sintering of textured $YBa_2Cu_3O_x$' *Appl. Phys. Lett.* **61** 492–494

[13] Arendt R H, Garbauskas M F, Lay K W and Tkaczyk J E 1991 'An alternate preparation for grain aligned structures of $(Bi,Pb)_2Ca_2Sr_2Cu_3O_z$' *Physica* **C176** 131–136

[14] Holloway A, McCallum R W and Arrasmith S R 1993 'Texture development due to preferential grain growth of Ho–Ba–Cu–O in 1.6-T magnetic field' *J. Mater. Res.* **8** 727–733

[15] Stassen S, Rulmont A, Vanderbemden Ph, Vanderschueren A, Gabelica Z, Cloots R and Ausoos M 1996 'Magnetic texturing of bulk samples of the superconductor $Bi_2Sr_2Ca_{0.8}Dy_{0.2}Cu_2O_{8-y}$' *J. Appl. Phys.* **79** 553–555

[16] Chen W P, Maeda H, Kakimoto K, Zhang P X, Watanabe K, Motokawa M, Kumakura H and Itoh K 1999 'Textured crystal growth of Bi(Pb)2212 bulk ceramics in high magnetic field' *J. Cryst. Growth* **204** 69–77
[17] Sazaki G, Yoshida E, Komatsu H, Nakada T, Miyashita S and Watanabe K 1997 'Effects of a magnetic field on the nucleation and growth of protein crystals' *J. Cryst. Growth* **173** 231–234
[18] Roberts B W 1956 'Neutron diffraction study of the structures and magnetic properties of manganese bismuthide' *Phys. Rev.* **104** 607–616
[19] Massalski T B *et al* (ed) *Binary Alloy Phase Diagrams* 2nd edition (ASM International)
[20] Yasuda H, Ohnaka I, Fukuda T, Watanabe K and Shimamura K 1999 'A magnetic alignment processing by using nonequilibrium structure of rapid solidification' *Proceedings of Japan–France Cooperative Science Program Seminar on New Aspects of Electromagnetic Processing of Materials* p 88
[21] Ohnaka I, Fukusako T and Tsutsumi H 1982 'Production of $Fe_{40}Ni_{40}B_{20}$ powder by the rotating water-atomization process' *J. Japan. Inst. Met.* **46** 1095–1102

Chapter 18

Direct observation of solidification and solid-phase transformations

Toshihiko Emi and Hiroyuki Shibata

Introduction

Theories have been developed to interpret crystal morphology changes in liquid to solid transformations and phase changes in peritectic (liquid + $\delta \to \gamma$) and high-temperature solid-state ($\delta \to \gamma$) transformations in simple metallic alloy systems. However, these theories have been tested only to a limited extent, either by dynamic observation in transparent low-temperature non-metallic analogues, such as succinonitrile and ammonium chloride, or low-melting-temperature alloys such as Sn–Pb and Sn–In, or by post mortem examination of quenched alloys such as Al–Zn and Fe–25Cr–20Ni. Furthermore, uncertainty in the physical properties necessary for the theories has made their validation more difficult, especially for metallic systems with high melting temperatures.

Steels are multi-component alloys based on the iron–carbon binary system, and are one of the most important groups of metallic materials, supporting the development of most industries. The properties of steel are much influenced by the segregation of impurities, including inclusions, and the crystallographic texture. The segregation of impurities can influence the solidification structure which in turn influences the crystallographic texture. The solidification structure determines the orientations of the crystals solidifying from the steel melt, and these crystals subsequently undergo solid-state transformations to give the resultant crystal structure, and hence determine the crystallographic texture.

In view of the lack of investigation of dynamic solidification and solid-state transformations in the Fe–C binary system, recent work has employed a combination of confocal scanning laser microscopy and an infrared imaging furnace for direct observation of the transformations in this system [1, 2]. Real-time progression of the transformations has been observed *in situ* on the surfaces of molten or solid Fe–C alloys. These observations have made it possible to reveal detailed characteristics of the dynamic changes in the morphology of crystals, their phase, type and the energy of the interphase boundaries in the Fe–C system. Dynamic sequences have been observed which reveal both engulfment and pushing of inclusions at advancing solid–liquid interfaces.

Direct observation at high temperature

Confocal laser scanning microscopy

A conforcal laser scanning microscope is shown in figure 18.1 [1]. The laser beam (1.5 mW helium–neon laser, wavelength 632.8 nm) is deflected and scanned in two dimensions (5.7 kHz and 60 Hz) by an acoustic optical device (AOD) and galvano-mirror, 45° polarized by a polarizing half mirror, and delivered through a long focal length objective lens on to the surface of the specimen. The reflected beam from the surface of the sample

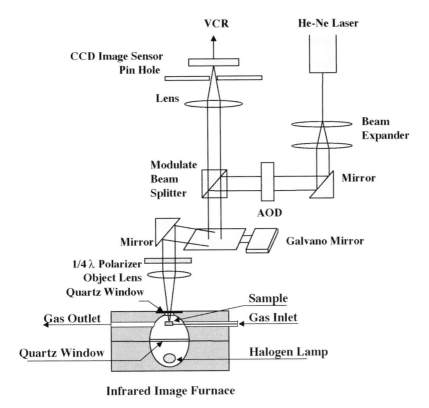

Figure 18.1. Optical system of a confocal scanning laser microscope with infrared imaging furnace.

enters a charge-coupled device (CCD) sensor after being further 45° polarized by the half mirror, and going through a modulated beam splitter and a pinhole. The pinhole serves to shut out any stray radiation from the unfocused area of the sample surface. The high intensity of the coherent laser beam (0.5 μm in diameter) is such that even the small difference in emissivity between the liquid and solid steel is clearly distinguishable. Signals from the charge-coupled device are displayed on a cathode ray tube recorded on video and stored in a computer for further processing. Signals can be obtained from planes located at different vertical distances from the specimen surface by vertically adjusting the microscope, and are stored and reproduced as a quasi-three-dimensional image. It is therefore possible to identify both concave and convex features on the sample. The resolution obtained is 0.5 μm with a maximum magnification of 2300 [1].

Infrared imaging furnace

A schematic diagram of an infrared imaging furnace is shown in figure 18.2 [1]. Specimens are placed at the upper focal point in a high-purity alumina

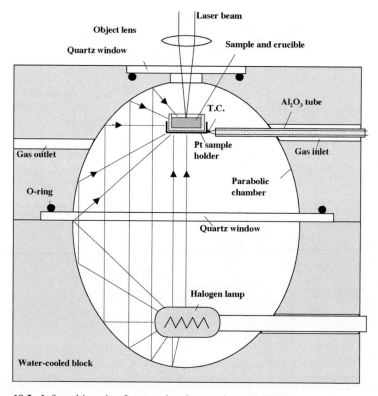

Figure 18.2. Infrared imaging furnace showing specimen position.

crucible in the upper half of the gold-plated ellipsoidal cavity of the imaging furnace. The top of the cavity has a viewing port covered by an airtight transparent quartz plate. At the lower focal point in the lower half of the cavity, separated from the upper part by a transparent quartz plate, a halogen lamp (1 kW) is used to heat the specimen (up to a maximum rate of 200 K min^{-1} up to 1923 K). Extra-high-purity argon gas, further deoxidized by passing over heated magnesium ribbons [2], is bled into the upper cavity to prevent oxidation of the specimen. The temperature of the furnace can be controlled to within ±1 K with a thermocouple welded to the platinum container that holds the alumina crucible.

The difference in the temperature between the surface of the specimen and the container was calibrated with type-B thermocouples (Pt–30%Rh/Pt–6%Rh, 0.1 mm in diameter) welded to two different parts of the surface of a pure iron specimen placed in the crucible. Extra care was taken to determine precisely the temperature distribution in the horizontal direction of the specimen by placing the thermocouple at the focal point on the specimen and changing the position of the thermocouple with a micrometer. This was used to determine the temperature gradient in the specimen for some experiments. The temperature difference between the surface and inside of the specimen was subsequently estimated by observing the solidification structure of melts quenched in the crucible under different cooling conditions. The temperature distribution in the depth-wise direction was generally found to be uniform [2].

Direct observation of Fe–C alloys

Cylindrical (4.5 mm inner diameter, 3 mm height) and rectangular (inner size, 3 mm width, 11 mm length, 3 mm height) crucibles were used, the latter specifically for studying instabilities in the liquid–solid interface during peritectic reactions and transformations [3]. During observations of the engulfment and pushing of inclusions by the growing solid–liquid interface in Fe–C alloys, care was taken not to melt the whole specimen, but instead to keep the peripheral part of the specimen solid. This was necessary to prevent the formation of a round sessile drop. The heating rate near to the melting point was reduced to about 10 K min^{-1} to allow melting to be observed whilst retaining a solid peripheral shell.

For more details of experimental conditions and chemistry of specimens specific to each of the following experiments, see references [1]–[3] and [11]–[13]. During solidification and inclusion engulfment experiments, the solid–liquid interface was set at the centre of the field of view. As the interface was continuously moving in the growth direction, it was occasionally necessary to reset the specimen position by controlling a micrometer connected to the specimen holder.

Instability of the solid–liquid interface during solidification

The relevant high-temperature, low-carbon portion of the Fe–C binary phase diagram is shown in figure 18.3. Dynamic observation of the solid–liquid interface of the alloys listed in table 18.1 reveals the following features:

(1) When an Fe–0.14% C melt is slowly cooled to 1773 K, stable planar liquid–δ interfaces form at a temperature gradient of 22 K mm^{-1}. The interface advances towards the left of the figure at a growth rate of 4.3 µm s^{-1}, as shown in figure 18.4. At a temperature gradient of 4.3 K mm^{-1}, the planar interface becomes unstable at a rate of 2.5 µm s^{-1}, as shown in figure 18.5, and exhibits an early indication of a transition into cellular configuration as shown in figure 18.6.

(2) For an Fe–0.22% C–0.80% Mn melt, the liquid–δ interfaces are unstable at a temperature gradient of 4.3 K mm^{-1} for a growth rate of 3.1 µm s^{-1}, resulting in the near sinusoidal perturbation shown in figure 18.7(a). The interval and amplitude of the perturbations were determined using commercial image analysis software (MacSALT) as the average distance between peaks and one half depth of the cavities at the interface. The interval and amplitude are 75 and 22 µm, respectively. The perturbations develop into a non-sinusoidal shape with time but without significant change in the interval and amplitude, as shown in figure 18.7(b).

(3) An Fe–0.42% C melt forms stable planar liquid–δ interfaces with cellular δ crystals for a temperature gradient of 22 K mm^{-1} and a growth

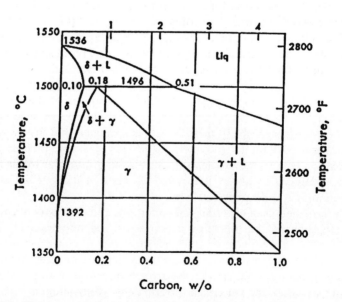

Figure 18.3. High-temperature–low-carbon portion of the Fe–C phase diagram.

Table 18.1. Composition of Fe–C alloy used for instability study (mass%).

Specimen	C	Si	Mn	P	S	Al	O	N
Fe–0.14C	0.14	0.01	0.01	0.001	0.001	0.003	0.0009	0.0011
Fe–0.22C	0.22	0.02	0.80	0.014	0.005	0.040	–	–
Fe–0.42C	0.42	0.01	0.01	0.001	0.001	0.002	0.0009	0.0010
Fe–0.81C	0.81	0.01	0.01	0.001	0.001	0.001	0.0007	0.0008

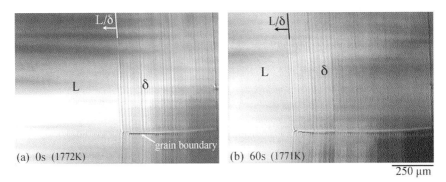

Figure 18.4. Planar liquid–δ crystal interfaces advancing towards the left of the figure in Fe–0.14 wt% C for a temperature gradient of 22 K mm^{-1} and a growth rate of 2.5 mm s^{-1} [2, 3].

rate of 2.0 µm s^{-1}, and also for a temperature gradient of 4.3 K mm^{-1} and a growth rate of 0.5–1.5 µm s^{-1}, as shown in figures 18.8(a) and 18.8(b) respectively. It can be seen that grain boundaries are present at the lower temperature gradient. The boundaries in figure 18.8(b) are found to be grooves, not ridges. A line along the groove is observed, fixed at the liquid–δ–δ triple

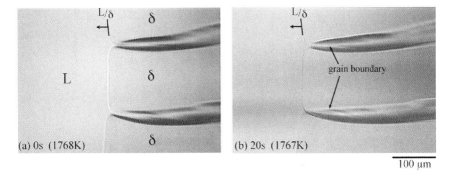

Figure 18.5. Early stage of instability at the liquid–δ crystal interface advancing towards the left of the figure in Fe–0.14 wt% C for a temperature gradient of 4.3 K mm^{-1} and a growth rate of 2.5 µm s^{-1} [2, 3].

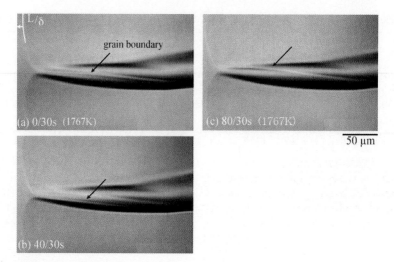

Figure 18.6. Early stage of planar to cellular transition of δ crystals advancing towards the left of the figure in Fe–0.14 wt% C for a temperature gradient of 4.3 K mm^{-1} and a growth rate of 2.5 μm s^{-1} [2, 3].

point, which moves up and down within the width of the groove over time. This phenomenon is often observed, but its nature is unknown.

(4) For an Fe–0.81% C melt, the liquid–γ interface is unstable for a temperature gradient of 4.3 K mm^{-1} and a growth rate of 0.6 μm s^{-1} with perturbations as shown in figure 18.9(a). The perturbations persist after 85 s, as shown in figure 18.9(b), without much development. When the specimen temperature is decreased from 1715 K to 1690 K, γ crystals grow from the bottom of the melt and appear on the surface of the melt, as shown in figure 18.9(c), inhibiting the horizontal unidirectional growth.

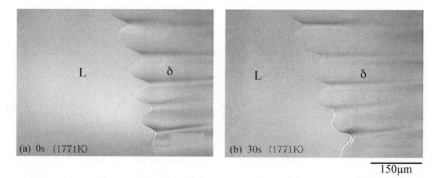

Figure 18.7. Near sinusoidal perturbation developing at the liquid–δ interface advancing towards the left of the figure in Fe–0.22 wt% C–0.80 wt% Mn for a temperature gradient of 4.3 K mm^{-1} and a growth rate of 3.1 μm s^{-1}.

Instability of the solid–liquid interface during solidification 305

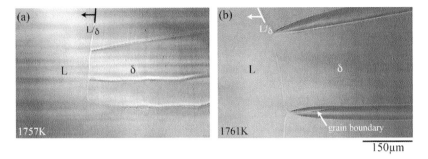

Figure 18.8. Planar liquid–δ interface advancing towards the left of the figure in Fe–0.42 wt% C (a) for a temperature gradient of 22 K mm^{-1} and a growth rate of 2.0 μm s^{-1} in a rectangular crucible, resulting in a macroscopically straight interface, and (b) for a temperature gradient of 4.3 K mm^{-1} and a growth rate of 1.2 μm s^{-1} in a cylindrical crucible, resulting in a macroscopically curved interface [2].

(5) The morphology of crystals solidifying from an Fe–C alloy melt is determined by the cooling rate of the melt, the concentration of carbon in the melt C_0, the temperature gradient in the melt G, and the growth rate of the crystals V. According to the detailed analyses by Mullins and

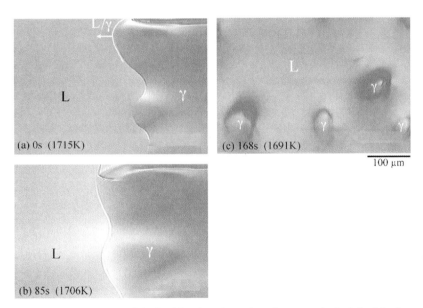

Figure 18.9. Perturbation of the liquid–γ interface advancing towards the left of the figure in Fe–0.81 wt% C for a temperature gradient of 4.3 K mm^{-1} and a growth rate of 0.6 μm s^{-1}. (a, b) Perturbations persist at 1706 and 1715 K respectively, whereas (c) the direction of solidification changes towards the top of the specimen as the melt is cooled to 1691 K and horizontal growth is hindered [2].

Sekerka [4], Sekerka [5] and Kurz and Fisher [6, 7], interfacial stability is given in a simplified form [7] by

$$\frac{1}{\varepsilon}\frac{\partial \varepsilon}{\partial t} = E = \frac{V}{mG_c}[-\Gamma\omega^2(\omega^* - VP/D) - G(\omega^* - VP/D) + mG_c(\omega^* - V/D)],$$

(18.1)

where ε is the amplitude of the perturbation, E is the relative growth rate of the amplitude of the perturbation, m is the slope of the liquidus line of the Fe–C binary phase diagram, G_c is the concentration gradient in the liquid at the liquid–solid interface, Γ is the Gibbs–Thomson parameter σ/S, σ is the specific solid–liquid interfacial energy, S is the entropy of melting, ω is the frequency $2\pi/\lambda$ of the perturbation, λ is the wavelength of the perturbation, P is the complementary distribution coefficient $(1 - k)$, k is the distribution coefficient, D is the diffusivity of carbon in the melt, $0.02\,\mathrm{mm^2\,s^{-1}}$, and

$$\omega^* = \left(\frac{V}{2D}\right) + \left[\left(\frac{V}{2D}\right) + \omega^2\right]^{1/2}.$$

(18.2)

A perturbation of the interface increases for $E > 0$, whereas a planar interface is stable for $E < 0$. Values calculated using equation (18.1) for Fe–0.22% C–0.80% Mn for a temperature gradient of $4.3\,\mathrm{K\,mm^{-1}}$ and a growth rate of $3.1\,\mathrm{\mu m\,s^{-1}}$ (see figure 18.7), are given in figure 18.10. The curve shown

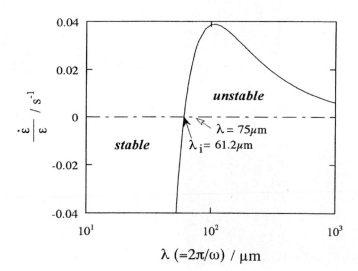

Figure 18.10. Calculated rate of development of a perturbation at the liquid–δ interface in Fe–0.22 wt% C–0.80 wt% Mn for a temperature gradient of $4.3\,\mathrm{K\,mm^{-1}}$ and a growth rate of $3.1\,\mathrm{\mu m\,s^{-1}}$. The calculated critical wavelength for developing the instability at the interface is 61 μm, whereas for the observed interface the perturbation becomes unstable at a wavelength of 75 μm [2].

in the figure indicates that when the wavelength is less than 61.2 μm, the amplitude will decrease to give a stable interface, whereas beyond this value for the wavelength the amplitude will increase making the interface unstable. This is in agreement with present observations whereby the interface becomes unstable, developing perturbations for a wavelength of 75 μm.

(6) The constitutional supercooling criterion predicts that a planar interface is stable if the following condition is met [6]:

$$V \leq -\frac{GDk}{mC_0(1-k)} = \frac{GD}{\Delta T_0} \qquad (18.3)$$

where ΔT_0 is the difference between the liquidus and solidus temperatures in the Fe–C phase diagram. In estimating ΔT_0 for Fe–0.22% C–0.80% Mn, allowance has been made for the influence of solutes (Mn, P and S), according to previous investigations [8, 9]. The observations given in sections (1)–(4) above are plotted in figure 18.11 together with lines calculated by equation (18.3). The theoretical prediction is found to be in reasonable agreement with the present observations for Fe–C alloys within the carbon concentration range examined.

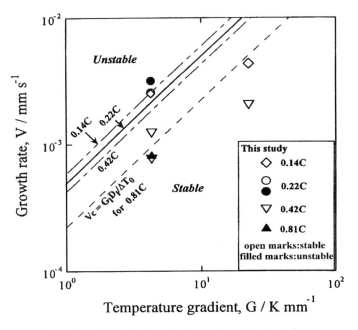

Figure 18.11. Comparison of the calculated and experimental criteria for the occurrence of the instability at the solid–liquid interface of Fe–C–(Mn) alloys (calculated on the basis of constitutional supercooling (equation (18.3)) [2].

(7) The Mullins and Sekerka stability criterion [4] has been simplified by Kurz and Fisher [6, 7] who neglected the thermal effect, such that

$$G > mG_c - \left(\frac{kV^2\Gamma}{D^2}\right). \quad (18.4)$$

Comparison of calculations using equation (18.4) with the present observations shows almost identical results [2] to those shown in figure 18.11, indicating that the effect of curvature (the second term on the right-hand side of equation (18.4)) has little influence on the stability under the present experimental conditions.

Transition from planar to cellular to dendritic solidification of steel melts

The sequences of transitions in crystal morphology have been dynamically observed during the solidification of a medium- and a high-carbon steel. Real time *in situ* observation has made the dynamic progress of the transition clearer then the behaviour previously observed by investigating quenched samples. The main characteristics of the observations made during solidification are summarized below.

For medium-carbon steel the melt solidifies to yield only cellular δ crystals of width 26–120 μm for temperature gradients in the range 4–5 K mm^{-1} and growth rates up to 12 μm s^{-1}. Coarsening and overgrowth of cells is often observed, with narrow cells often being overgrown by wider adjacent cells. The overgrowth increases with time up to 10 μm s^{-1} after 15 s [1]. Such a high rate of overgrowth may not be explained in terms of Ostwald ripening, and the underlying mechanism is still unknown.

For high-carbon steel the melt solidifies as γ phase at 1673–1675 K. When the growth rate is 0.5 μm s^{-1} for a temperature gradient of a few K mm^{-1}, planar γ crystals are formed, and the liquid–γ interface is stable. Increasing the growth rate clearly exhibits the following sequence, which is partly shown in figure 18.12. First, planar γ crystals grow towards the bulk of the liquid. As the growth rate increases to 1.6 μm s^{-1}, perturbation of the liquid–γ interface occurs, with a peak-to-peak distance of 30–65 μm, as shown in figure 18.12(a). At a growth rate of 2.1 μm s^{-1}, the interface becomes cellular, with some cells such as A and C growing preferentially over adjacent cells such as B, as shown in figure 18.12(b). Further increase in the growth rate to 2.3 μm s^{-1} results in a typical array of cellular crystals, as shown in figure 18.12(c). In the later stages of solidification, when the growth velocity has increased further, secondary perturbations are observed either side of the cellular crystals [1]. These perturbations grow into the secondary arms of dendrites as shown in figure 18.13. The observed coarsening of the secondary dendrite arm spacing is definitely not due to the

Transition from planar to cellular to dendritic solidification

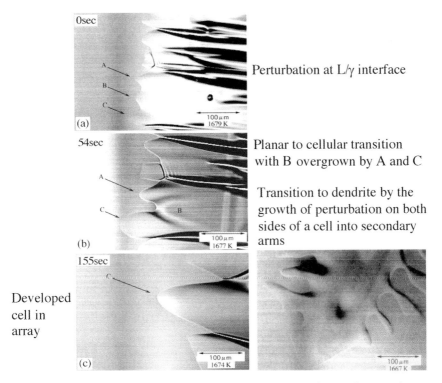

Figure 18.12. Planar to cellular interface transition in the high-carbon steel at growth rates of 1.5–2.5 μm s^{-1}; (a) perturbation occurs at planar liquid–γ interface, (b) cellular interface develops and overgrowth is observed, (c) formation of cellular array (only shown in part) [1].

thinning of the two tiny triangular arms observed in figure 18.13, but instead is clearly due to the preferred growth of the adjacent three thicker arms over the two smaller ones.

To examine the criterion for the dendritic transition, the radii of cell tips in the stabilized array just before the cells transformed into dendrites were observed and plotted (open circles) as a function of the growth rate in figure 18.14. A theoretical derivation of the criterion for the dendritic transition was made by Trivedi [10] who gave the relation between the tip radius R and growth velocity V for a temperature gradient of zero, as

$$R^2 V = \frac{2\sigma DL[1 - \phi(1-k)]}{\Delta S_m C_0 (k-1)}, \tag{18.5}$$

where L depends on the harmonic of the perturbation, and is estimated for spherical crystal tips to be between 10 and 28, ΔS_m is the entropy of fusion and ϕ is the Ivantsov function.

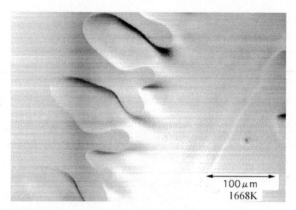

Figure 18.13. Development of secondary arms on the side of a primary arm in the high-carbon steel. Note that coarsening of the arm spacing is due to overgrowth [1].

A similar relation was given somewhat later, but independently, by Fisher and Kurz [7] as

$$V = \frac{2D(GR^2 + 4\pi^2\Gamma)}{R^3PG - 2R^2PCm + 4\pi^2\Gamma RP}. \tag{18.6}$$

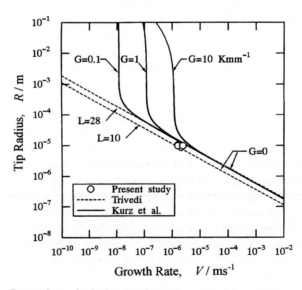

Figure 18.14. Comparison of calculated and experimental relations between the tip radius and growth rate of γ crystals in the high-carbon steel. The three open circles correspond to critical growth rates when cellular crystals are about to change into dendrites for temperature gradients of 4–5 K mm^{-1} [1].

Results calculated using equation (18.5) for $L = 10$ and 28 and those using equation (18.6) for temperature gradients in the range $0-10\,\text{K}\,\text{mm}^{-1}$, with data relevant to the present experimental conditions, are given in figure 18.14 as dotted and solid lines respectively [1].

The points at which the calculated solid curves in figure 18.14 branch off from the dotted lines indicate the conditions where a planar interface becomes unstable and planar–cellular–dendrite transitions occur. The calculated condition of branching for a temperature gradient of $10\,\text{K}\,\text{mm}^{-1}$ is in good agreement with the location of the open circles obtained in the present experiment for a temperature gradient in the range $4-5\,\text{K}\,\text{mm}^{-1}$, giving validation to the theories.

Peritectic reaction and transformation in Fe–C alloys

Fe–C alloys containing $0.10-0.51\,\text{wt\%}$ carbon may undergo the peritectic reaction liquid $+ \delta \rightarrow \gamma$ at the liquid–δ phase boundary. This is followed by a peritectic $\delta \rightarrow \gamma$ transformation during which the γ phase coarsens at the liquid–δ boundary. The equilibrium partition coefficients of common solutes are different for the δ and γ phases, and diffusion rates in γ are slower than in δ. Thus, the peritectic reaction liquid $+ \delta \rightarrow \gamma$ and the peritectic transformation $\delta \rightarrow \gamma$ cause complicated microscopic and macroscopic segregation patterns in peritectic medium-carbon and stainless steels.

The peritectic reaction and transformation are accompanied by a volume contraction that causes depression of the initially solidified shell in the mould. This phenomenon becomes marked for high-speed continuous casting of medium-carbon steel slabs of large width, generating surface cracks. In conventional slab casters, therefore, the maximum speed of casting peritectic steels is limited to about $1.8\,\text{m}\,\text{min}^{-1}$. In thin-slab casters where conventional steels are cast at speeds as fast as $5\,\text{m}\,\text{min}^{-1}$, peritectic steels are difficult to cast above $3.5\,\text{m}\,\text{min}^{-1}$, and hence are not usually cast. Controlling the rate of the peritectic reaction and transformation should provide an effective means of preventing surface cracks. For this reason, dynamic real time progression of the peritectic reaction and transformation has been observed *in situ* for peritectic Fe–$0.14\,\text{wt\%}$ C and Fe–$0.42\,\text{wt\%}$ C. Table 18.1 gives the full compositions of these alloys.

When a planar liquid–δ interface in the Fe–0.42% C alloy is cooled to $1754\,\text{K}$ for a temperature gradient of $4.3\,\text{K}\,\text{mm}^{-1}$, γ phase forms by the peritectic reaction liquid $+ \delta \rightarrow \gamma$ at the point where the liquid meets the grain boundary between two planar δ crystals. This can be seen by comparing figures 18.15(a) and (b). The γ phase propagates laterally from the triple point, liquid–γ–δ, along the liquid–δ crystal boundary. This lateral propagation of the γ phase along the liquid–δ interface, is very fast, approximately $3\,\text{mm}\,\text{s}^{-1}$ at a temperature gradient of $4.3\,\text{K}\,\text{mm}^{-1}$ and a cooling rate of

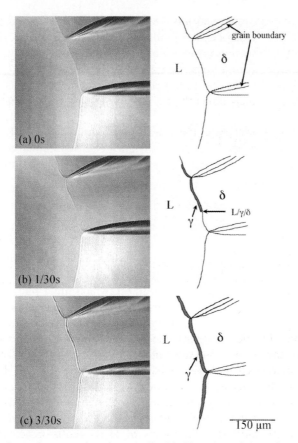

Figure 18.15. Lateral propagation along the liquid–δ interface of the γ phase formed at the liquid–δ–δ triple point due to the peritectic reaction in Fe–0.42 wt% C, while planar δ crystals grow towards the left of the figure, at a temperature of 1754 K and a temperature gradient of 4.3 K mm^{-1} [3].

1 K min^{-1}, covering the boundary with γ phase almost instantaneously, as shown in figure 18.15(c). This is followed, as shown in figures 18.16(a) to (c), by the peritectic transformation, i.e. growth (coarsening) of the γ phase in two opposite directions, one toward the liquid, and the other toward the δ phase. The growth rate toward the δ phase is faster than that toward the liquid, as clearly seen in figure 18.16(c), i.e. the γ–δ interface moves faster than the liquid–γ interface.

The growth rate of the γ phase during uni-directional solidification of Fe–0.42% C for a temperature gradient of 4.3 K mm^{-1} and cooling rate of 1 K min^{-1} is given by:

$$d = 40t^{1/2},$$

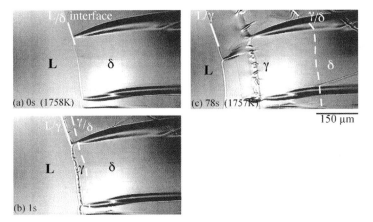

Figure 18.16. Peritectic transformation in Fe–0.42 wt% C for a temperature gradient of 4.3 K mm^{-1} and a cooling rate of 1 K min^{-1}. The γ–δ interface grows faster than the liquid–γ interface [(b) and (c)]. The original liquid–δ interface is left with wrinkles between the liquid–γ and γ–δ interfaces as the liquid–γ interface keeps growing towards the left while the γ–δ interface grows towards the right [3].

where d is the thickness of the γ phase in mm and t is the time in minutes. Figure 18.17 shows the observed γ phase growth rates toward the δ phase and liquid, compared with values calculated using a model [3], where the diffusion of carbon across the δ phase determines the growth rate toward δ and that across the γ phase controls the growth rate towards the liquid. This calculation, using reliable diffusivity data, confirms that the rate of the peritectic transformation is controlled by the diffusion of carbon in the γ phase on the liquid side of the interface. On the other hand, the diffusion of carbon in the δ phase controls the rate of peritectic transformation on the δ phase side of the interface [3].

The peritectic reaction was also investigated using δ crystals grown from the bottom of the melt to the surface as shown in figures 18.18–18.20. Initial formation of a rim of γ phase surrounds the δ phase very quickly within 0.2 s, as shown in figure 18.18(b), and is followed by γ growth to the centre of the δ crystals within 7 s. It is interesting to note that grooves are observed on the γ phase as the transformation proceeds, as shown in figure 18.18(d). This is obviously related to the volume change associated with the $\delta \rightarrow \gamma$ transformation. Similar characteristics, i.e. lateral progress of the liquid–γ–δ triple point by the peritectic reaction, growth of γ via the peritectic transformation, and faster growth of γ on the liquid side of the interface compared with that on the δ side, are also observed in Fe–0.14 wt% C, as shown in figures 18.19 and 18.20. However, the observed peritectic reaction is faster, approximately 4.5 mm s^{-1}, for Fe–0.14 wt% C compared with Fe–0.43 wt% C [2].

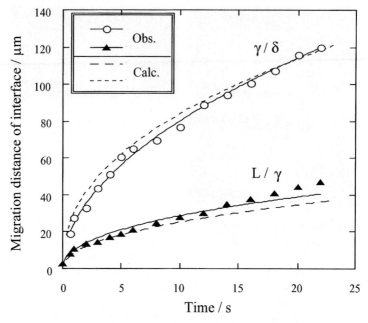

Figure 18.17. Growth of the γ phase towards the liquid and δ phase with time in Fe–0.42 wt% C at 1758 K. The sum of the two curves gives the total thickness of the γ phase. The dotted lines are calculated on the basis of a model given in reference [3].

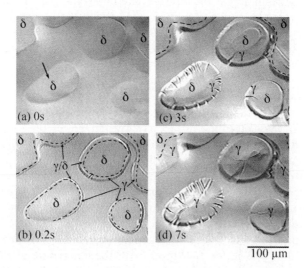

Figure 18.18. Progress of the peritectic reaction and transformation in Fe–0.42 wt% C; (a) δ crystals grow in the melt perpendicular to the meniscus, (b) γ phase rim forms around δ crystals, (c) γ phase grows with wrinkles towards the liquid, and (d) δ crystals transform completely into γ crystals [3].

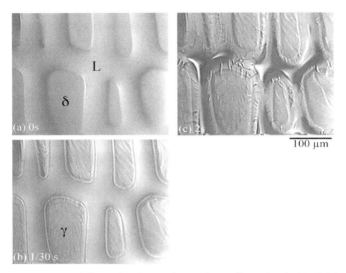

Figure 18.19. Progress of the peritectic reaction and transformation in Fe–0.14 wt% C at 1768 K with a temperature gradient of 4.3 K mm^{-1} and cooling rate of 20 K min^{-1}, (a) δ crystals grow in the melt perpendicular to the meniscus, (b) γ phase rim forms around δ crystals, and (c) δ crystals transform completely into γ crystals with wrinkle formation [3].

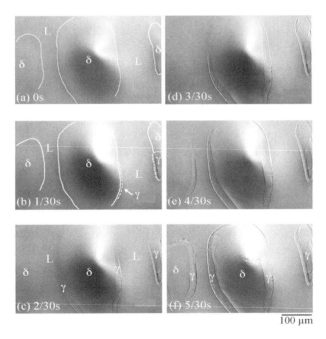

Figure 18.20. Progress of peritectic reaction [3] in Fe–0.14 wt% C at 1765 K with a temperature gradient of 4.3 K mm^{-1} and a cooling rate of less than 10 K min^{-1}.

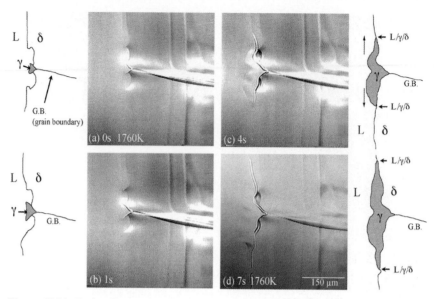

Figure 18.21. Peritectic reaction in Fe–5.1% Ni at 1760 K showing preferred lateral growth of liquid–γ–δ triple point along liquid–δ boundary with a temperature gradient of 4.3 K min^{-1}, supercooling of about 30 K, and solidification rate of 1.9 µm s^{-1} [2].

In contrast, Fe–5.10–5.25 wt% Ni also undergoes a peritectic reaction on solidification, but exhibits different behaviour. The peritectic reaction occurs in a manner similar to the medium-carbon Fe–C alloys, but in a more distinct way, as shown in figure 18.21 for Fe–5.1% Ni. At the triple point between the liquid and two adjacent δ crystals, a small shallow cavity is formed. The peritectic reaction initiates at this cavity, and propagates slowly, at 23–145 µm s^{-1}, in a lateral direction by preferred growth of another liquid–γ–δ triple point along the liquid–δ interface.

Surface energy of the δ ferrite–γ austenite interface

Only very limited data have been available on the surface energy of high-temperature δ–γ interfaces in steel. Nucleation and growth of both δ in a γ matrix and γ in a δ matrix have been observed dynamically. Static and dynamic dihedral angle measurements during the $\delta \rightarrow \gamma$ transformation of low-carbon steels exhibit the following features.

The surface energy of incoherent δ–γ interfaces in a low-carbon steel (0.007% sulphur) is calculated to be approximately 0.45 J m^{-2}. This is determined from the observed intrinsic dihedral angles at the triple point of one high-energy δ–δ grain boundary and two incoherent δ–γ interphase boundaries, as shown in figure 18.23. The intrinsic dihedral angle depends

Surface energy of the δ ferrite–γ austenite interface 317

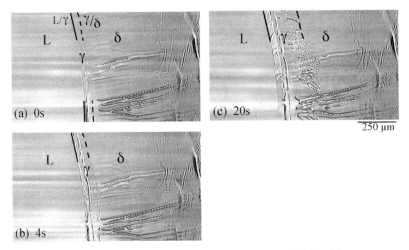

Figure 18.22. Peritectic transformation in Fe–5.25% Ni at 1760 K with a temperature gradient of 4.3 K min^{-1}, and solidification rate of 2.5 μm s^{-1}, showing characteristic cellular growth of γ crystals from original melt–δ crystal boundary into δ crystals [2].

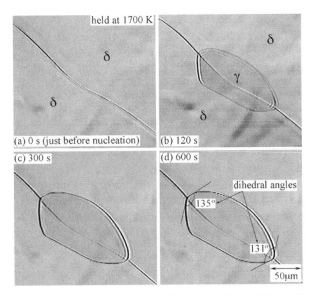

Figure 18.23. Dihedral angles at δ–γ interphase boundaries formed during δ → γ transformation of a low-carbon steel containing 0.007% sulphur, held at 1700 K. The γ crystal becomes discernible as it grows ≥0.5 μm after nucleation at a high-energy δ–δ crystal boundary. The angles at the incoherent δ–γ interphase boundaries do not change after holding for 300 s, indicating that they are in equilibrium [11].

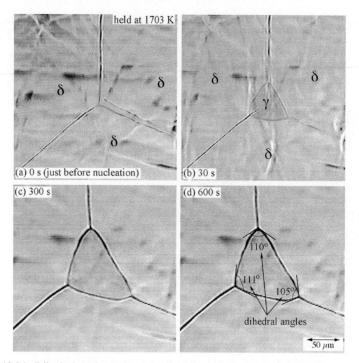

Figure 18.24. Dihedral angles at $\delta-\gamma$ interphase boundaries formed during $\delta \to \gamma$ transformation of a low-carbon steel containing 0.021% sulphur. The γ crystal becomes discernible as it grows $\geq 0.5\,\mu$m after nucleation at a high-energy $\delta-\delta$ crystal boundary. The angles at the incoherent $\delta-\gamma$ interphase boundaries do not change after holding for 300 s, indicating that they are in equilibrium [11].

strongly on the sulphur content, decreasing linearly from 131.5 to 107.3° as sulphur increases from 0.007 to 0.021%, as shown in figure 18.24. This indicates that the surface energy of incoherent $\delta-\gamma$ interfaces is reduced greatly, to 0.31 J m^{-2}, by the increase of sulphur to 0.021%. The dihedral angle is not affected by temperature in the narrow temperature range of $\delta-\gamma$ equilibrium, indicating that the energy of incoherent $\delta-\gamma$ interfaces remains the same at these temperatures. On the other hand, the intrinsic dihedral angle at the triple point is not obtained during the reverse $\gamma \to \delta$ transformation mainly due to either a shape effect or the absence of a suitable triple point [11].

Morphological instability of δ ferrite–γ austenite interfaces

The dynamic behaviour of $\delta-\gamma$ interfaces during $\delta-\gamma$ transformations has also been observed *in situ* to investigate the morphological instability of

Morphological instability of δ ferrite–γ austenite interfaces

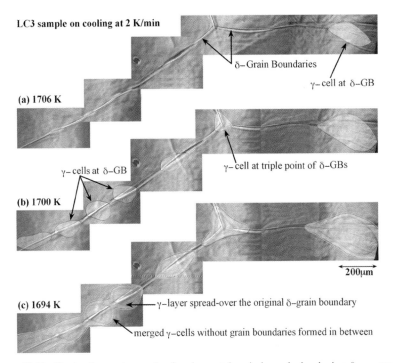

Figure 18.25. Nucleation and growth of γ phase at δ grain boundaries during δ → γ transformation in a low-carbon steel with 0.021% sulphur and a cooling rate of 2 K min^{-1} [12].

the boundary in the low-carbon steels, revealing the following interesting features.

During the δ → γ transformation, the γ phase frequently nucleates at the triple points of δ grain boundaries on the surface of the specimen, as shown in figure 18.25. The initial γ phase has a strong tendency to cover all δ grain boundaries completely, forming a thin layer on both sides of the original δ grain boundary. The δ–γ interfaces are always unstable having the finger-like morphology shown in figure 18.26. The finger-like γ phase develops along δ grain boundaries at low supercoolings and even grows into the δ matrix at high supercoolings, as shown in figure 18.27. During the reverse γ → δ transformation, however, the δ–γ interfaces are always planar or curved, without perturbations regardless of the degree of superheating, the size of the δ grains, the migration rate of the δ–γ interfaces or the diffusion field. Such a considerable difference in the characteristics of the δ–γ interfaces between δ → γ and γ → δ transformations indicates that the effect of thermal diffusion of carbon during the transformation may not be neglected. Classical theories of constitutional supercooling and instability during solidification have proved to be applicable to explain the morphological instabilities of the δ–γ interphase boundary in the δ → γ

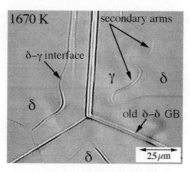

Figure 18.26. Instability at δ–γ interphase boundary during $\delta \to \gamma$ transformation in a low-carbon steel, with 0.021% sulphur at 1670 K, showing early stage development of secondary arms in the γ phase [12].

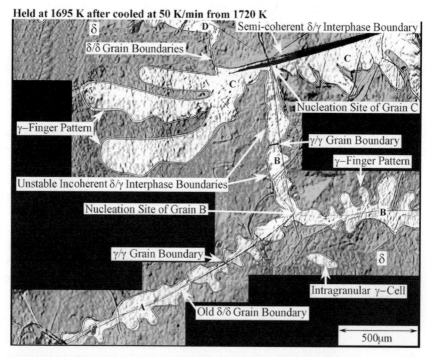

Figure 18.27. Morphological instability of δ–γ interphase boundaries during $\delta \to \gamma$ transformation of a low-carbon steel containing 0.021% sulphur held at 1695 K, i.e. at a supercooling of 11 K, after cooling at a rate of 50 K min^{-1} from 1720 K. Note the development of the finger-like γ crystals into the δ matrix caused by constitutional supercooling. A, B, C and D are all independent γ grains [12].

transformations. Calculations based on these theories agree well with the present observations [12].

Pinning of grain boundaries by inclusions

During the growth of γ crystals in low-carbon Al-killed steel, pinning of the grain boundary motion has been observed, as shown in figure 18.28, where B and C are traces of old γ and δ grain boundaries respectively. A new boundary moves to the right, as shown by a thin line marked A at 0 s in figure 18.28. After 1.5 s, the boundary contacts an inclusion and remains pinned until 12.0 s when it detaches, crosses the inclusion and continues to move freely further to the right.

Interaction of the solid–liquid interface with inclusions

The engulfment or pushing of inclusions by the advancing solid–liquid interface during the solidification of steel has important relevance to the production of clean steel. Many theories have been proposed to explain the factors influencing the pushing or entrainment of inclusions for metals and alloys. The factors include the inclusion size, solid–liquid interface velocity, melt viscosity, and interfacial energy at the inclusion–steel boundary. However, no experimental confirmation has as yet been made for steel.

Al_2O_3 clusters floating on a low-carbon Al-killed steel melt in front of an advancing solid–liquid interface were observed in order to determine the criteria for the engulfment or pushing of the clusters. Figure 18.29 shows that small clusters of fine Al_2O_3 inclusions (A and C) are pushed by the solid–liquid interface, while large clusters of Al_2O_3 inclusions (B) are engulfed by the solid–liquid interface. The aggregated Al_2O_3 inclusions A and C, slightly larger in apparent radius R_{incl}, are pushed by the interface for some distance, leaving a streak or groove behind, as shown in figure 18.29. The groove apparently forms due to insufficient supply of iron atoms behind the advancing aggregate. In the later stages of pushing, there are occasions where the liquid–δ interface near the aggregates A and C deforms in a concave fashion, surrounding the aggregate, and finally engulfing it.

Pushing is enhanced when the growth rate of the solid, i.e. the speed of advance of the liquid–δ interface V, is reduced. As shown in figure 18.30, an inverse relationship $V = 60/R_{incl.}$, where $R_{incl.}$ is the radius of the inclusion, is obtained as the criterion below which the engulfment of the aggregates will take place. The criterion was determined as the contour of the upper limit of half filled circles, each of which represents an event of pushing followed

322 Direct observation of solidification and solid-phase transformations

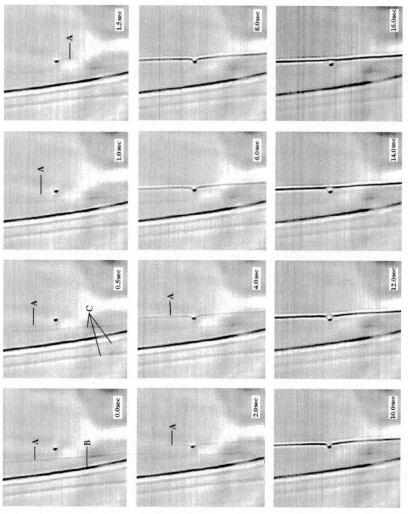

Figure 18.28. Pinning by an inclusion of γ grain boundary marked A moving towards the right, during grain growth in a low-carbon Al-killed steel.

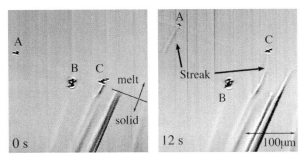

Figure 18.29. Pushing of small aggregates of Al_2O_3 inclusions by the advancing solid–liquid interface in a low-carbon Al-killed steel. Note that inclusions A and C are pushed, leaving grooves behind them, while inclusion B is engulfed [13].

by engulfment. The larger clusters of Al_2O_3 inclusions (14–29 μm) are engulfed at almost any growth rate of the solid faster than $8\,\mu m\,s^{-1}$. Just before the engulfment of such a cluster, the interface extends without exception in a convex fashion toward the cluster, as shown in figure 18.31. A similar phenomenon was also observed for round shaped liquid silicate inclusions an ultra-low-carbon Si-killed steel melt. The criterion for this steel is $V_1 = 23/R_{incl.}$. The convex extension was again observed, but to a lesser extent. A model calculation [13] shows that the convex extension of the interface is caused by inhibited heat transfer from the bulk of the melt to the interface. The heat transfer is reduced locally because of the low-thermal-conductivity large inclusions, and the reduced heat transfer

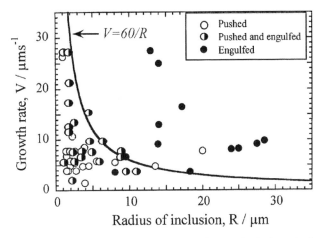

Figure 18.30. Critical condition for the engulfment of Al_2O_3 inclusions during solidification of low-carbon Al-killed steel at 1800 K [13].

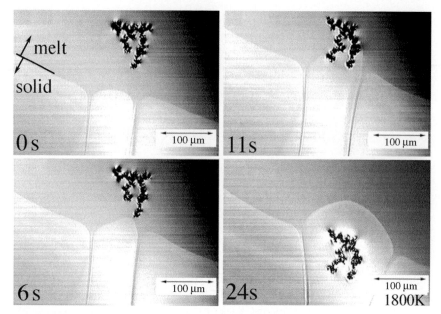

Figure 18.31. Preferred growth (convex extension) of the solid–liquid interface of a low-carbon Al-killed steel towards Al$_2$O$_3$ inclusions during engulfment of the inclusions at a temperature of 1800 K and a growth rate of 3 μm s^{-1} [13].

enhances the convex growth of the interface towards the intervening inclusion to accelerate the engulfment.

The characteristics of these observations together with computational analysis [13] indicate a number of practical implications for casting operation. Inclusions in the melt located close to the advancing solid–liquid interface are almost always engulfed in the solidifying steel shell at the growth rates normally encountered in continuous casting. To avoid engulfment, it is important to prevent inclusions being brought ahead of the solidifying shell. This can be achieved by sweeping or washing the melt–shell boundary by clean melt flow controlled by electromagnetic means.

Summary

In situ observation of the dynamics of phase transformations occurring in Fe–C alloys during solidification and cooling presents many interesting features, calling for some reconsideration of our current understanding. The use of this technique is stimulating further development of theory, and enhancing the progress of industrial processes on the basis of more realistic findings.

References

[1] Chikama H, Shibata H, Emi T and Suzuki M 1996 '*In-situ* real time observation of planar to cellular and cellular to dendritic transition of crystals growing in Fe–C alloy melts' *Mater. Trans., Japan. Inst. Met.* **37** 620–626

[2] Arai Y February 1998 '*In situ* observation and analysis of crystal growth and phase transformation of steel' Masters Thesis in Metallurgy, Tohoku University

[3] Shibata H, Arai Y, Suzuki M and Emi T 2000 'Kinetics of peritectic reaction and peritectic transformation in medium carbon steels' *Metall. Mater. Trans.* **B31**(5) in press

[4] Mullins W W and Sekerka R F 1964 'Stability of planar interface during solidification of a dilute binary alloy' *J. Appl. Phys.* **35** 444–451

[5] Sekerka R F 1965 'A stability function for explicit evaluation of the Mullins–Sekerka interface stability criterion' *J. Appl. Phys.* **36** 264–268

[6] Fisher D J and Kurz W 1980 'A theory of branching limited growth of irregular eutectics' *Acta Metall.* **28** 777–794

[7] Kurz W and Fisher D J 1981 'Dendrite growth at the limit of stability: tip radius and spacing' *Acta Metall.* **29** 11–20

[8] Esaka H and Ogibayashi N 1998 'Estimation of primary dendrite arm spacing and solidified interface morphology in low carbon steel by introducing quasi Fe–C binary alloy' *Tetsu-to-Hagane* **84** 49–54

[9] Nagata S, Matsumiya T, Ozawa K and Ohashi T 1990 'Estimation of critical strain for internal crack formation in continuously cast slabs' *Tetsu-to-Hagane* **76** 214–311

[10] Trivedi R 1980 'Theory of dendritic growth during the directional solidification of binary alloys' *J. Crystal Growth* **49** 219–232

[11] Yin H, Emi T and Shibata H 1998 'Determination of free energy of δ ferrite/γ-austenite interphase boundary of low carbon steels by *in-situ* observation' *Iron Steel Inst. Japan. Int.* **38** 794–801

[12] Yin H, Emi T and Shibata H 1999 'Morphological instability of δ-ferrite/γ-austenite interphase boundary in low carbon steels' *Acta Mater.* **47** 1523–1535

[13] Shibata H, Yin H, Yoshinaga S, Emi T and Suzuki M 1998 '*In-situ* observation of engulfment and pushing of non-metallic inclusions in steel melt by advancing melt/solid interface' *Iron Steel Inst. Japan. Int.* **38** 149–156

Chapter 19

Interfacial energy and structure in Al–Si alloys

Hideo Nakae

Introduction

This chapter presents a fundamental study into the role of interfacial energy in determining solidification structures. The Al–Si alloy system is considered because the interfacial energy plays such a significant role in the microstructure of the silicon phase [1, 2]. Modification of the silicon microstructure, i.e. refining it by adding sodium, strontium, antimony, etc., to Al–Si alloys, is a technique very widely used in foundries. The effects of solidification and cooling rates on the modification and microstructural refinement process are discussed in relation to the mechanisms operating. Overall nucleation, growth, grain refinement and phase selection effects are discussed in detail in chapters 12, 13, 10 and 15 respectively.

Attempts to explain the modification mechanism have involved various kinds of theory which can be separated into three main types [3]: the silicon growth restraint theory [4–6], the twin theory [5, 7–9] and the silicon nucleation restraint theory [10, 11]. However, a widely accepted theory is yet to be established. Calcium is in the same group as strontium in the periodic table, but there are only a few papers that discuss the effect of calcium on modification, such as Abdollahi *et al* [9, 12], despite the very many papers on the effects of sodium and strontium [3–11]. As a fundamental study to investigate the growth mechanism of the modification, we concentrate below on describing the change in silicon morphology due to the addition of strontium, strontium plus calcium, or just calcium, using the self-sealing unidirectional solidification method [13]. The self-sealing method is essential in order to prevent the strontium and calcium from oxidizing and evaporating during the experiments.

Gunduz and Hunt [14] have calculated the interfacial energy of Al–Si alloys by measuring the curvature of the solid–liquid interface, but commented that controlling the temperature was difficult. The interfacial energy of the Si/Al–Si melt has also been calculated using the dihedral angle method [15] and the sessile drop method [16]. These results show that the interfacial energy of the Si/Al–Si melt decreases by approximately 10–20% with the addition of strontium to the Al–Si melt. In order to explain these experimental results, we discuss below the equilibrated interfacial morphology of silicon at the solid–liquid interface for Al–Si melts to which strontium and calcium have been added.

Solidification rate of equiaxed grains

In order to discuss the effect of the solidification rate, R, on silicon modification, it is necessary to estimate the solidification rate of equiaxed grains. Very slow solidification rates of $0.5–16\,\text{mm}\,\text{h}^{-1}$ are used in order to correspond to typical macroscopic and microscopic solidification rates. Consider a plate casting of thickness 30 mm. If the solidification time is 10–15 min, the macroscopic solidification rate is $15\,\text{mm}\,(10–15\,\text{min})^{-1}$, which is approximately $75\,\text{mm}\,\text{h}^{-1}$. If we think about the microscopic solidification rate, however, solidification is complete when each eutectic cell has solidified. If a cell has diameter 2 mm and also solidifies in 10–15 min, then the solidification rate is $4–6\,\text{mm}\,\text{h}^{-1}$. Therefore, the range of solidification rates $0.5–16\,\text{mm}\,\text{h}^{-1}$ was used in the unidirectional solidification experiments.

Self-sealing unidirectional solidification

Al–Si–Sr and Al–Si–Ca alloy samples were prepared using 99.99 wt% Al, 99.999 wt% Si, Al–9.61 wt% Sr alloy and 99.8 wt% Ca. The calcium was supplied encapsulated in an argon-filled glass ampoule and was used immediately after breaking the ampoule. The sample alloys were melted in a 98 wt% Al_2O_3 crucible under an argon atmosphere and cast in a 4 mm inner diameter graphite mould. Table 19.1 shows the chemical compositions of the samples.

Unidirectional solidification was conducted by moving a stage equipped with an electric furnace at speeds set at six points between 0.5 and $16\,\text{mm}\,\text{h}^{-1}$. The temperature gradient at the solid–liquid interface was about $3.2\,\text{K}\,\text{mm}^{-1}$. Samples were inserted into 5 mm inner diameter 99.9 wt% Al_2O_3 crucibles. To prevent strontium and calcium from oxidizing and evaporating during melting, holding and solidification, a self-sealing method [13] was used, shown in figure 19.1, in which the melt in both the upper and lower parts was divided by a high-purity Al_2O_3 partition plate. Using this technique minimizes the losses of strontium and calcium. Samples were unidirectionally solidified to a length of 30 mm and then quenched in chilled water.

Table 19.1. Chemical compositions of Al–Si–Sr and Al–Si–Ca samples (wt%).

	Si	Sr	Mg	Ca	Fe	P
Al–Si	12.9	<0.001	<0.001	<0.001	<0.005	<0.001
0.004Sr	12.6	0.004	↓	↓	↓	↓
0.07Sr	13.1	0.069	↓	↓	↓	↓
0.34Sr	12.8	0.342	↓	↓	↓	↓
0.005Ca	12.8	<0.001	↓	0.0054	↓	↓
0.08Ca	12.9	↓	↓	0.08	↓	↓
0.32Ca	12.9	↓	↓	0.32	↓	↓

The structures of the samples, after etching using 0.1% hydrofluoric acid, were observed with an optical microscope, and samples, slightly etched using an NaOH aqua solution, were observed with a scanning electron microscope. The mean silicon interphase spacing was measured using the micrographs from an optical microscope. In some experiments, in order to observe the equilibrated interfacial morphology at the solid–liquid interface, the movement of the furnace was stopped and held for 5 h and then the samples were quenched in chilled water. In order to observe the continuity and discontinuity of silicon at the solid–liquid interface, the samples were deeply etched using NaOH to dissolve the aluminium matrix and allow extraction of the silicon phases from the samples. The silicon morphology was then observed in a scanning electron microscope.

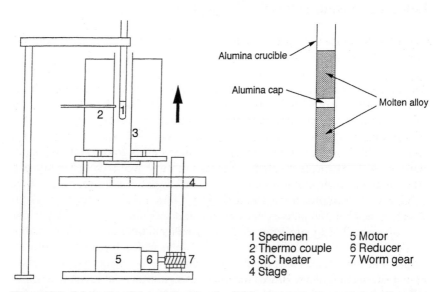

Figure 19.1. Schematic of the unidirectional solidification apparatus and the self-sealing method.

Table 19.2. Chemical compositions of Al–Si–Sr–Ca samples (wt%).

	Si	Sr	Mg	Ca	Fe	P
Al–Si–Sr	13.1	0.069	<0.005	<0.001	<0.005	<0.001
Al–Si–Sr*	14.2	0.058	<0.005	0.02	0.127	–
Sr–0.01Ca	14.4	0.06	<0.001	0.01	<0.01	<0.005
Sr–0.06Ca	14.4	0.06	↓	0.06	↓	↓
Sr–0.15Ca	14.3	0.05	↓	0.15	↓	↓
Sr–0.22Ca	14.4	0.06	↓	0.22	↓	↓

* Commercial grade.

To examine the interaction between strontium and calcium during silicon modification, a commercial grade Al–Si–Sr alloy and five Al–Si–Sr alloys that differed only in the calcium content were prepared by a method identical to that mentioned above, and cast in a graphite mould. The chemical compositions of the samples are shown in table 19.2. The strontium contents of these samples were nearly identical and they differed only in calcium content. These samples were remelted in 99.8 wt% Al_2O_3 and mullite Tanmann tubes using the self-sealing method, and solidified at cooling rates in the range 6–0.025 K min^{-1}, in the temperature range 1023–823 K. The morphology of the silicon was observed using an optical microscope.

Influence of strontium

The morphological changes in silicon as a function of strontium addition and solidification rate are shown in figure 19.2. In binary Al–Si, when the solidification rate is less than 2 mm h^{-1}, the silicon is aligned. However, when the solidification rate is greater than 4 mm h^{-1}, the silicon is disordered. With addition of 0.07 wt% strontium, as the solidification rate decreases below 2 mm h^{-1}, the silicon is again disordered and coarse compared with that in binary Al–Si. However, the silicon is modified and the shape of the solid–liquid interface becomes smooth when the solidification rate is greater than 4 mm h^{-1}. This means that strontium in Al–Si alloys has two kinds of effect, namely coarsening and modification. The solidification rate of the equiaxed grains in normal castings must be greater than 4 mm h^{-1} as otherwise coarse silicon will be formed in such castings.

The morphology of silicon in Al–Si–Sr alloys changes from coarse to fine at a solidification rate of 2–4 mm h^{-1}. The silicon is coarsened with strontium addition at low solidification rates, but is modified with strontium addition at high solidification rates. Based on these experimental results, Song *et al* [17] summarized the influence of strontium on the morphology of silicon in Al–Si alloys as shown in figure 19.3. In this figure, the range of normal silicon

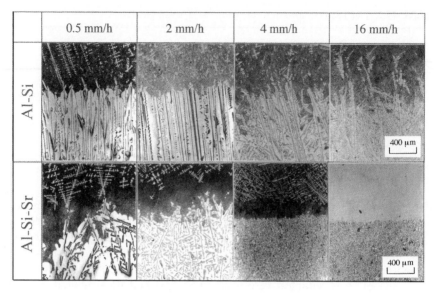

Figure 19.2. Influence of strontium on the structure of unidirectionally solidified Al–Si alloys at the solid–liquid interface.

becomes narrow with increasing strontium content, and disappears at 0.34 wt% Sr. This indicates that the transition in silicon morphology from coarsened to modified is similar to the graphite morphology change in cast iron [18, 19] and may explain why such a low solidification rate is needed to see the transition and the coarsening phenomenon.

The influence of solidification rate on the silicon interphase spacing, λ, is summarized for Al–Si and Al–Si–0.07 wt% Sr in figure 19.4. As can again be

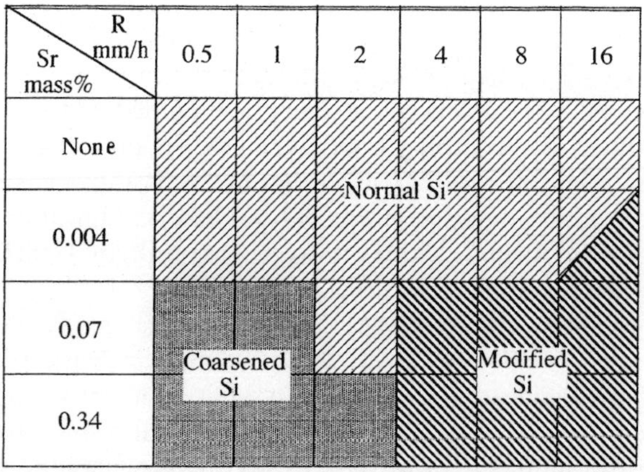

Figure 19.3. Influence of strontium content and solidification rate on silicon morphology.

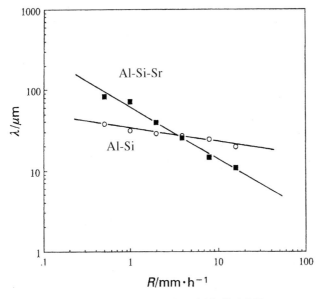

Figure 19.4. Silicon interphase spacing in Al–Si and Al–Si–0.07Sr.

seen clearly, the critical solidification rate is 4 mm h^{-1}, and at higher rates the silicon interphase spacing after modification with strontium becomes less than that in binary Al–Si. This value of the critical solidification rate is in the range of the solidification rates which lead to the formation of equiaxed

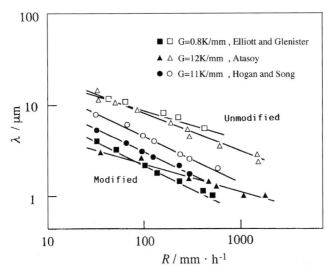

Figure 19.5. Interphase spacing reported by Hogan and Song [20], Elliott and Glenister [21] and Glenister and Elliott [23].

grains. Most of the previous published results [20–23] which describe the influence of solidification rate on interphase spacing are summarized in figure 19.5. Due to experimental difficulty, previous researchers have nearly always used much higher solidification rates for Al–Si–Sr than that required for equiaxed grains.

The solidified structure in front of the solid–liquid interface shows many α-dendrites in plain Al–Si. In the case of modified Al–Si–Sr, however, there is a dendrite-free zone and its width increases with solidification rate. The chemical composition of the dendritic free zone has become hypereutectic, indicating a difficulty of silicon growth in strontium-containing alloys, and a corresponding increase in the undercooling for solidification [21, 23, 24].

Influence of calcium

The influence of solidification rate on the morphological changes in silicon for Al–Si containing 0.08 wt% calcium is shown in figure 19.6. There are many papers which describe the weak modification effect of calcium, for example Abdollahi *et al* [12, 24], but this is not really confirmed in figure 19.6. The influence of solidification rate on the silicon interphase spacing, λ, is summarized for Al–Si and Al–Si–Ca alloys in figure 19.7. The critical solidification rate is more than $20\,\text{mm}\,\text{h}^{-1}$, much higher than that of the equiaxed grains, except in the final stages of solidification, such as at grain boundaries. Thus, unidirectional solidification experiments do not show any modification effect by calcium.

Silicon morphology at the solid–liquid interface

Scanning electron micrographs of solid–liquid interfaces are shown in figure 19.8 for Al–Si and Al–Si–Sr alloys solidified at a rate of $0.5\,\text{mm}\,\text{h}^{-1}$. The silicon in the Al–Si alloy is almost covered with α-aluminium except for a liquid channel at the tip of the silicon. The silicon in the 0.004 wt% strontium

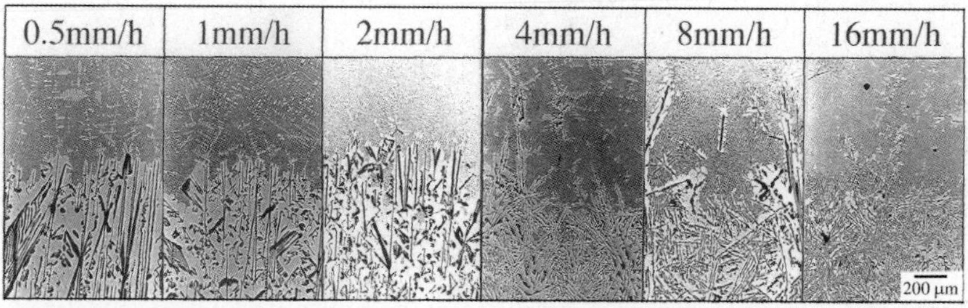

Figure 19.6. Influence of solidification rate on silicon morphology of Al–Si–0.08Ca.

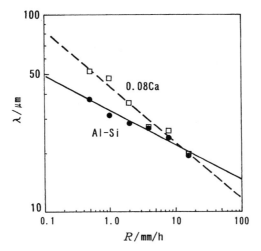

Figure 19.7. Influence of calcium content and solidification rate on silicon interphase spacing.

alloy is also covered by α-aluminium similar to the Al–Si sample, but the quenched silicon from the liquid phase is modified. In the case of the 0.07 wt% strontium sample, the silicon clearly protrudes into the liquid phase and the quenched silicon phase is modified. Moreover, most of the quenched silicon does not nucleate and grow from the protruded silicon, solidified at a rate of $0.5\,\mathrm{mm\,h^{-1}}$, as shown in figure 19.9. However, for samples solidified at rates of 4 and $16\,\mathrm{mm\,h^{-1}}$, the quenched silicon was

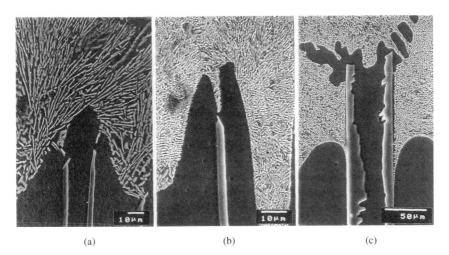

Figure 19.8. Influence of strontium content on interfacial morphology of samples solidified at $0.5\,\mathrm{mm\,h^{-1}}$. (a) Al–Si, (b) 0.004% strontium, (c) 0.07% strontium.

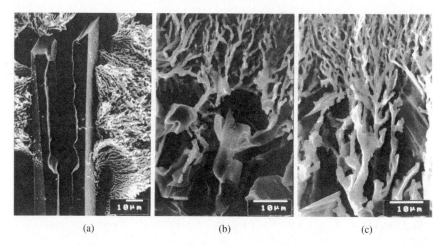

Figure 19.9. Influence of solidification rate on silicon morphology at solid–liquid interface. Al–Si–0.07Sr, (a) 0.5 mm h^{-1}, (b) 4 mm h^{-1}, (c) 16 mm h^{-1}.

nucleated and branched from the solidified silicon. This means that the silicon also protrudes into the liquid without an α-aluminium cover. Based on the observations in figures 19.8 and 19.9, we can postulate that the interfacial energy, γ, changes from $(\gamma_{Si/L} > \gamma_{\alpha/L})$ to $(\gamma_{Si/L} < \gamma_{\alpha/L})$, or from $(\gamma_{Si/L} > \gamma_{\alpha/L} + \gamma_{\alpha/Si})$ to $(\gamma_{Si/L} + \gamma_{\alpha/Si} < \gamma_{\alpha/L})$ by adding strontium. These phenomena are caused by a change in the silicon–liquid interfacial energy.

In the case of the 0.08 wt% calcium sample solidified at a rate of 0.5 mm h^{-1}, the silicon phase at the solid–liquid interface is covered with α-aluminium except for a liquid channel at the tip of the silicon, as shown in figure 19.10. This morphology is the same as that observed in binary Al–Si, as shown in figure 19.8. Nevertheless, the quenched silicon is modified

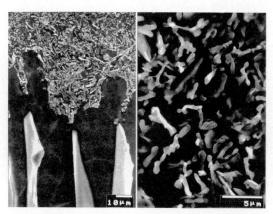

Figure 19.10. Influence of calcium on silicon morphology of samples solidified at 0.5 mm h^{-1}. (Al–Si–0.08Ca.)

due to the increased solidification rate. The solidification rate in the quenched region must be greater than 20 mm h^{-1}. This phenomenon agrees with the modification effect of calcium, as already reported [12, 24].

Equilibrated silicon morphology at the solid–liquid interface

To confirm a change in the interfacial energy with strontium and calcium additions, we can observe the equilibrated silicon morphology by stopping the movement of the furnace and allowing the silicon–liquid interface to equilibrate for 5 h. The changes in the interfacial morphology are shown in figure 19.11. In the case of the binary Al–Si, the silicon–liquid interface is completely covered with α-aluminium. The silicon in the 0.07 wt% strontium alloy clearly protrudes into the liquid phase. This means that the interfacial energies between the α-aluminium phase, silicon and liquid change with the addition of strontium as described in the previous section.

In the case of calcium addition, the morphology of the silicon–liquid interface is very similar to that with the strontium addition. This result shows that calcium is a surface active element similar to strontium, when the quenched silicon is modified with calcium addition as shown in figure 19.10. Calcium does not generally have a strong modification effect, because of the higher critical solidification rate as already described in an earlier section. We have reported [1, 17] that the modification mechanism is a branching of silicon to adjust the mass balance at the solid–liquid interface to maintain the solidification rate despite obstruction by adsorption of strontium on silicon which decreases its interfacial energy. The strontium adsorption layer controls the silicon growth, and the critical solidification rate is therefore dependent on the strontium content as shown in figure 19.3.

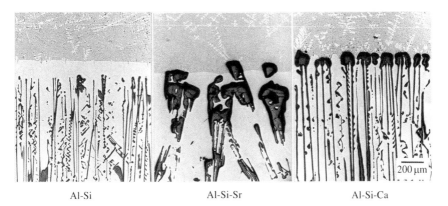

Figure 19.11. Influence of strontium or calcium on the equilibrated interfacial structure. (Solidified at 0.5 mm h^{-1} and held for 5 h.)

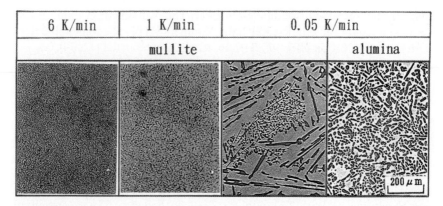

Figure 19.12. Influence of cooling rate and tube material on the morphology of silicon in commercial grade Al–Si–0.058Sr.

Critical cooling rate

The influence of cooling rate and tube materials on the silicon morphology for a commercial grade Al–Si–Sr alloy is shown in figure 19.12. The silicon is completely modified when solidified at a cooling rate of 0.05 K min^{-1} in an alumina tube, but is not modified when solidified at the same cooling rate in a mullite tube. This difference is due to an interaction between the tube and the melt [2, 25], with the strontium content decreased due to chemical reaction with silica in the tube. The silicon morphology in *commercial grade* Al–Si–Sr is not modified when solidified at a cooling rate of 0.025 K min^{-1} [2, 25] in an alumina tube. Nevertheless, the silicon morphology in *laboratory-prepared* Al–Si–Sr is completely modified at a cooling rate of 0.025 K min^{-1} in an alumina tube as shown in figure 19.13. This means that the critical cooling rate for the commercial grade Al–Si–Sr alloy is 0.05 K min^{-1}, and less than 0.025 K min^{-1} for the purer, laboratory prepared Al–Si–Sr alloy.

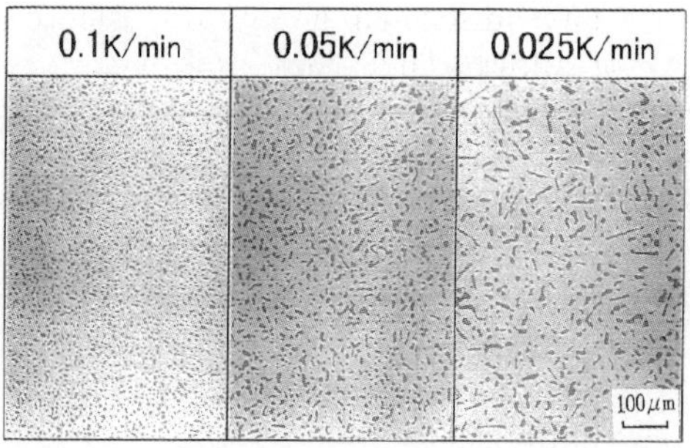

Figure 19.13. Morphology of silicon in Al–Si–0.069Sr solidified at 0.1, 0.05 and 0.025 K min^{-1} in alumina tubes.

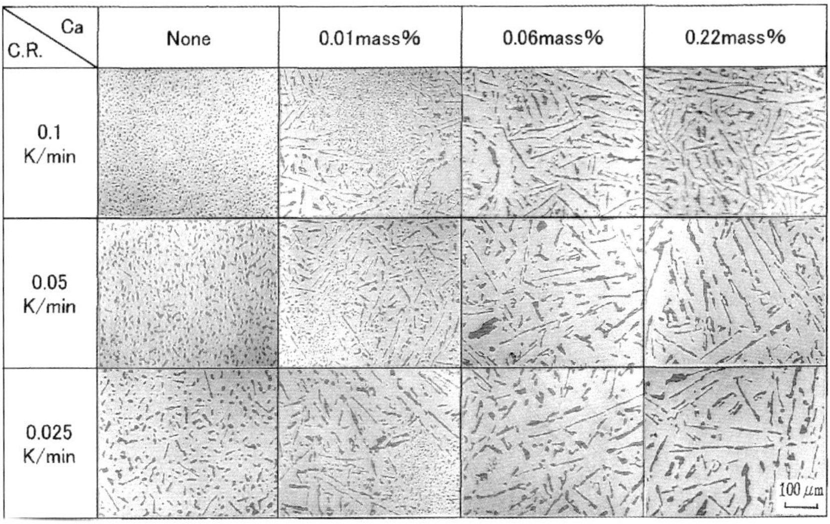

Figure 19.14. Influence of calcium content and cooling rate on silicon morphology for Al–Si–Sr–Ca alloys.

Commercial grade Al–Si–Sr contains a small amount of calcium, approximately 0.02 wt%, allowing investigation of the interaction between strontium and calcium in the melt as shown in table 19.2 and figure 19.14. The results obtained are in close agreement with those of the commercial grade Al–Si–Sr alloy for the silicon modification. This means that if the Al–Si–Sr alloys contain a small amount of calcium, the calcium nullifies the modification effect of strontium, perhaps due to the formation of a Ca–Sr silicide [2].

Summary

In this chapter, we have discussed growth aspects of the modification mechanisms in Al–Si alloys. The change in silicon morphology at the solid–liquid interface by the addition of strontium and calcium has been investigated using the self-sealing unidirectional solidification method. The modification effects of strontium and calcium are confirmed. When the solidification rate is greater than $4\,\text{mm}\,\text{h}^{-1}$, the eutectic silicon is modified in the strontium-containing samples. However, when the solidification rate is less than $2\,\text{mm}\,\text{h}^{-1}$, the eutectic silicon is not modified but is coarse. We have therefore defined the critical solidification rate as $4\,\text{mm}\,\text{h}^{-1}$ for the strontium-containing alloys. In the case of calcium additions, however, the critical solidification rate is greater than $20\,\text{mm}\,\text{h}^{-1}$. Therefore, the modification effect of calcium is very weak compared with that of strontium. The interfacial energy, $\gamma_{Si/L}$, between silicon and an Al–Si melt is decreased by the adsorption of strontium and calcium on silicon. These phenomena are confirmed by the observation of the equilibrated morphology of silicon at the solid–liquid interface. The modification mechanism is the branching of silicon to adjust the mass balance at the solid–liquid interface, because the silicon protrudes into the melt due to the adsorption of strontium to decrease its interfacial energy. The adsorption controls the silica growth. Therefore, the critical solidification rate depends on the strontium content as shown in figure 19.3. When strontium and calcium coexist in an Al–Si alloy, the strontium and calcium react to form a Ca–Sr silicide. The effect of modification of these elements is then nullified.

References

[1] Nakae H and Kanamori H 1997 *Proceedings of SP97* **7** 477-480
[2] Nakae H, Song K and Kanamori H 1995 *Proceedings 3rd AFC* 94–99
[3] Honma U 1969 *J. Japan. Inst. Light Met.* **19** 72
[4] Davies V de L and West J M 1963–64 *J. Inst. Met.* **92** 175–180
[5] Kobayashi K F and Hogan L M 1985 *J. Mater. Sci.* **20** 1961–1975
[6] Ransley C E and Neufeld H 1950 *J. Inst. Met.* **78** 26
[7] Day M G and Hellawell A 1968 *Proc. Roy. Soc.* **A305** 473–491
[8] Lu S-Z and Hellawell A 1985 *J. Cryst. Growth* **73** 316–328
[9] Lu S-Z and Hellawell A 1987 *Metall. Trans.* **18A** 1721–1733
[10] Plumb R C and Lewis J E 1957–58 *J. Inst. Met.* **86** 393
[11] Tiller W A 1958 *Liquid Metals and Solidification* (Cleveland, OH: ASM) p 276
[12] Abdollahi A and Gruzeleski J E 1998 *Int. J. Cast Met. Res.* **11** 145–155
[13] Nakae H, Song K and Fujii H 1992 *Mater. Trans. JIM* **11** 1051
[14] Gunduz M and Hunt J D 1985 *Acta Metall.* **33** 1651
[15] Song K, Kikuchi T, Yoshida M and Nakae H 1994 *J. Japan. Inst. Met.* **58** 1454
[16] Nakae H 1998 *Tetsu-To-Hagané* **84** 19–24
[17] Song K, Fujii H, Nakae H and Yamaura H 1993 *J. Japan. Inst. Light Met.* **43** 485
[18] Fredriksson H and Remaeus B 1975 'The metallurgy of cast iron' *Proceedings of the 2nd International Symposium on the Metallurgy of Cast Iron* (Georgi) p 315
[19] Nakae H and Shin H 1999 *Int. J. Cast. Met. Res.* **11** 345–350
[20] Hogan L M and Song H 1987 *Metall. Trans.* **18A** 707–713
[21] Elliott R and Glenister S M D 1980 *Acta Metall.* **28** 1489–1494
[22] Glenister S M D and Elliott R 1981 *Met. Sci. J.* **15** 181
[23] Atasoy O A 1984 *Aluminium* **60** 275
[24] Kim C B and Heine R W 1963–64 *J. Inst. Met.* **92** 367–376
[25] Song K, Park S and Nakae H 1994 *IMONO* **66** 822–826

SECTION 4

NEW MATERIALS AND PROCESSES

The ever-increasing demand for new products and new technology at reduced cost puts increasing requirements on conventional materials and processes. Improvements in all properties, be they mechanical, thermal, electrical, surface or the ability to withstand environmental degradation, are constantly in demand. In many cases these requirements can only be met by the introduction of new classes of materials with superior properties; in others, a change in the processing route used to manufacture more conventional materials can provide the required improvement in performance. This section describes a selection of recent developments in new materials and processes.

Chapter 20 describes the use of semi-solid processing to manufacture very fine-grained intermetallic alloys and composites. Chapter 21 discusses the effects of processing parameters on the ability to form bulk metallic glasses that have a range of novel properties. Chapter 22 describes the use of unidirectional solidification processing to improve the high temperature strengths and stabilities of eutectic ceramics, and Chapters 23 and 24 discuss the solidification processing of oxides containing superconducting phases.

Chapter 20

Rheocasting of TiAl alloys and composites

Kiyoshi Ichikawa and Yoshiji Kinoshita

Introduction

Semi-solid processing can be achieved by rotational stirring of an alloy during solidification, and is well-known as a promising process for the next generation of continuous casting, as it helps to homogenize the microstructure throughout the ingot [1]. Initial work on the processing of semi-solid alloys included the rheocasting process proposed by Flemings [2], and the casting process proposed by Hagiwara and Takahashi [3], in which the semi-solid zone at the solidification front is cyclically sheared. The main aims are the refinement of crystal grains and the elimination of segregation in an ingot. Semi-solid processes are often useful in controlling the macrostructure and decreasing the macrosegregation throughout the ingot, although irregular crystal forms such as rosette-shaped primary crystals are seen in the microstructures because of insufficient agitation during solidification [2, 3]. Ichikawa and co-workers have tried to homogenize and further refine the crystal grains [4–13], and produce highly ductile materials [5–12], by increasing both the stirrer rotation speed of and the cooling rate of the semi-solid alloy during rheocasting.

A national research and development programme, 'High-Performance Materials for Severe Environments', has been carried out in Japan in order to develop intermetallic compounds such as titanium aluminide, TiAl, and niobium aluminide, NbAl, for structural applications by improving their ductility at room temperature and strength at elevated temperatures. This chapter describes the use of a modified rheocasting process with high rotation speeds to manufacture intermetallic TiAl alloys and composites. Remarkable grain refinement and superior mechanical properties have been achieved in Ti–44at% Al and Ti–44at% Al–10vol% ZrC alloys.

The rheocasting process

A schematic diagram of the high-speed rheocasting apparatus used to manufacture intermetallic TiAl alloys and composites is shown in figure 20.1. Basically, the apparatus comprises an isothermal furnace in a vacuum or an inert gas atmosphere which holds the TiAl or composite sample, and a stirrer. A uniformly fine-grained microstructure is achieved by preventing formation of the normal lamellar structure and breaking up the primary crystals as the alloy solidifies by stirring at high speeds of up to $84\,\text{s}^{-1}$ (5000 rpm). After stirring the alloy for a given time, it is then transferred into a high-frequency furnace.

In more detail, the apparatus comprises a vacuum chamber with a hinged door. The vacuum chamber is segmented. In the top segment, the chamber holds a motor which rotates the stirrer that agitates the solidifying alloy or composite. In the middle segment, the isothermal chamber contains two concentric crucibles which hold the solidifying alloy with or without the

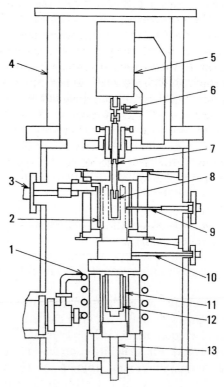

1, High-frequency induction coil. 2, Graphite heater. 3, Electrode. 4, Chamber. 5, Rotation motor. 6, Rotation detector. 7, Rotation axis. 8, Stirrer. 9, Thermocouple for temperature control. 10, Shutter. 11, Graphite crucible. 12, Calcia crucible. 13, Crucible support.

Figure 20.1. Experimental apparatus for vacuum rheocasting.

addition of ceramic particles to be stirred. In the bottom segment, the chamber contains a high-frequency furnace for melting the raw materials. The chamber is connected to a vacuum pump and a gas inlet to introduce an inert gas. The output shaft of the motor at the top of the chamber is supported by a bearing on a water-cooled mount on a partition between the upper and middle segments of the chamber. A rotation sensor is attached to the output shaft of the motor in order to measure the rotation speed of the stirrer. The apparatus was equipped with a 2 kW motor having the capacity to rotate the stirrer at a speed of up to $84\,\text{s}^{-1}$. If the motor is placed outside the chamber, its output shaft must be passed through the chamber wall in order to rotate the stirrer inserted in the molten alloy. For such a construction, it is difficult adequately to seal the hole through which the output shaft passes into the vacuum chamber. Our construction, whereby the motor is in the upper segment of the chamber itself, solves this problem.

The isothermal furnace is required to control the cooling rate of the alloy in order to produce a semi-solidified slurry with the desired solid fraction of not less than 0.5. The isothermal furnace is covered with heat insulators which contain a graphite heater. The heat insulator at the top of the isothermal furnace has a hole through which the output shaft of the motor passed into the centre of the furnace. The stirrer is attached to the bottom end of the output shaft in the isothermal furnace and is thus rotated by the motor. An illustration of this is shown in figure 20.2.

The bottom segment of the chamber contains the high-frequency furnace which heats the intermetallic alloy charge in two concentric crucibles, and whose side is covered with a heat insulator. A high-frequency coil is wound around the insulator. The crucible consists of an inner crucible of calcia and an outer crucible of graphite. A crucible cover that can be opened and closed from the outside of the chamber is mounted on the high-frequency furnace. Under the high-frequency furnace there is a lifting mechanism that selectively moves the crucible between the high-frequency furnace and the isothermal furnace. With the crucible cover kept in the

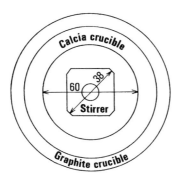

Figure 20.2. Transverse sections (mm) of crucible and stirrer used in the present work.

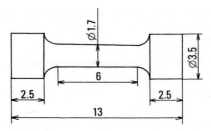

Figure 20.3. Size (mm) and shape of tensile test piece used in the present work.

open position, the lifting mechanism moves the crucible up into the isothermal furnace. The inner crucible has an inner diameter of 60 mm, an outer diameter of 83 mm and a depth of 130 mm. The stirrer, made of 99.7% purity titanium, has a downwardly tapering eight-sided cross-section: an upper cross-section of major axis 38 mm and minor axis 30 mm, and a lower cross-section of major axis 32 mm and minor axis 25 mm. The lateral dimensions are shown in figure 20.3. A new 99.7% Ti stirrer was used for each rheocasting experiment. Thus, even if the stirrer was eroded during the rheocasting process, the intermetallic alloy remained uncontaminated, with a composition which could be appropriately controlled by adjustment of the charge compositions. The actual temperature of the charge was derived from the known relationship between the temperature measured at the centre of the outer wall of the outer graphite crucible with a radiation pyrometer and the temperature determined by the direct observation of the charge through the peephole, using a radiation pyrometer.

Rheocast alloys

To prepare the intermetallic TiAl alloys and the necessary composites, mixtures of the elemental materials were placed in the crucible. Granular sponge titanium of not lower than 99.5% purity, small aluminium pieces of 99.99% purity, fragments of chromium, manganese, vanadium, silicon and niobium of not less than 99% purity, and 500 g of ceramic particles were used to prepare Ti–44, 49 and 54 at% Al binary alloys, Ti–44, 46 and 48 at% Al–2% Cr, Mn or V ternary alloys, Ti–44, 46 and 48 at% Al–2% Cr–2% Si or Nb quaternary alloys, and Ti–44, 46 and 49 at% Al–0.5–15 vol% TiC, ZrC, TaC, NbC, VC, TaB_2 and TiN alloy composites.

After placing the crucible containing the mixed raw materials in the high-frequency furnace, the vacuum chamber was evacuated to 1×10^{-3} MPa or below. Argon gas was then introduced until the argon gas pressure in the chamber reached the desired level of 0.1 MPa. The raw metallic elements were readily melted in a short time in the argon gas atmosphere by the high-frequency heating. At this point, the crucible cover above the high-frequency furnace was opened in order to confirm that the metallic charge had melted. The lifting mechanism was then actuated to raise the crucible containing the charge, consisting of a mixture of molten titanium aluminide with or without solid ceramic particles, into the isothermal furnace where the stirrer was inserted into the charge. The ascent of the crucible was stopped automatically when the bottom end of the stirrer reached a point of 10 mm above the inner floor of the crucible.

The charge was continuously cooled under conditions of no heating of the isothermal furnace to achieve the maximum cooling rate of the charge. The stirrer which had begun to rotate immediately after insertion into the molten charge, continued to rotate for a maximum period of 30 s, and was then removed from the semi-solid alloy. The rotation speed of the stirrer was relatively low, $15\,\mathrm{s}^{-1}$, until the start of solidification, and was then gradually increased at a steady rate to prevent spatter of the solidifying charge out of the crucible. By continuing the agitation at a constant high speed, the primary crystals which formed in the molten charge were broken up, and a fine-grained microstructure was obtained. The constant rotation speed of the stirrer during solidification varied from 15 to $70\,\mathrm{s}^{-1}$. After continuing the constant-speed agitation for a given period of time, which gave a solid fraction greater than 0.5, the lifting mechanism was actuated to lower the crucible into the high-frequency furnace, thereby preventing deposition and solidification of the semi-solid charge on the stirrer.

Microscopic observation was made of the top, middle and bottom transverse sections cut from ingots rheocast or solidified without stirring. Tensile test pieces were also machined from the ingots as illustrated in figure 20.3. Tensile tests were carried out in air at room temperature and in an argon atmosphere at elevated temperatures using a strain rate of $1.1 \times 10^{-3}\,\mathrm{s}^{-1}$ and room temperature, 1073, 1173 and 1273 K for the intermetallic TiAl alloys and strain rates of $1.1 \times 10^{-4}\,\mathrm{s}^{-1}$ at room temperature and $1.1 \times 10^{-3}\,\mathrm{s}^{-1}$ at 1373 K in the intermetallic TiAl alloy composites.

Rheocast Ti–44% Al

The microstructures of intermetallic TiAl alloys prepared by rheocasting at high stirrer speeds were observed by scanning electron microscopy. For the purpose of comparison, reference intermetallic TiAl alloys were prepared by allowing the molten charge to solidify spontaneously in the isothermal furnace without using the stirrer. As an example of the effect of rheocasting, the microstructures of Ti–44% Al alloys solidified without stirring and rheocast in an argon gas atmosphere at 0.1 MPa are shown in figure 20.4. Figure 20.4(a) shows a micrograph of the microstructure of a Ti–44% Al alloy melted and solidified naturally without stirring. Conventional lamellar structures of alternating TiAl and Ti$_3$Al are observed. The formation of this

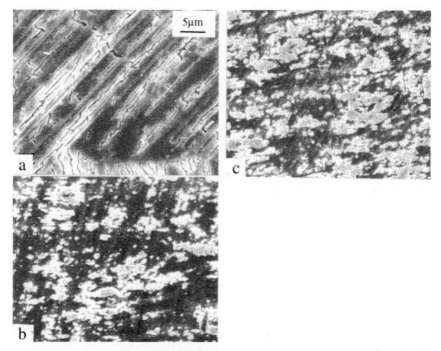

Figure 20.4. Microstructures of Ti–44% Al ingots (a) conventionally solidified without stirring, (b) rheocast at a rotation speed of $15\,\text{s}^{-1}$, and (c) rheocast at a rotation speed of $70\,\text{s}^{-1}$, under an argon pressure of 0.1 MPa.

conventional lamellar structure is via a solid-state transformation at the eutectoid temperature, and can cause cracking of the material.

Figure 20.4(b) shows a micrograph of the microstructure of a Ti–44% Al alloy melted and allowed to solidify while being agitated with a stirrer rotating at a speed of $15\,\text{s}^{-1}$ for a period of 30 s. The lamellar structure seen in the alloy solidified without stirring is destroyed completely, and an extremely refined microstructure containing dispersed aggregates of flaky crystal grains is formed instead. The average size of the aggregates is $6.6 \pm 3.3\,\mu\text{m}$ and $2.3 \pm 1.0\,\mu\text{m}$ along their major and minor axes, respectively. The average size of individual crystal grains in these aggregates is $1.6 \pm 1.4\,\mu\text{m}$ and $0.7 \pm 0.5\,\mu\text{m}$ in their major and minor axes, respectively. Figure 20.4(c) shows a micrograph of the microstructure of a Ti–44% Al alloy melted and allowed to solidify while being agitated with a stirrer rotating at a faster speed of $70\,\text{s}^{-1}$ for a period of 21 s, in which an extremely refined microstructure with dispersed aggregates composed of flaky crystal grains was again formed. The average size of these aggregates is $6.4 \pm 3.7\,\mu\text{m}$ and $2.0 \pm 1.1\,\mu\text{m}$ in the major and minor axes, respectively.

The normal lamellar structure as seen in the Ti–44% Al alloy solidified without stirring is destroyed completely and an extremely refined microstructure is formed in the rheocast Ti–44% Al alloy. The shape of crystal grains in the rheocast Ti–44% Al alloy was flaky and not as spherical as those seen in an intermetallic CuAl alloy produced by the stirring synthesis process [15]. In the Ti–44% Al alloy shown in figure 20.4(b), the observed fine-grained microstructure forms as a result of the complete break-up of the lamellar structure even at low speed. Figure 20.4(c) also shows a refined microstructure from crystals through high-speed stirring for a shorter period. These extremely refined microstructures with dispersed aggregates are a special morphology which cannot be formed by conventional casting or powder metallurgy processes. The special morphology of dispersed aggregates has a strong effect on the improvement of mechanical properties in Ti–44% Al alloy, as described later in this chapter.

Other rheocast TiAl alloys

Intermetallic TiAl alloys which usually undergo a solid state transformation into a lamellar structure at the eutectoid temperature are in the range 44–54 at% Al. The alloys Ti–49% Al and Ti–54% Al were therefore selected in order to investigate the effect of the chemical composition of the intermetallic TiAl alloy on the destruction of a lamellar structure by rotational stirring. First of all, two reference Ti–49% Al and Ti–54% Al ingots were prepared by allowing the molten alloys to solidify in the isothermal furnace without stirring. Unstirred Ti–49% Al exhibited a coarse lamellar microstructure containing many cavities and other casting defects, while the unstirred Ti–54%Al exhibited a lamellar substructure both in and between dendritic crystals.

The microstructures of rheocast Ti–49% Al and Ti–54% Al are shown in figure 20.5. Figures 20.5(a) and (b) show micrographs of the microstructure of Ti–49% Al and Ti–54% Al prepared by rheocasting with stirrer

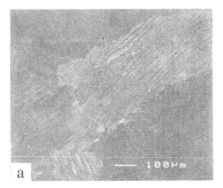 

Figure 20.5. Microstructures of (a) Ti–49% Al rheocast at a rotation speed of $33\,\text{s}^{-1}$, and (b) Ti–54% Al rheocast at a rotation speed of $37\,\text{s}^{-1}$, under an argon pressure of 0.1 MPa.

speeds of 33 and 37 s^{-1} respectively. Solidification was complete in both cases 20 s after insertion of the stirrer. Rheocast Ti–49% Al exhibited a dense lamellar microstructure, while rheocast Ti–54% Al exhibited lamellar substructures both in and between spheroidal crystals, which resulted from the break-up of dendritic crystals by the mechanical stirring [14]. Similarly, lamellar structures could not be destroyed in the ternary and quaternary TiAl alloys by the rheocasting process.

Rheocast TiAl composites

Optical microstructures of rheocast Ti–44% Al–ZrC alloy composites are shown in figure 20.6. Figures 20.6(a), (b) and (c) show the microstructures of Ti–44% Al composite ingots with 2, 5 and 10 vol% ZrC respectively, allowed to solidify while being agitated with stirrer speeds of 43–50 s^{-1}, for periods of 20 s in each case. For the composites containing 5 and 10 vol% ZrC, zirconium-rich lamellar grains and titanium-rich and carbon-rich precipitates were formed in the lamellar grains, with more being formed in the 10 vol% ZrC-containing composite. The decomposition of zirconium carbide particles and the formation of titanium-rich and carbon-rich precipitates will be explained later in this chapter.

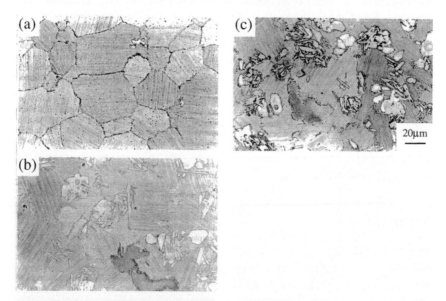

Figure 20.6. Microstructures of (a) Ti–44% Al–2 vol% ZrC rheocast at a rotation speed of 45 s^{-1}, (b) Ti–44% Al–5 vol% ZrC rheocast at a rotation speed of 43 s^{-1}, and (c) Ti–44% Al–10 vol% ZrC rheocast at a rotation speed of 50 s^{-1}, under an argon pressure of 0.1 MPa.

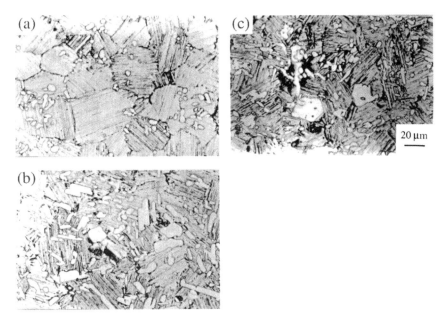

Figure 20.7. Microstructures of (a) Ti–44% Al–10 vol% TiC rheocast at a rotation speed of $50\,s^{-1}$, (b) Ti–46% Al–10 vol% TiC rheocast at a rotation speed of $44\,s^{-1}$ and (c) Ti–49% Al–10 vol% TiC rheocast at a rotation speed of $50\,s^{-1}$, under an argon pressure of 0.1 MPa.

Optical microstructures of rheocast TiAl–10 vol% TiC alloy composites are shown in figure 20.7. Figures 20.7(a), (b) and (c) show microstructures of TiAl–10 vol% TiC alloy composite ingots, containing 44, 46 and 49 at% Al respectively, allowed to solidify while being agitated with stirrer speeds in the range $44–50\,s^{-1}$ for period in the range 22–25 s. For each of the 44, 46 and 49 at% Al-containing composites, spheroidal titanium-rich and carbon-rich precipitates were formed in the lamellar grains, with more being formed in the 46 at% Al-containing composite than either the 44 or 49 at% Al-containing composite.

Microanalysis

The distributions of titanium and aluminium were measured in rheocast and unstirred intermetallic TiAl alloys and composites and microanalysis in the scanning electron microscope on transverse sections from the top, middle and bottom of ingots. Secondary electron and X-ray images, using an electron probe microanalyser beam of 50 nm in diameter, from the middle section of a Ti–44% Al ingot solidified without stirring are shown in figure 20.8. Figure 20.8(a) shows the secondary electron image, while figures

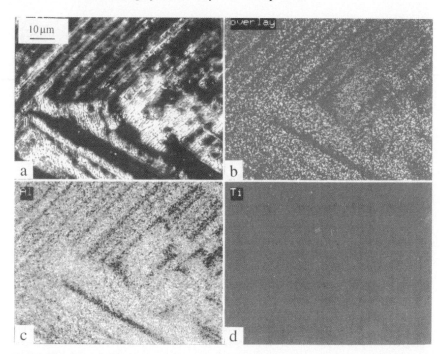

Figure 20.8. Microanalysis of (a) secondary electron image, and X-ray images of (b) titanium and aluminium, (c) aluminium and (d) titanium, for the middle section of a Ti–44% Al solidified without stirring under an argon pressure of 0.1 MPa.

20.8(b), (c) and (d) show X-ray images using a combination of titanium and aluminium, just aluminium, and just titanium, respectively. Segregation of aluminium and titanium was observed to be consistent with the lamellar structure shown in figure 20.8(a).

Figure 20.9 shows equivalent secondary electron and X-ray images from the middle section of a Ti–44% Al ingot rheocast at a rotation speed of $70\,\text{s}^{-1}$. A uniform distribution of aluminium and titanium elements was observed, as opposed to the lamellar segregation seen in the alloy solidified without stirring. The secondary electron and X-ray images from the middle section of a Ti–44% Al ingot rheocast at a rotation speed of $15\,\text{s}^{-1}$ also showed lamellar segregation of aluminium and titanium. Hence, homogenization of chemical composition in combination with refinement of the microstructure of an intermetallic TiAl alloy can only be achieved by rheocasting at high rotation speeds.

As an example of microsegregation in a rheocast composite ingot, figure 20.10 shows the microstructure and the X-ray images from the middle section of Ti–44% Al–5 vol% ZrC rheocast at a rotation speed of $43\,\text{s}^{-1}$. Figures 20.10(a) and (b) show the secondary electron image and optical microstructure, respectively. Figures 20.10(c), (d), (e) and (f) show the X-ray

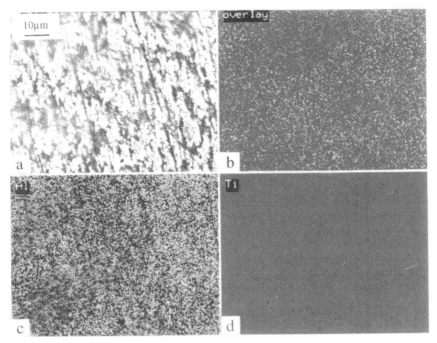

Figure 20.9. Microanalysis of (a) secondary electron image, and X-ray images of (b) titanium and aluminium, (c) aluminium and (d) titanium, from the middle section of a Ti–44% Al ingot rheocast at a rotation speed of $70\,\text{s}^{-1}$ under an argon pressure of 0.1 MPa.

images of carbon, titanium, zirconium and aluminium, respectively. These images confirm that the zirconium carbide particles decompose during the melting and solidification of the composite, and that zirconium-rich lamellar grains, as seen in the upper right of figure 20.10(a), and titanium-rich and carbon-rich precipitates, as seen in the bottom part of figure 20.10(a), form in the comparatively small lamellar grains.

Effect of cooling rate

In order to examine the possible influence of cooling rate during rheocasting and solidification on the microstructure of the TiAl alloys, two solidification experiments were performed. The first was conventional solidification of the ingot, i.e. with no stirrers in the melt. The second was solidification with the stirrer inserted in the melt, but not rotating. The microstructure of the conventionally solidified Ti–44% Al ingot is shown in figure 20.4(a). The microstructure of the Ti–44% Al ingot solidified with the stirrer inserted into the melt, but with no rotation, is shown in figure 20.11. Almost the same lamellar spacings were observed with and without the stirrer inserted

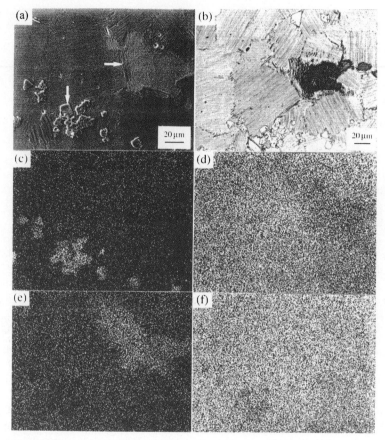

Figure 20.10. Microanalysis of (a) secondary electron image, (b) optical microstructure, and X-ray images of (c) carbon, (d) titanium, (e) zirconium and (f) aluminium, from the middle section of a Ti–44%Al–5vol%ZrC composite ingot rheocast at a rotation speed of $43\,\text{s}^{-1}$ under an argon pressure of 0.1 MPa.

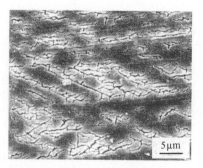

Figure 20.11. Microstructure of Ti–44% Al solidified with no rotation of a stirrer inserted into the melt under an argon pressure of 0.1 MPa.

into the melt, confirming that there is no great difference in cooling rate during solidification with or without the stirrer inserted. Hence the drastic refinement of primary crystals in the rheocast alloys is not due to a difference in cooling rate caused by the presence of the stirrer, but is directly caused by fragmentation of the lamellar structure with rotational stirring.

Tensile properties of Ti–44% Al

Tensile testing at room and elevated temperature was carried out on the rheocast and unstirred intermetallic TiAl alloys and composites. The maximum values of strength were taken to represent the inherent mechanical properties of test pieces without porosity formed by forced entrapment of gas during solidification with stirring in an argon atmosphere.

The maximum tensile strength and elongation in conventionally cast and rheocast Ti–44% Al alloys are shown in table 20.1. The tensile elongations and ultimate tensile strengths measured at room temperature of the rheocast Ti–44% Al alloys were larger than those of alloys solidified without stirring [14]. The elongation of the rheocast alloy at a high rotation speed exceeded 3% while its tensile strength exceeded 460 MPa. Considerable improvement in tensile strength and ductility at room temperature was therefore achieved by rheocasting the Ti–44% Al alloy.

The tensile elongation at elevated temperature increased with temperature from 1073 to 1273 K in the Ti–44% Al alloy solidified without

Table 20.1. Tensile test data on Ti–44% Al alloys solidified without stirring and rheocast at rotation speeds of 15 and 17 s^{-1}.

Casting condition	Temperature (K)	Tensile strength (MPa)	Elongation (%)	Reduction of area (%)
Solidification without stirring	Room	245	2.0	2.3
	1073	320	2.5	3.4
	1173	363	1.7	7.0
	1273	389	4.3	4.7
Rheocasting at 15 s^{-1}	Room	359	2.7	2.3
	1073	475	3.3	3.5
	1173	309	1.7	3.4
	1273	323	20.7	27.7
Rheocasting at 70 s^{-1}	Room	468	3.3	2.4
	1073	406	2.3	3.5
	1173	538	5.2	3.5
	1273	439	4.2	3.4

stirring. The tensile strengths of the rheocast Ti–44% Al alloys were higher than those of the alloy solidified without stirring, although the values did fluctuate. Such fluctuations in the high-temperature tensile properties in the rheocast Ti–44% Al alloys were due to porosity formed by forced gas entrapment between primary crystals during solidification with rotational stirring in an argon atmosphere. The Ti–44% Al alloy rheocast at $15\,\mathrm{s}^{-1}$ showed a tensile elongation of 20.7% at 1273 K [14]. It is considered that this large elongation was achieved by dynamic recrystallization during deformation during testing at 1273 K, and that such a large elongation would be easily reproducible under similar conditions.

Tensile properties of other rheocast TiAl composites

The tensile testing data on the intermetallic TiAl binary, ternary and quaternary alloys produced by rheocasting are summarized in figure 20.12. This figure shows the relationship between the tensile elongation at room temperature and the ultimate tensile strength at 1273 K. Tensile test data obtained by other researchers [15] are also shown in figure 20.12. This figure confirms that the combination of elongation at room temperature with tensile strength at high temperature for rheocast Ti-44% Al ingots is much superior to those obtained by other workers on unstirred alloys [16].

A comparison was also made of the specific strength in the range room to high temperature of the Ti–44% Al ingots rheocast in the present work

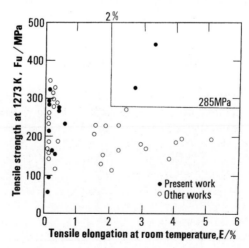

Figure 20.12. Relationship between the tensile elongation at room temperature and the ultimate tensile strength at 1273 K for rheocast intermetallic TiAl alloys.

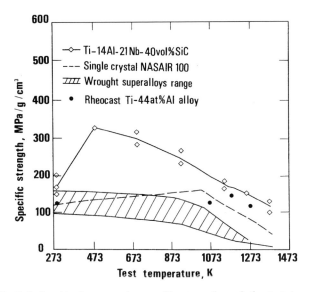

Figure 20.13. Relationship between the specific strength and the test temperature for rheocast Ti–44% Al alloy and other heat-resistant materials taken from the literature.

and other heat-resistant materials taken from the literature. The relationship between the specific strength and the test temperature for 40 vol% SiC fibre-reinforced Ti–14% Al–21% Nb [17], single-crystal NASAIR 100 [17], a range of wrought superalloys [17], and the rheocast Ti–44%Al ingots are shown in figure 20.13. The specific strengths in rheocast Ti–44% Al were much superior to single-crystal NASAIR 100 and approaching those of 40 vol% SiC fibre-reinforced Ti–14% Al–21% Nb at 1173 and 1273 K.

Tensile properties of rheocast TiAl composites

The relationship between the tensile elongation at room temperature and the ceramic particle content in the intermetallic TiAl composites is shown in figure 20.14. Elongation generally increased with increase in ceramic particle content, although the highest elongation of 5% was measured in a low-volume-fraction, 1 vol% TaB_2 particle-dispersed Ti–44% Al ingot. Figure 20.15 shows the relation between the ultimate tensile strength at 1373 K and the ceramic particle content in the intermetallic TiAl composites. The ultimate tensile strength at 1373 K again generally increased with increase in ceramic particle content. In particular, the ultimate tensile strength at 1373 K was 280 MPa in a 10 vol% ZrC particle-dispersion-strengthened Ti–44% Al composite ingot. Finally, the relationship between the tensile elongation at room temperature and the ultimate tensile strength at 1373 K

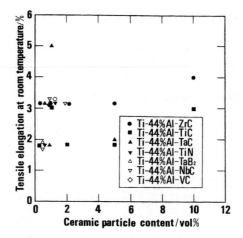

Figure 20.14. Relationship between the tensile elongation at room temperature and the ceramic particle content of rheocast intermetallic TiAl alloy composites.

in the intermetallic TiAl composites is summarized in figure 20.16, which also includes tensile test data obtained by other researchers [16]. The combination of elongation at room temperature and tensile strength in the rheocast Ti–44% Al–10 vol% ZrC composite material was much superior to those of other workers. Intermetallic TiAl alloys and composites with excellent

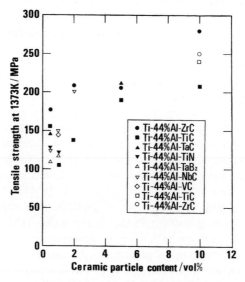

Figure 20.15. Relationship between the ultimate tensile strength at 1373 K and the ceramic particle content of the rheocast intermetallic TiAl alloy composites.

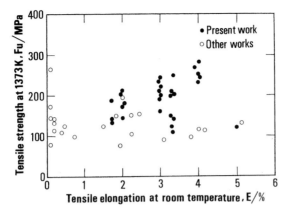

Figure 20.16. Relationship between the tensile elongation at room temperature and the ultimate tensile strength at 1373 K of rheocast intermetallic TiAl alloy composites.

room-temperature and high-temperature properties can clearly be produced by rheocasting at high rotation speed.

Summary

In the microstructures of rheocast Ti–44% Al the conventional lamellar structure seen in alloys solidified without stirring is completely destroyed, and an extremely refined microstructure composed of dispersed aggregates is formed. Oval lamellar grains are seen in the microstructure of rheocast Ti–49% Al and a lamellar substructure observed in the oval primary crystal grains in the microstructure of rheocast Ti–54% Al. Microanalysis of Ti–44% Al rheocast at high rotation speed shows that a uniform distribution of Al and Ti is achieved by refinement by fragmentation of the lamellar structure with rotational stirring. The room-temperature elongations and tensile strengths of Ti–44% Al rheocast at rotation speeds of 15 and $70\,s^{-1}$ are higher than those of the alloy solidified without stirring. The elevated-temperature elongation and ultimate tensile strength of rheocast Ti–44% Al are also improved compared with those of the alloy solidified without stirring.

Zirconium carbide particles are distributed homogeneously in the lamellar grains of a rheocast Ti–44% Al–2 vol% ZrC composite, whilst zirconium carbide particles decompose to give zirconium-rich lamellar grains and titanium-rich and carbon-rich precipitates in the lamellar grains in rheocast Ti–44% Al–5 vol% ZrC composite. The precipitates increase with increased zirconium carbide content from 5 to 10 vol% or with increased aluminium content from 44 to 46 at%, but decrease with an increased aluminium content from 46 to 49 at%. The rheocast Ti–44% Al–10 vol% ZrC composite has an excellent combination of mechanical properties.

References

[1] Courtois J A and Koch A A *Proceedings of the Third International Conference on Processing of Semi-Solid Alloys and Composites* ed M Kiuchi (Institute of Industrial Science, University of Tokyo, Tokyo) p 213
[2] Flemings M C 1974 *Metall. Trans.* **5** 2121
[3] Hagiwara I and Takahashi T 1965 *J. Japan. Inst. Met.* **29** 637
[4] Ichikawa K, Kinoshita Y and Shimamura S 1985 *Trans. Japan. Inst. Met.* **26** 513
[5] Ichikawa K, Ishizuka S and Kinoshita Y 1987 *Trans. Japan. Inst. Met.* **28** 135
[6] Ichikawa K and Ishizuka S 1987 *Trans. Japan. Inst. Met.* **28** 145
[7] Ichikawa K, Ishizuka S and Kinoshita Y 1988 *Trans. Japan. Inst. Met.* **9** 598
[8] Ichikawa K and Ishizuka S 1989 *Trans. Japan. Inst. Met.* **30** 431
[9] Ichikawa K 13 January 1987 US Patent No. 4 636 355
[10] Ichikawa K and Ishizuka S 1989 *Mater. Trans., Japan. Inst. Met.* **30** 915
[11] Ichikawa K and Ishizuka S 1990 *Mater. Trans., Japan. Inst. Met.* **31** 75
[12] Ichikawa K, Ishizuka S, Achikita M, Kinoshita Y and Katou M 1990 *Mater. Trans., Japan. Inst. Met.* **31** 730
[13] Ichikawa K and Ishizuka S 1987 *Trans. Japan. Inst. Met.* **28** 434
[14] Ichikawa K and Kinoshita Y 1996 *Mater. Trans., Japan. Inst. Met.* **37** 1311
[15] Ichikawa K and Kinoshita Y 1993 *Mater. Trans., Japan. Inst. Met.* **34** 467
[16] Hashimoto K, Hanamura T and Mizuhara Y 1990 *Proceedings of Basic Technologies for Future Industries, The 6th Symposium on High Performance Materials for Severe Environments, RIMCOF/JITA* p 67
[17] Stoloff N S 1989 *International Compounds; Structure, Physical Metallurgy and Applications* TMS Fall Meeting 1 October 1989 p 325

Chapter 21

Solidification of metallic glasses

Nobuyuki Nishiyama and Akihisa Inoue

Introduction

For the past three decades, metallic glasses have attracted great interest because of useful properties such as high mechanical strength, high corrosion resistance, high viscous flowability and good soft magnetism, resulting from their unique atomic configuration. However, metallic glasses found before 1988 do not have large glass-forming ability and these glasses require rather high cooling rates above 10^4 K s^{-1} for iron-, cobalt- and nickel-based systems and 10^3 K s^{-1} for Pd–Ni–P and Pt–Ni–P systems [1] to freeze into glasses. Therefore, the resulting sample thickness for glass formation is usually below 100 μm for the former metallic glasses and less than several millimetres for the latter metallic glasses. The critical cooling rates for glass formation of these metallic glasses are higher by several orders of magnitude than those for oxide or fluoride glasses. Finding new metallic glass systems with much lower critical cooling rates is strongly desired for the future development of basic science and the application of metallic glasses.

Recently, we have succeeded [2] in finding new multicomponent metallic glasses with much lower critical cooling rates ranging from 0.1 K s^{-1} to several hundred K s^{-1}, as shown in figure 21.1 [3]. It should be noted that the new metallic glasses, identified by underlining, have much larger supercooled liquid regions before crystallization, defined by the difference $\Delta T_x = T_x - T_g$ between crystallization temperature T_x and glass transition temperature T_g. The higher limits of cooling rate for typical sample preparation methods such as melt-spinning ($<10^5$ K s^{-1}) or casting ($<10^2$ K s^{-1}) and thermal analysis ($<10^0$ K s^{-1}) are also shown in figure 21.1. It is said that the new metallic glasses can be cast into bulk form with ease. In addition, a new Pd–Cu–Ni–P metallic glass [4–7] enables us to investigate *in situ* undercooling behaviour using ordinary thermal analysis techniques because of its outstandingly low critical cooling rate of 0.1 K s^{-1} [6]. It is generally accepted that such results are useful for understanding the glass-forming ability of metallic glasses.

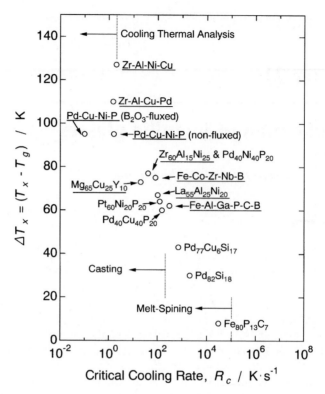

Figure 21.1. Relationship between the critical cooling rate for glass formation and the supercooled region before crystallization for metallic glasses. The underlining indicates systems investigated by the present author and co-workers.

This chapter presents the outstandingly large glass-forming ability of the new $Pd_{40}Cu_{30}Ni_{10}P_{20}$ alloy through the investigation of *in situ* undercooling behaviour using a thermal analysis technique. The results are discussed within the framework of classical nucleation theory.

Optimization of composition for large glass-forming ability

In order to find a new metallic glass system with a large glass-forming ability, we examined the possibility of improving the glass-forming ability of traditional metallic glasses by increasing the number of constituent elements [5]. The ternary $Pd_{40}Ni_{40}P_{20}$ alloy was chosen as the fundamental system for the study because it had the largest reported glass-forming ability at the time and was known to be stable against oxidation. The high stability against oxidation was necessary in order to investigate the undercooling behaviour by an ordinary thermal analysis method.

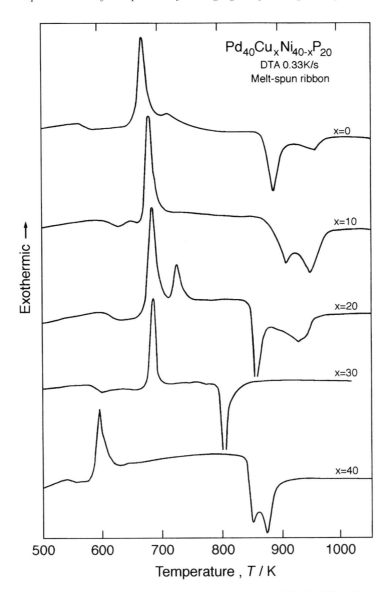

Figure 21.2. Differential thermal analysis curves of melt-spun $Pd_{40}Cu_xNi_{40-x}P_{20}$ ($x = 0$, 10, 20, 30 and 40 at%) glassy ribbons.

Figure 21.2 shows differential thermal analysis curves on heating of melt-spun $Pd_{40}Ni_{40-x}Cu_xP_{20}$ glassy ribbons. As seen in this figure, on heating all alloys the first reaction observed was an endothermic reaction due to the glass transition, followed by a supercooled liquid region preceding an exothermic crystallization reaction and finally an endothermic reaction

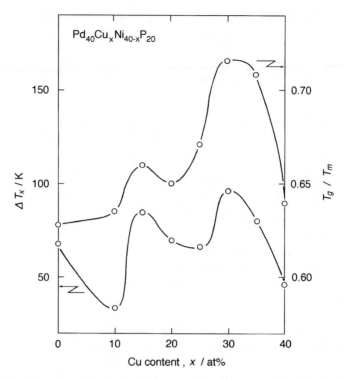

Figure 21.3. The temperature interval of the supercooled liquid region (ΔT_x) and reduced glass transition temperature (T_g/T_m) as a function of copper content for melt-spun $Pd_{40}Cu_xNi_{40-x}P_{20}$ glassy ribbons.

due to melting. Here, it is to be noted that the $Pd_{40}Cu_{30}Ni_{10}P_{20}$ ($x = 30$) glassy ribbon has the widest supercooled liquid region of all of the alloys, and is then followed by an apparent single-stage crystallization peak and single melting peak, with melting occurring at around 800 K, the lowest of any of the alloys investigated. Based on the differential thermal analysis curves shown in figure 21.2, the temperature range of the supercooled liquid region ΔT_x and the reduced glass transition temperature defined by the ratio of the glass transition temperature divided by the onset of melting temperature (T_g/T_m) for the Pd–Ni–Cu–P glassy ribbons are plotted as a function of copper content in figure 21.3. It is seen that the $Pd_{40}Cu_{30}Ni_{10}P_{20}$ glassy ribbon exhibits the highest values of 95 K for ΔT_x and 0.72 for T_g/T_m. The values for the extent of the supercooled region before crystallization and the reduced glass transition temperature are much higher than the largest values ($\Delta T_x = 63$ K and $T_g/T_m = 0.66$) for previously reported Pd–Ni–P ternary metallic glasses, indicating that the present new palladium-based quaternary metallic glass has a much larger glass-forming ability. This remarkable extension of the supercooled liquid region and the increment in the value of the reduced glass transition temperature allow us to expect a significant decrease in the critical cooling rate for glass formation.

Preparation of bulk metallic glass

In order to demonstrate the large glass-forming ability of the new palladium-based quaternary metallic glass, the production of a bulk metallic glass was tried by water quenching of the molten alloy in a quartz tube [6]. It has been reported [8–10] that flux treatment using a B_2O_3 medium is very effective at increasing the maximum sample thickness obtainable in Pd–Ni–P metallic glasses, therefore this treatment was applied to the preparation of samples. As an example, figure 21.4 shows the outer surface appearance of a cylindrical bulk sample with a diameter of 72 mm and a length of 75 mm. The bulk sample has a smooth surface and shows a good metallic lustre. The amorphicity of the sample was examined by X-ray diffraction, differential scanning calorimetry (DSC) and transmission electron microscopy. Figure 21.5 shows the differential scanning calorimetry curve taken from the central region of a bulk sample with a diameter of 72 mm, together with data from a melt-spun glassy ribbon with a thickness of about 20 μm. The characteristic temperatures of the bulk sample such as glass transition crystallization and onset of melting temperatures agree with those for the melt-spun glassy ribbon, suggesting that there are no distinct differences in the disordered structure and thermal stability of the two supercooled liquids despite their very different sample thicknesses. In addition, the X-ray diffraction patterns consist of only a broad peak, and no appreciable contrast is revealed in transmission electron microscopy of precipitation of a crystalline phase, even in

Figure 21.4. Outer surface appearance of the cylindrical $Pd_{40}Cu_{30}Ni_{10}P_{20}$ bulk metallic glass, 72 mm in diameter and 75 mm in length, prepared by quenching a fluxed molten alloy in water.

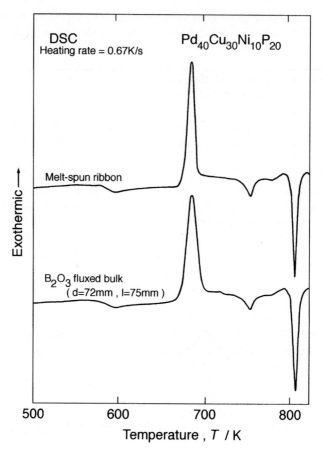

Figure 21.5. Differential scanning calorimetry curve of the cylindrical $Pd_{40}Cu_{30}Ni_{10}P_{20}$ bulk metallic glass, 72 mm in diameter, taken from the central region. The data from the melt-spun ribbon are also shown for comparison.

the central region of the bulk sample. These results allow us to conclude that the bulk sample of 72 mm in diameter is composed of only a glassy phase and that the new quaternary Pd–Cu–Ni–P alloy has an outstandingly large glass-forming ability. The supercooled liquid region before crystallization and reduced glass transition temperature are thought to be the origin for the large glass-forming ability of the new quaternary Pd–Cu–Ni–P alloy.

Undercooling behaviour

The outstandingly large glass-forming ability of the alloy enables us to investigate the undercooling behaviour by ordinary thermal analysis [7].

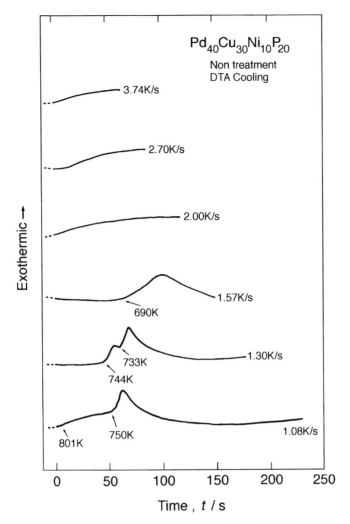

Figure 21.6. Differential thermal analysis curves of non-fluxed $Pd_{40}Cu_{30}Ni_{10}P_{20}$ measured at various cooling rates.

Consequently, we tried to determine the critical cooling rate for glass formation of the $Pd_{40}Cu_{30}Ni_{10}P_{20}$ alloy by using a differential thermal analyser (DTA) under continuous cooling conditions. Figure 21.6 shows the differential thermal analysis curves of the alloy during continuous cooling at rates from 3.74 to 1.08 K s^{-1}. Although no appreciable exothermic peak is seen on the differential thermal analysis curves at the cooling rates above 2.00 K s^{-1}, the curves at the lower cooling rates include an exothermic peak. The data from the differential thermal analysis curves indicate clearly that the critical cooling rate lies between 2.00 and 1.57 K s^{-1}.

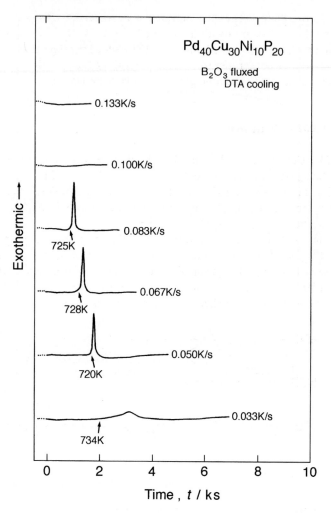

Figure 21.7. Differential thermal calorimetry curves of fluxed $Pd_{40}Cu_{30}Ni_{10}P_{20}$ measured at various cooling rates.

With the aim of further improving the glass-forming ability, we examined the critical cooling rate of the alloy subjected to B_2O_3 flux treatment. Figure 21.7 shows the differential thermal analysis curves of the molten alloy cooled at rates from 0.133 to 0.033 K s^{-1}. As seen from the figure, no appreciable exothermic peak was seen in the curves cooled at 0.133 and 0.1 K s^{-1}, while the curves at rates of 0.083 to 0.033 K s^{-1} include an exothermic peak due to solidification into a crystalline phase. This result indicates that the critical cooling rate of fluxed samples is as low as about 0.1 K s^{-1}, which is lower than that (1.57 K s^{-1}) for the same alloy without

Theoretical calculation

B_2O_3 flux treatment. The flux treatment is therefore effective at further increasing the glass-forming ability of the Pd–Cu–Ni–P alloy. It is therefore concluded that the flux treated alloy has an outstandingly large glass-forming ability, larger than other metallic glasses, and the R_c of the alloy (0.1 K s^{-1}) is the lowest value in the metallic glasses reported up to date.

Theoretical calculation

Figure 21.8 shows the incubation times to crystallize a volume fraction X of 10^{-6}, X being obtained from the data in Figures 21.6 and 21.7 and plotted as a function of temperature [11]. By fitting these experimental results to the following theoretical calculations, the nucleation and growth mechanism can be clarified.

Figure 21.9 shows a typical differential scanning calorimeter curve from the $Pd_{40}Cu_{30}Ni_{10}P_{20}$ glassy alloy at a heating rate of 0.67 K s^{-1}. It can be clearly seen that the glass transition and crystallization temperatures are 575 and 670 K, respectively. Furthermore, the onset of melting temperature T_m and the heat of fusion ΔH_m^f are determined as 804 K and 4840 J mol^{-1}

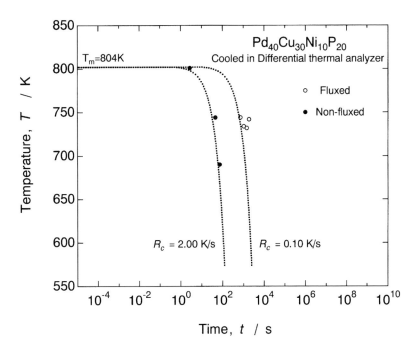

Figure 21.8. Relationship between the incubation time to crystallize a volume fraction and the crystallization temperature for fluxed and non-fluxed molten $Pd_{40}Cu_{30}Ni_{10}P_{20}$ alloys under continuous cooling.

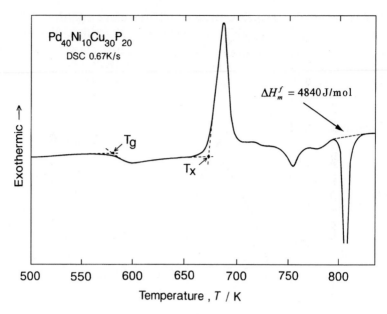

Figure 21.9. Typical differential scanning calorimetry curves of the $Pd_{40}Cu_{30}Ni_{10}P_{20}$ glassy alloy at a heating rate of $0.67\,K\,s^{-1}$.

from near-isothermal differential scanning calorimetry scans. In the construction of the time–temperature transformation and continuous cooling transformation diagrams through the calculation of the nucleation frequency and crystal growth rate, the following parameters are used. The value of mean atomic diameter obtained from the weighted average of the constituent elements is 2.55×10^{-10} m. The number of single molecules per unit volume is calculated to be 7.60×10^{28} atom m^{-3} using the measured density 9.30×10^3 g m^{-3} of the alloy. The work of nucleation ΔG^* is taken as $60kT$, as suggested by Onorato and Uhlmann [12]. The temperature dependence of the viscosity was estimated using Kissinger's method and the results of thermomechanical analysis under compressive stress [13].

The temperature dependences of homogeneous nucleation frequency I_v^{hom} and crystal growth rate u_c calculated using the classical nucleation theory are shown in figure 21.10. The homogeneous nucleation frequency increases rapidly with increasing undercooling below the onset of melting temperature, and shows a maximum of 10^7 m^{-3} s^{-1} at 593 K near the glass transition temperature. On the other hand, the crystal growth rate decreases monotonically with increasing undercooling, which is mainly dominated by the atomic mobility.

Figure 21.11 shows the time–temperature transformation and continuous cooling transformation curves constructed by using the calculated

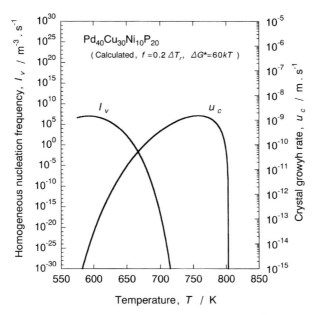

Figure 21.10. The calculated homogeneous nucleation frequency I_v^{hom} and crystal growth rate u_c as a function of temperature for the $Pd_{40}Cu_{30}Ni_{10}P_{20}$ glassy alloy using the classical nucleation theory (in the case of homogeneous nucleation).

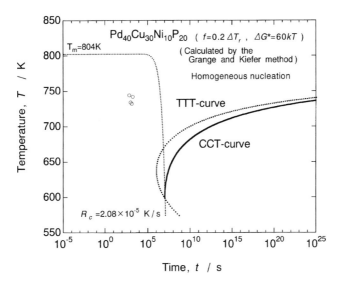

Figure 21.11. Time–temperature transformation and continuous cooling transformation curves constructed using the calculated homogeneous nucleation frequency and crystal growth rate, together with the experimental results obtained by differential thermal analysis.

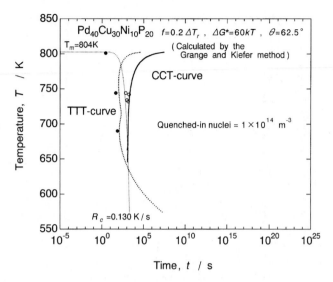

Figure 21.12. Modified continuous cooling transformation curve based on the assumption of taking the density of quenched-in nuclei as 10^{14} m^{-3}.

homogeneous nucleation frequency and crystal growth rate together with the experimental results obtained by differential thermal analysis. The nose temperature, incubation time and critical cooling rate are determined to be 599 K, 9.86×10^6 s and 2.08×10^{-5} K s^{-1}, respectively, by using the calculated continuous cooling transformation curve. In comparison with the continuous cooling transformation curve, the experimental results lie towards higher temperatures and shorter times. Thus, the calculated critical cooling rate is underestimated by four orders of magnitude compared with the experimentally determined value. Alternatively, it is assumed that heterogeneous nucleation cannot be suppressed completely by using the flux technique and actually solidification into crystals is dominated by heterogeneous rather than homogeneous nucleation.

To clarify the actual nucleation mechanism, heterogeneous nucleation is considered. Drehman *et al* have pointed out [14] that there are crystal growth centres in the $Pd_{40}Ni_{40}P_{20}$ melt, so-called quenched-in nuclei. By taking the density of such quenched-in nuclei as 10^{14} m^{-3}, the modified continuous cooling transformation curve can represent the experimental results for the fluxed sample, as shown in figure 21.12. This value for the density of quenched-in nuclei gives the best fit between the theoretical and experimental results for the fluxed samples and is almost consistent with the result of cold-crystallization studies during isothermal annealing [15]. However, it seems that the value 10^{14} m^{-3} for the density of quenched-in nuclei is much higher than the 10^9 to 10^{10} m^{-3} reported by Drehman and Greer [14]. The atmosphere of each of the cooling equipments was almost the same (evacuated to less than 3×10^{-2} torr and refilled with inert gas). The difference in the values for the density of quenched-in nuclei was probably caused by a difference in the atmosphere during the alloy preparation and the chemical etching treatment. On the other hand, it is difficult to obtain modified continuous cooling transformation curves, which correctly describe the experimental results for non-fluxed samples. This difficulty is probably related to surface nucleation due to oxidation, and currently is under investigation.

Summary

This chapter describes a new metallic glass system with a large glass-forming ability, investigates *in situ* undercooling behaviour and discusses glass-forming ability with regard to classical nucleation theory. The results obtained are summarized as follows:

- The new $Pd_{40}Cu_{30}Ni_{10}P_{20}$ alloy exhibits a large supercooled liquid region before crystallization of 95 K and large reduced glass transition temperature of 0.72. This value for the reduced glass transition temperature is believed to be largest for a metallic glass reported to date.
- Cylinders of the Pd–Cu–Ni–P glass can be formed with diameters up to 72 mm by quenching the molten alloy with water in a quartz tube.
- The fluxed Pd–Cu–Ni–P alloy has a critical cooling rate of 0.1 K s^{-1} for glass formation. This is a much slower cooling rate than that required for the same alloy without flux treatment (1.57 K s^{-1}).
- Heterogeneous nucleation in the alloy cannot be suppressed completely by using the flux technique and solidification into crystals seems to be dominated by growth of quenched-in nuclei rather than homogeneous nucleation.

References

[1] Davies H A 1983 *Amorphous Metallic Alloys* ed F E Luborsky (Butterworths) pp 8–25
[2] Inoue A 1995 *Mater. Trans., Japan. Inst. Met.* **36** 866
[3] Inoue A 1997 *Proc. Japan. Acad.* **B73** 19
[4] Inoue A, Nishiyama N and Matsuda T 1996 *Mater. Trans., Japan. Inst. Met.* **37** 181
[5] Nishiyama N and Inoue A 1996 *Mater. Trans., Japan. Inst. Met.* **37** 1531
[6] Inoue A, Nishiyama N and Kimura H M 1997 *Mater. Trans., Japan. Inst. Met.* **38** 179
[7] Nishiyama N and Inoue A 1997 *Mater. Trans., Japan. Inst. Met.* **38** 464
[8] Kui H W, Greer A L and Turnbull D 1984 *Appl. Phys. Lett.* **45** 615
[9] Kui H W and Turnbull D 1985 *Appl. Phys. Lett.* **47** 796
[10] Willnecker R, Wittmann K and Görler G P 1993 *J. Non-Cryst. Solids* **156–158** 450

[11] Nishiyama N and Inoue A 1999 *Acta Mater.* **47** 1487
[12] Onorato P I K and Uhlmann D R 1976 *J. Non-Cryst. Solids* **22** 367
[13] Nishiyama N and Inoue A 1999 *Mater. Trans., Japan. Inst. Met.* **40** 64
[14] Drehman A J and Greer A L 1984 *Acta Metall.* **32** 323
[15] Nishiyama N and Inoue A 1998 *Proceedings of the International Conference on MRS Fall Meeting*, Boston, to be published

Chapter 22

New eutectic ceramics

Itsuo Ohnaka, Yoshiharu Waku, Hideyuki Yasuda and Yoshiki Mizutani

Introduction

Ceramics and ceramic-matrix composites are promising as structural materials with excellent heat resistance, oxidation resistance and abrasion resistance, and are therefore being researched and developed on a global scale. However, the strength of nearly all ceramic polycrystalline materials drops off rapidly with increasing temperature [1]. On the other hand, Viechnicki et al [2] showed that the microstructure of eutectic ceramics could be controlled by unidirectional solidification based on microstructural studies on an $Al_2O_3/Y_3Al_5O_{12}$ (yttrium aluminium garnet, YAG) system using the Bridgman method, suggesting the possibility of better properties. Recently Waku et al [3–7] found that unidirectionally solidified eutectics such as Al_2O_3/YAG and Al_2O_3/GdAlO_3$ (gadolinium aluminium perovskite, GAP) have superior high-temperature strength characteristics and excellent thermal stability up to around 1900 K. These eutectics consist of a single-crystal Al_2O_3 phase and a single-crystal YAG or GAP phase without grain and colony boundaries. This discovery has opened a new field of solidification and casting research, because there are still many challenging problems to use such materials in the real world. This chapter reviews such work including new work on phase selection in the Al_2O_3/YAG eutectic system, and discusses the future for these new ceramics and solidification problems. Peritectic ceramic systems are discussed in chapters 23 and 24. Eutectic solidification theory is discussed in chapter 10.

Al_2O_3/YAG and Al_2O_3/GAP unidirectionally solidified eutectic ceramics

Unidirectional solidification of Al_2O_3/YAG and Al_2O_3/GAP eutectics was carried out as described below, with the results presented and compared

Table 22.1. Mechanical properties of unidirectionally solidified eutectic ceramics.

System	Strength (MPa) F: flexural B: buckling	Temperature (K)	Processing and structure	Ref.
ZrO_2–MgO	159 (F)	room temp.	Bridgman: fibrous	[9]
Al_2O_3–$ZrO_2(Y_2O_3)$	~510 (F) ~450	room temp. 1848	fibrous	[14]
Al_2O_3–$ZrO_2(Y_2O_3)$	1.13–2.45 GPa (B)	room temp.	laser heated floating zone: highly oriented fibrous	[15]
$Al_2O_3Y_3$–$Y_3Al_5O_{12}$ (alumina–YAG)	373 (F) 272 265 4.2 MPa m$^{1/2}$	room temp. 1648 1858 1859	Bridgman: cellular	[16]
$Al_2O_3Y_3$–$Y_3Al_5O_{12}$ (alumina–YAG)	Comp. creep flow stress $\sigma = 10^{3.24}\varepsilon^{0.2}$	1923	Bridgman: elongated grains	[17]
$Al_2O_3Y_3$–$Y_3Al_5O_{12}$ (alumina–YAG)	360–450 (F) comp. creep flow stress $\sigma = 10^{3.16}\varepsilon^{0.171}$	up to 1873 1973	Bridgman: single crystal	[5, 6]
$Al_2O_3Y_3$–$GAlO_3$ (alumina–GAP)	Yielding ~700 (F)	1873	Bridgman: single crystal	[5, 6]

with conventional results in table 22.1 [3–7]. Commercially available Al_2O_3, Y_2O_3 and Gd_2O_3 powders were mixed in molar ratios of $Al_2O_3/Y_2O_3 = 82/18$ and $Al_2O_3/Gd_2O_3 = 77/23$, and ball-milled using ethanol. After removing the ethanol and drying the slurry, the mixture was melted into an ingot by using high-frequency induction heating in a molybdenum crucible for Al_2O_3/YAG and by arc melting for Al_2O_3/GAP. Unidirectional solidification of Al_2O_3/YAG was performed by using Bridgman-type equipment. The ingot was inserted into a molybdenum crucible (50 mm outside diameter, 200 mm length, 5 mm thickness) placed in a vacuum chamber, and a graphite susceptor was heated by high-frequency induction heating. After keeping the melt at 2223 K (about 100–120 K above the melting point) for 30 min, the molybdenum crucible was lowered at 1.4 μm s^{-1}. Al_2O_3/GAP powder, obtained by crushing the arc-melted ingots, was put into a molybdenum crucible (12 mm outside diameter, 120 mm length, 1 mm thickness) and placed in a chamber. Melting was performed in the molybdenum crucible heated by high-frequency induction heating under a pressure of 10^{-5} mmHg of argon, and then, after holding for 1800 s at 2123 K, unidirectional solidification by lowering the molybdenum crucible at a speed of 1.4 μm s^{-1}. For accurate control of crystal growth, a mini-crucible (2 mm in diameter and 17 mm in length) for growing suitable seed crystals was set at the bottom of the main crucible.

Mechanical properties

Specimens for three-point flexural tests, tensile tests and creep tests were selected so that their axial direction was parallel to the direction of solidification. The dimensions of the flexural test specimen were 3 mm × 4 mm × 36 mm with a 30 mm support span. The three-point flexural strength was measured from room temperature to 2073 K in an argon atmosphere at a cross-head speed of 0.5 mm min^{-1}. The tensile tests were also conducted at a strain rate of 10^{-4} s^{-1} in an argon atmosphere from room temperature to 2023 K. Strain rate control compressive creep tests were carried out on 3 mm × 3 mm × 6 mm specimens. Specimens were heated by heating the graphite susceptor using high-frequency induction heating. Tests were conducted in an argon gas atmosphere.

Figure 22.1 compares the flexural strength of Al_2O_3/YAG eutectic composites made by sintering and by unidirectional solidification. It is interesting that the strength of the unidirectionally solidified eutectic ceramics is constant up to 2000 K, while the sintered composite strength decreases rapidly above 1000 K. Figure 22.2 shows the relationship between compressive creep stresses and strain rates in the unidirectionally solidified eutectic ceramics and a sintered composite at test temperatures of 1773, 1873 and

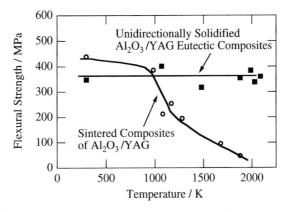

Figure 22.1. Temperature dependence on flexural strength of unidirectionally solidified eutectic composites compared with sintered composites.

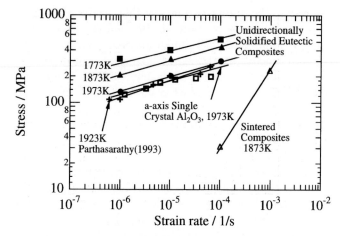

Figure 22.2. Comparison of compressive flow stress of the unidirectionally solidified eutectic composite and a sintered composite as a function of strain rate and temperature.

1973 K. While the unidirectionally solidified eutectic ceramics and the sintered composite share the same chemical composition and contain the same structural phases, their creep characteristics are markedly different. That is, at a strain rate of $10^{-4}\,\text{s}^{-1}$ and test temperature of 1873 K, the sintered composite showed a creep stress of 33 MPa, while the unidirectionally solidified eutectic ceramic's creep stresses were approximately 13 times higher at 433 MPa. Moreover, as can be seen from figure 22.2, the unidirectionally solidified eutectic Al_2O_3 composites have creep characteristics that surpass those of *a*-axis sapphire fibres [8] and display overall excellent creep resistance. Their creep strength is higher than that reported by Parthasarathy *et al* [17] for similar unidirectionally solidified eutectic ceramics.

Figure 22.3 compares typical stress–displacement curves of the unidirectionally solidified eutectic ceramics with sintered composites obtained from the flexural test at 1873 K. The Al_2O_3/GAP unidirectionally solidified eutectic ceramic shows a yielding phenomenon under high stress, with a flexural yield stress of about 695 MPa. Further, the relative fracture energy of the Al_2O_3/GAP unidirectionally solidified eutectic ceramic, which is calculated from the area below the stress–strain curve, is much larger than that of the Al_2O_3/YAG unidirectionally solidified eutectic ceramic which undergoes brittle fracture at this temperature but shows plasticity at higher temperatures above approximately 1900 K. At 1873 K the sintered Al_2O_3/GAP undergoes plastic deformation at lower stress, with less fracture energy. The Al_2O_3/GAP unidirectionally solidified eutectic ceramic did not fracture even after the flexural test, showing its substantial plasticity.

The thermal stability of the microstructures of the unidirectionally solidified eutectic ceramics is also superior. That is, even after 1000 h of heat

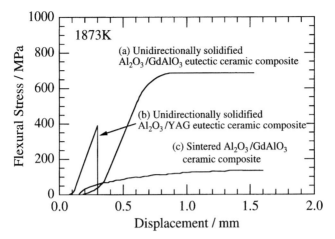

Figure 22.3. Typical stress–displacement curves from three-point flexural tests at 1873 K of (a) unidirectionally solidified Al_2O_3/GAP eutectic composites compared with (b) unidirectionally solidified Al_2O_3/YAG eutectic composites and (c) sintered Al_2O_3/GAP composites.

treatment at 1973 K in an air atmosphere, no grain growth is observed. Further, more interestingly, their machinability is rather good, and they can be machined by a diamond wheel or high-frequency vibration more easily than S_3N_4 or SiC.

Solidification structures

The solidification structures were examined in order to explain the excellent high-temperature characteristics of the unidirectionally solidified eutectic ceramics. X-ray diffraction patterns showed only strong diffraction from the Al_2O_3 (110) plane, indicating that the as-grown unidirectionally solidified eutectic ceramics consisted of a single-crystal YAG or GAP and a single-crystal Al_2O_3 (sapphire). As shown in figure 22.4, the eutectic is shaped very much like hieroglyphs, consisting of single-crystal Al_2O_3 with an irregular distribution at the microscopic level, and single-crystal GAP. At the macroscopic level, the unidirectionally solidified eutectic ceramics have a uniform microstructure with no colonies or pores. High-resolution transmission electron micrographs showed that an amorphous phase exists at the grain boundaries of the sintered composites, while none was found in the unidirectionally solidified eutectic ceramics. Further, transmission electron microscopy revealed dislocation structure in the Al_2O_3 and GAP phases at the tensile-stressed side of the plastically deformed specimens after the three-point flexural test at 1873 K of the Al_2O_3/GAP unidirectionally solidified eutectic ceramic.

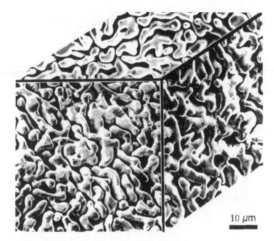

Figure 22.4. Scanning electron micrographs showing the three-dimensional configuration of single-crystal GAP in a unidirectionally solidified Al_2O_3/GAP eutectic composite.

These facts together with data from existing literature can explain the plastic behaviour and high-temperature strength of the unidirectionally solidified eutectic ceramics. First, the unidirectionally solidified eutectic ceramic high-temperature strengths are superior because they have much lower impurities than the sintered ceramics, which have an amorphous phase at the grain boundaries, deteriorating the high-temperature strength. The reason why the current unidirectionally solidified eutectic ceramics are better than conventional unidirectionally solidified eutectic ceramics may be that they consist of single crystals of Al_2O_3 and YAG or GAP with few defects, and specifically no colonies. The ceramic strength is affected by the colony size, especially the width of the colony boundaries which are weak [1, 9, 10, 12, 13]. The unidirectionally solidified eutectic ceramic high-temperature stability may be because of the semicoherent interface structure between the matrix and strengthening phase resulting in a smaller interfacial energy as well as thermodynamic stability of the constituent phases of the single-crystal Al_2O_3 and the single-crystal YAG. Parthasarathy et al [18] examined the creep strength of an Al_2O_3/YAG unidirectionally solidified eutectic ceramic and expected that this material may satisfy the design criteria for turbine blades at 1530°C.

Solidification of eutectic ceramics

The control of solidification structures in eutectic ceramics is essential for getting better properties, as mentioned above. Furthermore, the structure of our present unidirectionally solidified eutectic ceramics is different from conventional ones. Table 22.2 lists the results of conventional work on eutectic solidification of ceramics.

Table 22.2. Solidification conditions and structures of some eutectic ceramics.

System	Processing	Conditions	Structure	Ref.
Al_2O_3–$Y_3Al_5O_{12}$	Bridgman	$G = 19\,\mathrm{K\,mm^{-1}}$ $R = 5.6$–$33\,\mathrm{\mu m\,s^{-1}}$ $G/R = (0.6$–$13) \times 10^9\,\mathrm{K\,s\,m^{-2}}$	• colony type structure • rod and platelet YAG	[2]
Al_2O_3–ZrO_2	Bridgman	$G = 19\,\mathrm{K\,mm^{-1}}$ $R = 3.6$–$43.3\,\mathrm{\mu m\,s^{-1}}$ $G = 120\,\mathrm{K\,mm^{-1}}$ $R = 0.2$–$2.8\,\mathrm{\mu m\,s^{-1}}$ $G/R = (0.4$–$5.3) \times 10^9\,\mathrm{K\,s\,m^{-2}}$	• cellular structure with rodlike monoclinic ZrO_2 • non-cellular structure with lamellar or platelike eutectics	[24]
Al_2O_3–$Y_3Al_5O_{12}$ (alumina–YAG)	Bridgman	$G \simeq 20\,\mathrm{K\,mm^{-1}}$ $R \simeq 17\,\mathrm{\mu m\,s^{-1}}$ $G/R = 1.2 \times 10^9\,\mathrm{K\,s\,m^{-2}}$	• elongated grains and colonies aligned with growth direction • hieroglyphic shape c-axis of Al_2O_3 perpendicular to growth direction • elongated grains and colonies • rod or lamellar eutectics	[25]
Al_2O_3–$YAlO_3$ (alumina–YAP)				
$ZrO_2(Y_2O_3)$ 65.07 mol% Al_2O_3–33.69 ZrO_2–1.24 Y_2O_3	unidirectional solidification with laser zone melting	$R = 2.8$–$41.7\,\mathrm{\mu m\,s^{-1}}$ $R = 2.8\,\mathrm{\mu m\,s^{-1}}$ $R = 41.7\,\mathrm{\mu m\,s^{-1}}$	• coupled growth • well-aligned fibrous structure • lamellar structure	[15]
Al_2O_3–$Y_3Al_5O_{12}$	floating zone method using image furnace	$R = 0.28\,\mathrm{\mu m\,s^{-1}}$	• elongated grain along growth direction • hieroglyphic eutectics	[26]
Al_2O_3–$Y_3Al_5O_{12}$	Bridgman method with Mo	$G = 20\,\mathrm{K\,mm^{-1}}$ $R = 1.4$–$1.5\,\mathrm{\mu m\,s^{-1}}$ $G/R = 1.3 \times 10^{10}\,\mathrm{K\,s\,m^{-2}}$	• columnar grains irregular and elongated eutectics in shape	[16]
Al_2O_3–$Y_3Al_5O_{12}$	h.f. Bridgman method	$R = 1.4\,\mathrm{\mu m\,s^{-1}}$	• single crystals of eutectic phases tangled with each other • macroscopically uniform Waku	

Faceted and non-faceted growth

It has been reported that the Hunt–Jackson criteria [20] can be rather well applied to ceramics. This is also the case for our experiments. Cockayne *et al* [21] have reported that YAG solidifies by non-faceted growth if planar solidification occurs. It is interesting to examine the mechanical properties of unidirectionally solidified eutectic ceramic which solidified with a non-faceted phase.

Lamellar or fibrous structure

Although lamellar and fibrous structures have been observed previously, the present unidirectionally solidified eutectic ceramics have no such structure. Minford *et al* [22] reported that oxide eutectic microstructures can be predicted from the volume per cent of the minor phase combined with minimization of the interfacial surface area per unit volume. Probably, however, we should also consider the effect of cooling rate [23] and impurity content, as in metallic systems. For example, Viechnicki *et al* [2] reported that both rod and platelet morphologies appear in the Al_2O_3/YAG system depending on cooling rate.

Elimination of colonies

Because colony boundaries are weak, we should try to eliminate them. This requires a high ratio of temperature gradient to growth rate. The critical value of this ratio is estimated as 6×10^7 K s m^{-2} for many oxide ceramics [13], which is one order of magnitude lower than that for metallic systems such as Co–Cr_7C_3 or Ni_3Al–Ni_3Nb. The critical value does not vary greatly with composition, because the diffusion coefficient does not change much at high temperature. As seen from table 22.2, two orders of magnitude higher values of this ratio, say 10^{10} K s m^{-2}, were used in these experiments. Probably, a lower growth rate should be used with increasing diameter of samples in order to realize a uniform temperature distribution in the radial direction due to the smaller thermal conductivities of ceramics. Furthermore, we should take into account the low accuracy of the available diffusion coefficient data. In addition, although purification can theoretically help to prevent the formation of cellular structures, there is no experimental evidence of this effect.

Structural refinement

Although it is not clear that finer structures always have higher strengths, finer structures at least have the possibility of improving room-temperature strength. It has been reported that the lamellar spacing is proportional to the

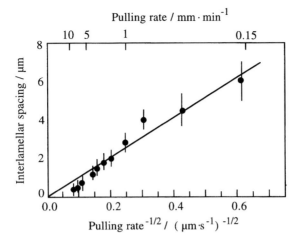

Figure 22.5. Interlamellar spacing for $Al_2O_3/Y_3Al_5O_{12}$ eutectic structures versus pulling rate and the corresponding microstructures.

inverse square root of the growth rate for many oxide systems [22, 27], just like metallic systems. Although measuring the spacing for the present unidirectionally solidified eutectic ceramics is not easy, Epelbaum et al [28] reported that the spacing of Al_2O_3/YAG unidirectionally solidified eutectic ceramics follows the rule as shown in figure 22.5, suggesting that they behave in a similar manner to metallic systems.

Crystallographic relationships

High-temperature stability is also related to the crystallographic relationship at the eutectic phase boundary. Minford et al [22] reported that the interface and orientation of directionally solidified phases are determined by:

- minimization of the lattice misfit between the phases, and
- neutralization of ionic charge across the interface plane.

Waku et al [5] found that the boundary consists of the single-crystal Al_2O_3 and single-crystal YAG with $\langle 110 \rangle$ Al_2O_3 parallel to $\langle 420 \rangle$ YAG, suggesting the minimization of the thermal strain.

Selection of eutectic systems in Al_2O_3–Y_2O_3 ceramics

As mentioned above, it is important to understand solidification behaviour in order to control the eutectic structure. However, there are still ambiguities in the understanding of solidification behaviour. For example, it is still not clear which phases appear during cooling of Al_2O_3–Y_2O_3 mixtures

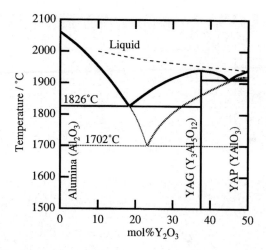

Figure 22.6. Al$_2$O$_3$-rich portion of the phase diagram of the Al$_2$O$_3$–Y$_2$O$_3$ system.

with compositions corresponding to YAG, YAlO$_3$ (YAP) or Y$_4$Al$_2$O$_9$ compounds. The pseudo-binary Al$_2$O$_3$–Y$_2$O$_3$ system is shown in figure 22.6 [29]. Caslavsky et al [29] performed optical differential thermal analysis and reported that melts heated to temperatures below 2263 K followed the equilibrium phase diagram, while melts cooled down from temperatures above 2263 K followed the metastable phase diagrams. However, in general, the selection of the solidified structure is determined not by the melt temperature before cooling but by a combination of nucleation and the growth velocity of the interface. Therefore, we examined the eutectic selection in terms of cooling rate, nucleation temperature and maximum holding temperature of the melt as described below.

Specimens were prepared from 99.99% α-Al$_2$O$_3$ powder and 99.9% Y$_2$O$_3$ powder. Wet ball milling with addition of ethanol was used in order to obtain a homogeneously mixed slurry. The composition of the slurry was Al$_2$O$_3$–18.5 mol% Y$_2$O$_3$ which corresponded to the eutectic composition of α-Al$_2$O$_3$/YAG. The slurry was dried in a vacuum. A molybdenum crucible was used to melt specimens and to measure the cooling curves. Specimens were 8 mm in diameter and 5 mm in height. The molybdenum crucible was heated by radio-frequency induction under a vacuum of 10^{-2} torr. Temperature measurement was carried out by two-colour pyrometers. The pyrometers were calibrated so that the temperature at which the specimen was completely melted coincided with the equilibrium eutectic temperature (1826 K). The crystal structures of the specimens were identified by X-ray diffraction at room temperature to discriminate between the formation of Y$_3$Al$_5$O$_{12}$, yttrium aluminium garnet (YAG) and YAlO$_3$, yttrium aluminium perovskite (YAP). The microstructures were observed by backscattered electron imaging in the scanning electron microscope.

Selection of eutectic systems in Al_2O_3–Y_2O_3 ceramics 383

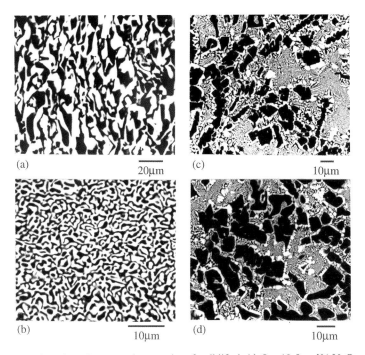

Figure 22.7. Scanning electron micrographs of solidified Al_2O_3–18.5 mol% Y_2O_3. (a) α-Al_2O_3/YAG eutectic, cooled from 2173 K, cooling rate 0.2 K s^{-1}, nucleation temperature 2073 K, (b) α-Al_2O_3/YAG eutectic, cooled from 2173 K, cooling rate 1.5 K s^{-1}, nucleation temperature 2018 K, (c) α-Al_2O_3 and α-Al_2O_3/YAP eutectic, cooled from 2173 K, cooling rate 20 K s^{-1}, nucleation temperature 1893 K, (d) α-Al_2O_3 and α-Al_2O_3/YAP eutectic, cooled from 2273 K, cooling rate 20 K s^{-1}, nucleation temperature 1898 K.

Figure 22.7 shows scanning electron micrographs of solidified structures. A faceted–faceted eutectic structure is clearly observed in the specimen cooled at 0.1 K s^{-1} and nucleated at 2073 K, as shown. The black phase is α-Al_2O_3, while the white phase is the YAG. The lamellar spacing in figure 22.7(a) is about 10 µm. The specimen is cooled at 1.5 K s^{-1} and nucleated at 2018 K exhibits a finer faceted–faceted eutectic structure, as shown in figure 22.7(b). The lamellar spacing is between 2 and 5 µm. On the other hand, primary α-Al_2O_3 and rodlike eutectic structure are observed when the specimen is cooled from about 2173 K at a cooling rate of 20 K s^{-1} and nucleated at 1892 K as shown in figure 22.7(c). The α-Al_2O_3/YAP eutectic structure is observed between the primary α-Al_2O_3. Figure 22.7(d) shows the Al_2O_3/YAP eutectic structure in the specimen cooled from 2273 K at 20 K s^{-1} and nucleated below 1898 K. There is no difference in the solidified structure between the specimen cooled from 2173 K and the specimen cooled from 2273 K when the Al_2O_3/YAP system was selected.

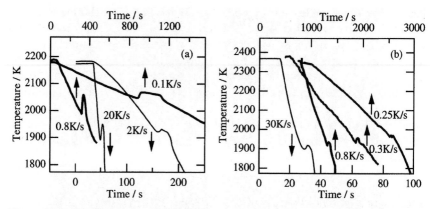

Figure 22.8. Cooling curves from Al_2O_3–18.5 mol% Y_2O_3. (a) Cooling curves from temperatures below 2000 K and (b) cooling curves from temperatures above 2000 K.

The selection of the eutectic systems is clearly determined by the nucleation temperature; the Al_2O_3/YAG (equilibrium eutectic) system is selected when nucleation occurs above 1973 K, while the Al_2O_3/YAP (metastable) system is selected when the melt is undercooled below 1973 K, as shown in figure 22.8. If the melt is cooled from about 2173 K and at cooling rates below $2\,\mathrm{K\,s^{-1}}$, exothermic heat due to nucleation and growth is detected above 1973 K and results in formation of the Al_2O_3/YAG eutectic. If the melt is cooled at cooling rates greater than $2\,\mathrm{K\,s^{-1}}$, exothermic heat is always detected below 1973 K, resulting in formation of the Al_2O_3/YAP eutectic. This can be explained by the proposed phase diagram by Caslavsky [29], where the metastable Al_2O_3/YAG eutectic temperature is 1975 K. Nucleation of the Al_2O_3/YAP system is intrinsically easier than that of the Al_2O_3/YAG eutectic system, because the former has a much larger undercooling from the liquidus temperature and YAG is difficult to nucleate due to its complicated structure.

The maximum melt temperature before cooling significantly affects nucleation of YAG. The melt heated up to temperatures above 2273 K never nucleates above 1973 K and consequently the metastable system is selected as shown in figure 22.8(b). Melts heated above 2273 K are difficult to solidify. Even when the melt is kept at 1993 K for 1.8 ks, solidification does not occur. Caslavsky *et al* [29] reported a possibility of two immiscible liquids above 2273 K, because the melt was opaque for the YAG composition. If a miscibility gap exists above 2273 K and the melt is kept above this temperature, coarsening should occur, but this was not the case in our experiments. We [31] have developed equipment for optical differential thermal analysis which has been used to measure the differential thermal analysis curves of Al_2O_3–18.5 mol% Y_2O_3. However, no exothermic or endothermic heat was detected during heating up to 2473 K, although the heat of fusion was clearly detected. Thus, the change in the melt around 2273 K, if it exists, will cause only a small enthalpy change in comparison with the heat of fusion. Coordination of oxygen around aluminium may have an important role in the melt [29]. Fourfold coordination exists in the YAG structure, while only Al–O octahedra exist in the α-Al_2O_3 and YAP structures. Coordination in the melt may change during heating and affect the nucleation behaviour. There is no effect of the initial structure of the ingot on nucleation. For example, the Al_2O_3/YAG system appears even when specimens solidified as the metastable Al_2O_3/YAP system are re-melted and slowly cooled.

The future of unidirectionally solidified eutectic ceramics

Applications

For the practical application of unidirectionally solidified eutectic ceramics, not only their strength but also their weaknesses should be considered. Their weaknesses include higher melting point, lower toughness especially near room temperature, lower productivity and higher raw materials cost, compared with conventional superalloys. Therefore, their applications should be developed where conventional superalloys cannot be used. The following applications might be possible, though further work is necessary:

- new aero-gas turbines and power generation systems with non-cooled turbine blades for very high temperatures,
- fastening parts such as bolts, nuts and springs for high-temperature uses such as solid-electrolyte fuel cells, and
- parts for high-temperature burners, etc.

Relationship between solidification structure and properties

It is necessary to improve the mechanical properties, especially near room temperature, and optimize the properties in all useful temperature ranges at minimum cost. Although there are some data on mechanical properties, they are mostly for cellular structures and it is hoped to get more data for the unidirectionally solidified eutectic ceramics with fewer defects and a more detailed knowledge of the mechanisms of cracking and fracture. For example, it is reported that the lamellar spacing or the growth rate hardly affects the strength [9, 12–14], unlike metallic systems. This should be checked with the unidirectionally solidified eutectic ceramics which have fewer defects. In comparison, greater growth rate increases the hardness for MgO–$MgAl_2O_4$ [19]. There are few data on fatigue strength or creep.

Physical properties

Not enough accurate data are available regarding, for example, thermal conductivity, latent heat, specific heat, thermal radiation properties, temperature dependence of density, viscosity of the liquid eutectics, phase diagrams, diffusion coefficients, solid–liquid interfacial energies, mechanical properties at high temperatures including Young's modulus, thermal expansion coefficient depending on crystallographic orientation, etc. Those are necessary not only for the basic understanding of the solidification and fracture phenomena but also for computer simulation.

Solidification phenomena

To make clear the relationship between structure and solidification conditions and to understand the solidification mechanisms for various unidirectionally solidified eutectic ceramics, more work is necessary as mentioned above, including effects of growth condition, impurities and additions of third elements, fluid flow, electromagnetic force, gravity and centrifugal force, and the use of seed crystals on the solidification structure. Furthermore, a detailed understanding of the structure of the eutectic interface and its formation mechanism, and an explanation of the effect of phase boundary morphology and consistency on mechanical properties are also desired.

Computer simulation

Computer simulation is very important not only for understanding physical phenomena but also to develop production processes and control the solidification structure, for example, lamellar spacing, phase morphology, macrosegregation and other defects, because physical experiments are usually very expensive, especially if the scale becomes large. One of the difficulties in the simulation of solidification is the treatment of radiative heat transfer, because very high temperatures are used and many ceramics are rather transparent. Stress analysis is also important to avoid cracking and requires mechanical property data at high temperatures.

Control of solidification structures

It is very important to control the ratio of temperature gradient to the growth rate throughout the product in order to produce a single-crystal structure. This is especially difficult for products with complicated shape and larger sizes. If conventional processes are difficult to apply, adjustment of chemical composition, application of electromagnetic and/or centrifugal forces may also have to be considered. In any case, computer simulation is essential to decrease the cost and time of development.

Increasing productivity

Although the conventional Bridgman method can be applied for products with simple shapes, the temperature gradient is relatively low, of the order of 20 K m^{-1}, resulting in a growth rate of the order of 1 μm s^{-1} being required to eliminate cellular structures, giving very low productivity. The zone melting method can increase the temperature gradient but is not easy to apply for products with complicated shapes. Other processes may have to be developed such as, for example, the Stepanov method.

Manufacturing processes

New processes should be developed to produce products with complicated shapes. If the lost wax process is applied, the following should be developed: mould materials preventing reaction with the melt at high temperatures, methods for making and heating the mould, and controlling the solidification. New, alternative processes may have to be developed. This might include the use of rapidly solidified powders containing metastable phases, resulting in undercooled liquids and easier shaping at lower temperatures.

Summary

Al_2O_3/YAG and Al_2O_3/GAP have better high-temperature strengths than conventional ceramics. Eutectic composites have new microstructures, in which single-crystal Al_2O_3 and single-crystal complex oxide compounds (YAG or GAP) are three-dimensionally, continuously connected and finely entangled without grain boundaries. There are many problems to be solved in the solidification of these eutectic ceramics. Phase selection in the Al_2O_3–Y_2O_3 system has been examined and it is found that when the melt is cooled from 2373 K at any cooling rate, it always nucleates below the Al_2O_3/YAP eutectic temperature, selecting the metastable Al_2O_3/YAP eutectic. The equilibrium Al_2O_3/YAG eutectic is always selected when the melt is cooled from the lower temperature of 2173 K at a cooling rate less than 1 K s^{-1}. Although these new eutectic ceramics have fascinating high-temperature properties and are promising for high-temperature applications such as turbine blades, there are still many problems to be addressed, including:

- the relationship between solidification structure and properties,
- basic physical property data,
- solidification phenomena,
- computer simulation,
- control of solidification structure,
- increasing productivity, and
- manufacturing processes.

References

[1] Hilling W B 1986 'Prospects for ultrahigh-temperature ceramic composites' *Tailoring Multiphase and Composite Ceramics* ed R E Tressler, G L Messing, C G Pantano and R E Newnham, *Mater. Sci. Res.* **20** 697–712

[2] Viechnicki D and Schmid F 1969 'Eutectic solidification in the system $Al_2O_3/Y_3Al_5O_{12}$' *J. Mater. Sci.* **4** 84–88

[3] Y Waku, N Nakagawa, H Ohtsubo, Y Ohsora and Y Kohtoku 1995 'High temperature properties of unidirectionally solidified Al_2O_3/YAG composites' *J. Japan. Inst. Met.* **59** 1, 71-78

[4] Waku Y, Otsubo H, Nakagawa N and Kohtoku Y 1996 'Sapphire matrix composites reinforced with single crystal YAG phases' *J. Mater. Sci.* **31** 4663–4670

[5] Waku Y, Nakagawa N, Wakamoto T, Ohtsubo H, Shimizu K and Kohtoku Y 1997 'A ductile eutectic ceramic composite with high strength at 1873 K' *Nature* **389** 49–52

[6] Waku Y, Nakagawa N, Wakamoto T, Ohtsubo H, Shimizu K and Kohtoku Y 1998 'High-temperature strength and thermal stability of a unidirectionally solidified Al_2O_3/YAG eutectic composite' *J. Mater. Sci.* **33** 4943

[7] Waku Y, Nakagawa N, Wakamoto T, Ohtsubo H, Shimizu K and Kohtoku Y 1998 'The creep, oxidation resistance, and thermal shock characteristics of a unidirectionally solidified Al_2O_3/YAG eutectic composite' *J. Mater. Sci.* **33** 1217

[8] Kotchick D M and Tressler R E 1980 'Deformation behavior of sapphire via the prismatic slip system' *J. Am. Ceram. Soc.* **63** 429–434

[9] Kennard F L, Bradt R C and Stubican V S 1974 'Directional solidification of the ZrO_2–MgO eutectic' *J. Am. Ceram. Soc.* **57** 428–431

[10] Ashbrook R L 1977 'Directionally solidified ceramic eutectics' *J. Am. Ceram. Soc.* **60** 435–428

[11] Stubican V S and Bradt R C 1981 'Eutectic solidification in ceramic systems' *Annu. Rev. Mater. Sci.* **11** 267–297

[12] Stubican V S, Bradt R C, Kennard F L, Minford W J and Sorrel C C 1986 'Ceramic eutectic composites', *Tailoring Multiphase and Composite Ceramics, Materials Science Research* (Plenum) **20** 103–114

[13] Lakiza S N 1990 'Directional solidification of eutectics: new prospects for refractory oxide ceramics (review)', *Sov. Powder. Metall. Met. Ceram. (USA)* **28** 637–646 [1989 *Poroshk. Metall. Ukrainian SSR* **28** 8, 58–69]

[14] Hulse C O and Batt J A May 1974 'Effect of eutectic microstructure on the mechanical properties of ceramic oxides' Final Technical Report VARL-N910803-10, NTIS AD-781995/6GA, 140 pp

[15] Haggerty J S, Sigalovsky J and Courtright E L 1993 'Processing and properties of unidirectionally solidified $ZrO_2(Y_2O_3)$–Al_2O_3 eutectic reinforcement fibers' Presented at Symposium on Processing and Fabrication of Advanced Materials at 1993 TMS Fall Meeting, 17–21 October, Pittsburgh, PA, pp 93–104

[16] Mah T, Parthasarathy T A and Matson L E 1990 'Processing and mechanical properties of $Al_2O_3/Y_3Al_5O_{12}$ (YAG) eutectic composite' *Ceram. Eng. Sci. Proc.* **11**(9–10) 1617–1627

[17] Parthasarathy T A, Mar T and Matson L E 1993 'Deformation behavior of an Al_2O_3–$Y_3Al_5O_{12}$ eutectic composite in comparison with sapphire and YAG' *J. Am. Ceram. Sci.* **76**(1) 29–32

[18] Parthasarathy T A, Mah T and Matson L E 1990 'Creep behavior of an Al_2O_3–$Y_3Al_5O_{12}$ eutectic composite' *Ceram. Eng. Sci. Proc.* **11** 1628–1638

[19] Kennard F L, Bradt R C and Stubican V S 1976 'Mechanical properties of the directionally solidified MgO–$MgAl_2O_4$ eutectic' *J. Am. Ceram. Soc.* **56** 160–163

[20] Hunt J D and Jackson K A 1966 'Binary eutectic solidification' *Trans. AIME* **236** 843–852

[21] Cockayne B, Chesswas M and Gasson D B 1968 'The growth of strain-free $Y_3Al_5O_{12}$ single crystals' *J. Mater. Sci.* **3** 224–225

[22] Minford W J, Bradt R C and Stubican V S 1979 'Crystallography and microstructure of directionally solidified oxide eutectics' *J. Am. Ceram. Soc.* **62** 154–157

[23] Croker M N, Fidder R S and Smith R W 1973 'The characterization of eutectic structure' *Proc. Roy. Soc.* **335A** 15–37

[24] Schmid F and Viechnicki D 1970 'Oriented eutectic microstructures in the system Al_2O_3/ZrO_2' *J. Mater. Sci.* **5** 470–473

[25] Matson L E, Hay R S and Mah T 1990 'Characterization of alumina/yttrium-aluminum garnet and alumina/yttrium-aluminum perovskite eutectics' *Ceram. Eng. Sci. Proc.* **11** 995–1003

[26] Kawakami S 1991 'Directional solidification of the Al_2O_3–$Y_3Al_5O_{12}$ eutectic by the floating zone method' *Reports of the Government Industrial Research Institute, Nagoya* **40** 1, 54–57

[27] Sorrell C C, Beratan H R, Bradt R C and Stubican V S 1984 'Directional solidification of (Ti,Zr) carbide–(Ti,Zr) diboride eutectics' *J. Am. Ceram. Soc.* **67** 190–194

[28] Epelbaum B M, Yoshikawa A, Shimamura K, Fukuda T, Suzuki K and Waku Y 1999 'Microstructure of $Al_2O_3/Y_3Al_5O_{12}$ eutectic fibers grown by μ-PD method' *J. Crystal Growth* **198/199** 471–475

[29] Caslavsky R L and Viechnicki D J 1980 'Melting behaviour and metastability of yttrium aluminum garnet (YAG) and AlO_3 determined by optical differential thermal analysis' *J. Mater. Sci.* **15** 1709–1718

[30] Cockayne B 1985 'The use and enigmas of the Al_2O_3–Y_2O_3 phase system' *J. Less-Common. Met.* **114** 199–206

[31] Mizutani Y, Yasuda H and Ohnaka I 2002 in preparation

Chapter 23

Rapid solidification of peritectics

Kuzuhiko Kuribayashi, Yuzuru Takamura and Kosuke Nagashio

Petitectic solidification

Figure 23.1 shows a typical peritectic phase diagram, where β and γ are the primary phase and peritectic phase, respectively. In this phase diagram, if a sample, the chemical composition of which is that of γ, is cooled from the completely molten state, the β phase is first precipitated at the liquidus temperature T_L. When the temperature is cooled further, the chemical composition of the remaining melt is changed from the initial composition toward C_p along the liquidus line. At the peritectic temperature T_p, the incongruent melting point of γ, the γ phase is precipitated by the reaction

$$L + \beta \rightarrow \gamma,$$

where L represents a liquid with the bulk composition C_p. For this reaction to be continued under the condition of unidirectional solidification, three distinct elemental processes, the dissolution of β into L, the transport of the solute to γ, and an interface reaction which incorporates solute onto the lattice sites of γ, must occur successively at the triple junction of β, γ and L under an undercooling below T_p [1]. If a particle of β that is relatively large remains undissolved, it will be engulfed by the γ phase. The engulfed β phase gradually diminishes in size. Since the diminishing rate is controlled by solute diffusion from β to the L/γ interface through the surrounding γ phase, it will take an extremely long time for the reaction to be completed.

Single crystals of peritectic phases such as γ are usually grown by solution growth at temperatures below T_p, because in melt growth at temperatures above T_p it is difficult to inhibit precipitation of the β phase. Even in solution growth, it takes a long time for a bulky γ crystal to be grown, and in order to finish crystal growth within a reasonable time, precipitation of the β phase must still be suppressed.

Deep undercooling is one of the most promising methods for precipitating the peritectic γ phase directly from the melt, because the driving force for forming the peritectic phase vanishes at the congruent

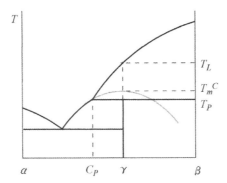

Figure 23.1. Typical peritectic phase diagram for functional oxide materials.

melting temperature, T_m^c. Direct solidification of peritectic γ thus becomes possible at undercoolings greater than $T_L - T_m^c$.

Containerless processing and electromagnetic levitation

Containerless processing, which is free from heterogeneous nucleation on the crucible wall and, therefore, can deeply undercool the melt, has been established for studying undercooling-related phenomena, such as nucleation, dendrite growth dynamics, nonequilibrium phase formation and thermophysical property measurements of molten materials.

Electromagnetic levitation (EML), in particular, has contributed significantly to the research of undercooling-related phenomena in metallic materials [2, 3]. For nonmetallic materials, however, fewer studies have been reported, because the electrical resistivities are too high for the materials to be levitated by electromagnetic levitation. The aero-acoustic levitator has been developed for levitating and heating electrically non-conductive materials, such as glass, ceramics and other chemically reactive materials [4].

Oxide peritectics

Superconducting oxides, $REBa_2Cu_3O_{7-x}$ (RE123) where RE can be, e.g., Y, Nd or Sm, and magneto-optic materials for magnetic bubble memories, $Y_3Fe_5O_{12}$ (yttrium iron garnet, YIG), are typical examples of oxides which show a peritectic phase diagram. Several trials for direct growth of the superconducting phase from the undercooled melt have been carried out on $YBa_2Cu_3O_{7-x}$ (Y123) by aero-acoustic levitation [5] and a drop-tube apparatus [6]. However, they resulted entirely in eutectic microstructures of Y_2BaCuO_5 (Y211) and $BaCuO_2$. The congruent growth of Y123 from the undercooled melt has never been reported.

Recently, Nagashio et al [7] carried out containerless solidification of $NdBa_2Cu_3O_{7-x}$ (Nd123) using aero-acoustic levitation, and observed a huge amount of the superconducting phase in the as-processed sample, the volume fraction of which was more than 90%. Their result is very successful for the purpose of forming the peritectic phase in a short time. The reason, however, why the Nd123 phase could be grown directly from the undercooled melt, when the Y123 phase could not, remains unclear.

This chapter describes the containerless processing of Nd123 and yttrium iron garnet by aero-acoustic levitation. Nucleation and phase selection in the deeply undercooled melt have been investigated in order to establish the conditions for direct growth of the peritectic phases by containerless processing. This chapter concentrates on solidification of the peritectic phase in levitated droplets. Chapter 11 discusses unidirectional solidification of peritectic metallic alloys, and chapter 24 discusses solidification of two-phase peritectic oxides consisting of primary-phase particles distributed throughout the peritectic phase as a matrix.

Aero-acoustic levitation

The aero-acoustic levitation experimental setup combined with a CO_2 laser irradiation system was used for the undercooling experiments. Figure 23.2 shows a schematic view of the experimental arrangement. Pre-sintered Nd123 and yttrium iron garnet (YIG) were initially melted in a water-cooled laser hearth to be shaped as small drops. The samples with a mean diameter of 3 mm or less were levitated by aero-acoustic levitation and heated by CO_2 laser irradiation up to temperatures above their equilibrium melting points under ambient air. The sample temperatures were calibrated by measuring the liquidus temperature or the peritectic temperature, monitored by a high-speed two-colour pyrometer with a spot size of 1 mm diameter. Shutting out the laser beam, the samples were cooled down rapidly. The cooling rate was typically 400 K s^{-1} in Nd123. The rapidly solidified samples were characterized by X-ray diffraction, scanning electron microscopy, and energy dispersive X-ray microanalysis.

In some cases, solidification was allowed to occur without any intervention, i.e. nucleation was allowed to take place spontaneously in the melt during cooling, followed by recalescence and complete solidification. In other cases, solidification was forced to occur by seeding the melt by inserting a single crystal at a controlled undercooling, followed again by recalescence and complete solidification. This was in order to reveal the relation between the microstructure of the as-solidified material and the undercooling at which nucleation was initiated.

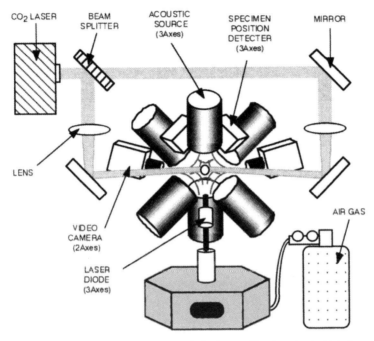

Figure 23.2. Schematic view of aero-acoustic levitator. Heating the levitated sample is carried out by CO_2 laser irradiation.

Cooling curves

Cooling curves of samples showing nearly maximum undercooling are shown in figure 23.3, where the left hand figure is for Nd123 at an undercooling of 149 K below T_p and the right hand figure is for yttrium iron garnet at an undercooling of 158 K below T_p. Almost all samples were successfully undercooled more than 100 K below T_p. After spontaneous nucleation, the temperatures increased rapidly up to near T_p, with pronounced recalescence, as shown in figure 23.3.

Solidified phases

Figure 23.4 shows the corresponding X-ray diffraction pattern for the sample of Nd123 spontaneously nucleated at high undercooling, as shown in figure 23.3, compared with a standard sample synthesized by conventional sintering. In the sintered specimen, diffraction from the Nd123 phase is clearly observed. The diffraction peaks in the spontaneously nucleated sample are quite similar to those of the standard specimen, and diffraction

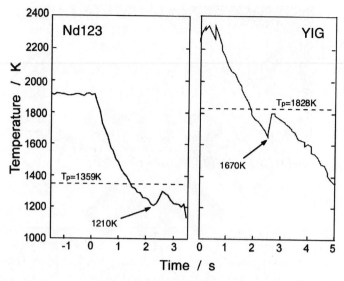

Figure 23.3. Cooling curves of samples showing undercooling and recalescence.

peaks from other phases, such as $Nd_4Ba_2Cu_2O_{20}$ (Nd422) and $BaCuO_2$ are not seen. Figure 23.5 shows a scanning electron micrograph from a sample whose chemical composition is slightly deviated from stoichiometric Nd123 on the copper-rich side. The bright region consisting of superconducting phase shows a mixed pattern of cells and dendrites as in stoichiometric Nd123 [7]. The volume fraction of superconducting phase shows a maximum of more than 98%, in the undercooled and spontaneously nucleated samples.

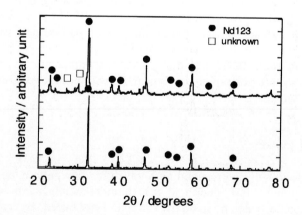

Figure 23.4. X-ray diffraction patterns from Nd123. The upper pattern is from a spontaneously nucleated sample, and the lower pattern is from a standard sample made by conventional sintering.

Figure 23.5. Scanning electron microscope photograph of a cross-section of spontaneously nucleated Nd123.

X-ray diffraction patterns from samples of yttrium iron garnet are shown in figure 23.6 [8]. The upper pattern is from the sample seeded at 1750 K and the lower one is from the sample nucleated spontaneously at nearly the maximum undercooling. In the sample seeded at 1750 K, medium intensity diffraction peaks are observed from the yttrium iron garnet phase together with high intensity diffraction peaks from $YFeO_3$ (orthoferrite) and FeO_x. No yttrium iron garnet phase is observed in the sample spontaneously nucleated at the maximum undercooling. Figure 23.7 shows a sketch of the cross section of the 1750 K nucleated sample, where the garnet phase grown from the seed crystal is located at the

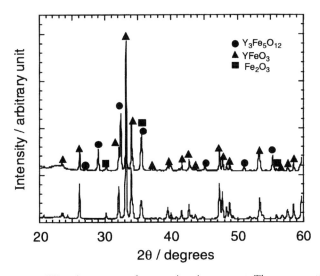

Figure 23.6. X-ray diffraction patterns from yttrium iron garnet. The upper pattern is from a sample seeded at 1750 K, and the lower pattern is from a spontaneously nucleated sample.

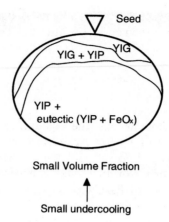

Figure 23.7. Sketch of the cross-section in the seeded sample of yttrium iron garnet.

mantle area near the nucleation point. Away from the nucleation point, the volume fraction of yttrium iron garnet was reduced and $YFeO_3$ and the eutectic between $YFeO_3$ and FeO_x become dominant.

Melt synthesis versus sintering

In earlier studies on the containerless processing of $YBa_2Cu_3O_{7-x}$ (Y123) [4], post-processing annealing was performed in two steps, the first step at a rather high temperature and the second at a relatively low temperature, such as 14 h at 1200 K and then 40 h at 690 K. If the Y123 phase can be grown congruently from the undercooled melt, the first-step annealing is unnecessary. Such high-temperature annealing destroys the advantage of containerless processing, since the Y123 phase can be synthesized at around 1200 K directly by sintering the mixed constituent oxides Y_2O_3, BaO and CuO. The sintering synthesis of RE123 is usually carried out at around 1200 K.

Nucleation rates

According to the classical theory of nucleation, the activation energy required to form a nucleus of critical size is given by

$$\Delta G^* = \frac{16\pi\sigma^3}{3\Delta G^2}, \tag{23.1}$$

where σ is the solid–liquid interface energy and ΔG is the driving force for congruent solidification, i.e. the Gibbs free energy difference between liquid and solid. In the case of congruent solidification from the melt of a stoichiometric compound, such as Nd123, it can be assumed that the driving force ΔG is given by

$$\Delta G = \frac{\Delta H_f \Delta T}{T_E}, \qquad (23.2)$$

similar to pure materials, where ΔH_f, ΔT and T_E are respectively the heat of fusion, the total undercooling and the equilibrium melting temperature. Since ΔH_f is the product of T_E and the entropy of fusion ΔS_f, ΔG is proportional to ΔS_f. There are no data for the entropy of fusion of Nd123. However, the entropy of fusion of Nd422 is expected to be smaller than that of the primary phase $Nd_4Ba_2Cu_2O_{20}$ because its atomic configuration is simpler. This assumes that the entropy of fusion is principally configurational. In addition, lattice site substitution between barium and neodymium can also reduce the entropy of fusion in Nd123. The solid–liquid interface energy σ of Y123 is equal to or slightly larger than that of Y211 [9], and a similar relation is expected for the solid–liquid interface energy of Nd123 and Nd422.

Based on the above discussion, the nucleation of the primary phase $Nd_4Ba_2Cu_2O_{20}$ is expected to dominate over that of the peritectic phase, even below the peritectic temperature. Nevertheless, the Nd123 superconducting phase can be grown directly from undercooled melts.

Growth rates

In the framework of the free-dendrite growth theory in undercooled melts, the total undercooling ΔT consists of four components,

$$\Delta T = \Delta T_t + \Delta T_r + \Delta T_c + \Delta T_k, \qquad (23.3)$$

where ΔT_t, ΔT_r, ΔT_c and ΔT_k are thermal undercooling, curvature undercooling, constitutional undercooling and kinetic undercooling, respectively. The kinetic undercooling ΔT_k is usually very small and can be neglected for an atomically rough interface which is typically observed in metallic materials. In nonmetallic materials, however, which have an atomically smooth interface showing a faceted pattern, atoms coming to the interface must migrate across the interface before being incorporated at a suitable site. In this case, the kinetic undercooling ΔT_k is the dominant factor that controls the growth rate. The constitutional undercooling ΔT_c is the secondary rate-controlling factor for peritectic oxide systems, particularly in the growth of the primary phase whose chemical composition is far from that of the bulk liquid.

Figure 23.8 shows growth rate V versus undercooling ΔT for a peritectic system schematically, where T_q is the temperature below which the growth

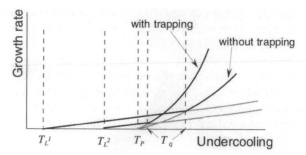

Figure 23.8. Schematic relation between growth rate and undercooling for primary and peritectic phases.

rate of the peritectic phase is greater than that of the primary phase. The superscript i denotes the following: $i = 1$ is when the difference between T_L and T_p is large, and $i = 2$ is when the difference between T_c and T_p is small. It is assumed that $dV/d\Delta T$ of the peritectic phase is larger than that of the primary phase, because long-range diffusion of solute rejected from the moving primary phase interface makes its motion sluggish. At very low temperatures, however, non-equilibrium solute and disorder trapping are expected to make the interface more mobile.

When melts are undercooled sufficiently below T_q, the peritectic phase dominates the microstructure. The reason why less of the peritectic phase is observed in Y123 can be rationalized if T_q is lower than the maximum undercooling. The phase diagram of Nd123 is quite different from that of Y123 as shown in figure 23.9, where bold and fine lines correspond to Nd123 and Y123, respectively. The peritectic temperature T_p of Nd123 is

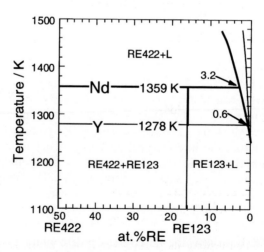

Figure 23.9. Phase diagrams of Nd123 (bold lines) and Y123 (fine lines).

81 K higher than that of Y123, and the slope of the liquidus line is smaller than that of Y123. It means that T_q of Y123 is much lower than that of Nd123 provided that $dV/d\Delta T$ of primary and peritectic phases in Nd123 are equal to those in Y123. In addition, the kinetic barrier for growth of Nd422 is expected to be higher than that of Y211, and non-equilibrium trapping in Nd123 derived from the nonstoichiometry of the phase, $Nd_{1+y}Ba_{2-y}Cu_3O_{7-x}$, unlike Y123, reduces $dV/d\Delta T$ of the primary phase and increases $dV/d\Delta T$ of the peritectic phase.

Seeding

The above hypothesis can be verified by forced nucleation by seeding a single crystal into the melt at a controlled undercooling. Figure 23.10 shows the volume fractions of the Nd123 and Nd422 phases in as-solidified samples as a function of the nucleation temperature. The sum of Nd123 and Nd422 is less than unity because the volume fraction of post-recalescence phases showing peritectic and eutectic microstructures has been eliminated. As shown in the figure, the volume fraction of Nd422 decreases while that of Nd123 increases drastically below the peritectic temperature T_p. The seeding experiment suggests that T_q is nearly equal to T_p in Nd123.

In the case of yttrium iron garnet, the microstructures of samples nucleated spontaneously consist of eutectic $YFeO_3$ and Fe_2O_3, and no garnet phase. As already shown in figure 23.7, seeding with a single crystal of yttrium iron garnet induces growth of the garnet phase, located at the

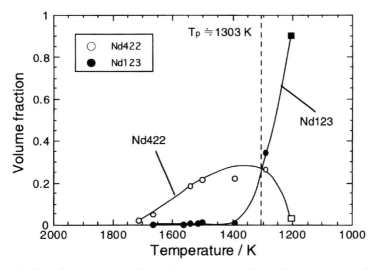

Figure 23.10. Relation between the volume fraction and seeding temperature for the Nd123 and Nd422 phases.

mantle area near the seeding point. Away from the seeding point, the volume fraction of yttrium iron garnet decreases and $YFeO_3$ and eutectic $YFeO_3$ and FeO_x become dominant. It is expected that T_q is near T_p in yttrium iron garnet, because the difference between liquidus and peritectic temperatures, T_c and T_p, which are 1970 K and 1823 K respectively, is small compared with that in Nd123. Nevertheless, the volume fraction of the peritectic phase in as-solidified yttrium iron garnet samples is much less than in Nd123. One reason for this difference is that the kinetic barrier for growth of $YFeO_3$ is lower than that of yttrium iron garnet. It is well known that the kinetic coefficient for growth of the garnet phase is very large [10]. A second reason for the difference is that the energy barrier for nucleation of the garnet phase is large because of the huge unit cell consisting of hundreds of atoms [11]. As a result, the free energy barrier for nucleation and the kinetic barrier for growth of the perovskite phase are expected to be low enough that the metastable eutectic between perovskite and the constituent oxide could be formed even below the peritectic temperature. A similar metastable eutectic structure is observed in $Y_3Al_5O_{12}$ (yttrium aluminum garnet, YAG), that shows a congruent melting temperature [12].

Summary

Superconducting oxides, $NdBa_2Cu_3O_{7-x}$ (Nd123), and magneto-optic materials, $Y_3Fe_5O_{12}$ (yttrium iron garnet, YIG), levitated by the aero-acoustic levitator can be deeply undercooled below their peritectic temperatures. Yttrium iron garnet nucleates spontaneously to solidify with a eutectic structure of $YFeO_3$ and Fe_2O_3, while Nd123 nucleates spontaneously to solidify mostly as the superconducting Nd123 phase with a typical dendrite structure. Seeding with a single crystal at a controlled undercooling reveals that the driving force for congruent growth of the peritectic phase becomes non-vanishing when the melt is undercooled below the peritectic temperature, and the peritectic phase becomes dominant when the melt is undercooled sufficiently below T_q, the temperature below which the growth rate of the peritectic phase is greater than that of the primary phase.

References

[1] Kerr H W and Kurz W 1996 *Int. Mater. Rev.* **41** 129
[2] Herlach D M, Cochrane R F, Egry I, Fecht H J and Greer A L 1993 *Int. Mater. Rev.* **38** 273
[3] Koseki T and Flemings M C *Metall. Mater. Trans.* 1995 **26A** 2991, 1996 **27A** 3226 and 1997 **28A** 2385
[4] Weber J K R, Hampton D S, Merkley D R, Rey C A, Zatarski M M and Nordine P C 1994 *Rev. Sci. Instrum.* **65** 456

[5] Weber J K R, Zima W P, Nordine P C, Goretta K C and Poeppel R B 1993 in: *Containerless Processing* ed W Hofmeister (TMS) p 123
[6] Olive J R, Hofmeister W H, Bayuzick R J, Carro G, McHugh J P, Hopkins R H and Vlasse M 1993 in: *Containerless Processing* ed W Hofmeister (TMS) p 111
[7] Nagashio K, Takamura Y, Kuribayashi K and Shiohara Y 1999 *J. Crystal Growth* **200** 118
[8] Takamura Y, Nagashio K and Kuribayashi K Proc. Drop Tower Days 1998 *Hokkaido* in press
[9] Shiohara Y unpublished
[10] Goernert P 1981 *J. Crystal Growth* **52** 88
[11] Bennema P, Giess F A and Weidenborner J E 1983 *J. Crystal Growth* **62** 41
[12] Nagashio K, Takamura Y and Kuribayashi K *Proceedings of the 3rd International Conference on Solidification and Gravity* ed R Roosz and M Rettenmayr (Transtech) in press

Chapter 24

Peritectic solidification of superconducting oxides

Yuh Shiohara, Makoto Kambara, Teruo Izumi and Yuichi Nakamura

Introduction

The melt growth process [1–3] is effective for attaining a high critical current density (J_c) for practical applications of $REBa_2Cu_3O_x$ oxide superconductors (RE123, RE = Y, Nd, Sm, ...), since it enables the growth of large and highly oriented textures. It is also effective for introducing particles of the high-temperature stable phase (RE_2BaCuO_5, RE211) into the RE123 phase matrix as strong magnetic flux pinning sites, through peritectic solidification of the RE123 crystals. RE123 superconducting oxides solidify through a peritectic reaction between the higher temperature stable solid phase RE211 and the liquid phase, governed by solute diffusion in the liquid phase. Steady-state solidification models based on this mechanism were suggested by different research groups at almost the same time [4–6], and explain the basic phenomena about the solidification behaviour in RE123 superconducting materials. In addition, however, several other important concepts, such as pushing/trapping of the high-temperature stable phase RE211 particles by the solidification front [7], need to be considered to understand the growth mechanism of the system in further detail.

In this chapter, more recent progress in studying the solidification behaviour of Y123 superconducting oxide, especially non-steady-state growth, is reported. Experimental investigations into the growth of bulk materials [8] are described, and a simple solidification model for non-steady-state growth is developed.

Experimental procedure

Bulk growth of Y123 crystals was carried out by the undercooling and seeding method. Precursor powder mixtures with compositions Y:Ba:Cu = 1.4:2.2:3.2

and 1.8 : 2.4 : 3.4 with 0.5 wt% Pt were selected and uniaxially pressed into pellets of 20 mm diameter and 7–10 mm thickness. The sample pellets were placed on an MgO single-crystal substrate in a furnace and were heated up to a maximum temperature of 1150°C, above the peritectic temperature of 1010°C. After holding at 1150°C for 1 h, the samples were cooled to the seeding temperature and a piece of seed crystal was placed on the sample surface. After seeding, the samples were rapidly cooled down to the growth temperature T_h, below the peritectic temperature, and held for predetermined holding times t_h. Finally, the samples were quenched by dropping into a coolant. The growth length, microstructure and compositional distributions in the samples were observed by optical microscopy and electron probe microanalysis (EPMA). Since the crystal size in the *ab* plane is important for the superconducting current loop in bulk applications, the growth characteristics were studied mainly along the *a* direction. More details about these experiments are reported elsewhere [8].

Growth rates

Figure 24.1 shows a typical top view of a melt grown sample. The Y123 crystal grew epitaxially from the Nd123 seed crystal. The growth lengths of Y123 crystals grown under different conditions can be defined as the distance from the edge of the seed crystal to the 123/liquid interface. Figure 24.2 shows the relation between the Y123 growth length along the *a* axis of the crystal and holding time t_h under different growth conditions, i.e. for different growth temperatures T_h and Y211 contents. In the early stages of growth, the growth length increases proportionally with increasing holding time at all growth temperatures and Y211 contents. However, the Y123 growth rate begins to decrease under all conditions with a holding time longer than about 10^5 s. The Y123 growth rate with 40 mol% excess Y211 decreases earlier and the crystal size is also smaller than with 20 mol% excess Y211, even though the growth rates were similar in the

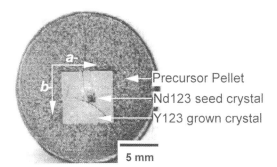

Figure 24.1. Typical top view of a solidified Y123 crystal grown from an Nd123 seed.

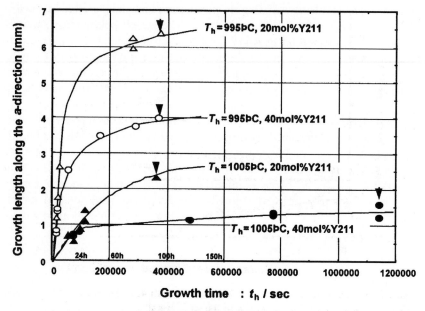

Figure 24.2. Relation between crystal growth length and growth time.

early stages of growth. These results suggest that Y123 growth depends not only on undercooling but also on the amount of the excess Y211.

Distribution of Y211 particles

To clarify the effect of the Y211 particles on the growth of Y123, volume fractions of Y211 particles were evaluated in both the Y123 solid and the liquid regions for samples quenched during solidification, using quantitative electron probe microanalysis. Figure 24.3 shows the relation between the evaluated Y211 volume fraction and distance from the seed crystal. The Y211 volume fraction in the Y123 crystal region is lower than expected from the initial material composition and the phase diagram, and gradually decreases during growth of the Y123 crystal. On the other hand, the Y211 volume fraction in the liquid is higher than expected from the phase diagram. This phenomenon can be explained by particle pushing behaviour. The Y211 particles are pushed by the growing Y123 interface and accumulate in the liquid [7].

Figure 24.4 shows distributions of Y211 particles in the liquid, evaluated from quenched samples (shown by arrows in figure 24.2). The Y211 volume fractions in the liquid are again higher than expected from the initial material composition and the phase diagram, with a similar value of about 0.6 near

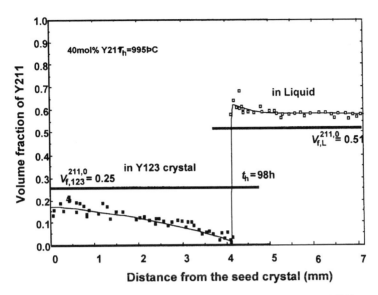

Figure 24.3. The volume fraction of Y211 particles with 40 mol% excess Y211 grown at 995 °C as a function of the distance from the seed crystal.

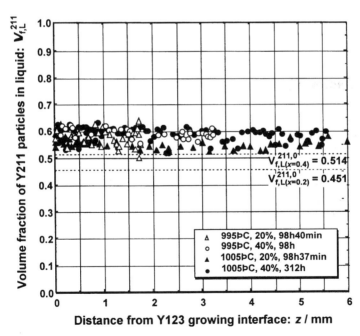

Figure 24.4. The volume fraction of Y211 particles in the liquid after quenching when the Y123 crystal growth rate decreased sharply.

the solid–liquid interface, irrespective of the initial material composition and growth conditions. These results indicate that the Y211 particles are pushed by the growing Y123 crystal, and accumulate ahead of the Y123 solid–liquid interface during solidification and pile up in the liquid just at the point when the crystal growth rate decreases sharply.

Y123 crystal growth

The Y123 crystals show non-steady-state growth behaviour. In the RE123 system, the growth of the RE123 crystals is known to proceed by solute diffusion through the liquid phase between RE211 particles dispersed in the liquid and the RE123/liquid interface. The compositional difference between RE211/liquid and RE123/liquid interfaces is the driving force for the solute diffusion. This mechanism is shown schematically in figure 24.5, assuming local equilibrium at the Y123/liquid and Y211/liquid interfaces for simplicity except in the peritectic reaction boundary layer.

When the initial material composition is on the line connecting Y123 and Y211, the average compositions in the liquid and in the crystal are also on this line, even though the pushing of Y211 particles occurs. In this case, the compositions of the Y123/liquid interface (point A in figure 24.5) and the Y211/liquid interface (point B in figure 24.5) keep the same value during solidification at a constant undercooling temperature. The growth rate of the Y123 crystal, therefore, should not change. However if the average composition is out of the Y123–Y211 tie-line (e.g. CuO rich), the liquid composition near the two interfaces will be A' and B' in figure 24.5, and will

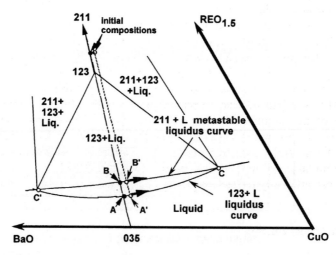

Figure 24.5. Schematic diagram of the ternary phase diagram in the Y system showing the compositions at the Y123/liquid and Y211/liquid interfaces.

gradually shift towards the Y123–Y211–liquid three-phase equilibrium point C in figure 24.5. For BaO-rich material, the liquid compositions near the two interfaces will gradually shift towards the other three-phase equilibrium point C′ in figure 24.5. In this case, the compositional difference between the interfaces, which is the driving force for solute diffusion, decreases to zero and crystal growth stops. From conventional solidification modelling, this compositional shift caused by a final transient will occur only in the final stage of solidification when the deviation from the tie-line is not large.

Effect of Y211 particles

Piling up of Y211 particles near the Y123 interfaces as shown in figure 24.4 suggests that such a high volume fraction of Y211 particles may also play a role in the non-steady-state growth process. In diffusional solidification models for Y123 growth [4–6], interaction among Y211 particles is ignored and Y211 particles are considered only as sources of solute. However, interactions such as coarsening do take place, and the effective area through which solute diffusion can take place is determined by the size and volume fraction of the Y211 particles in the liquid. In peritectic Y123 growth, the Y211 particles dissolve in the liquid near the Y123 interface. Consequently, the volume fraction of Y211 should decrease gradually near the Y123/liquid interface. At the same time, residual Y211 particles are pushed ahead of the Y123/liquid interface and accumulate in the liquid. There should therefore be a drastic change in the Y211 particle volume fraction near the interface, rising rapidly from a low value in the dissolution region to a high value in the particle accumulation region.

The large volume fraction of Y211 particles caused by pushing will effectively impede and decrease solute diffusion away from the Y123/liquid interface, because of an effective decrease in the diffusion area in the liquid. There is, therefore, a local boundary region near the interface where the compositional shift to the three-phase equilibrium point C on figure 24.5 is enhanced easily because of its small volume compared with the size of the sample. This explains why the Y123 crystal growth rate decreases sharply with crystal size, as shown in figure 24.2. Any slight compositional shift from the Y123–Y211 tie-line will be enhanced in the local boundary region of Y211 particle accumulation, and will cause a final transient locally to decrease the Y123 crystal growth rate.

Solidification model

Based on the ideas described in the previous section, we consider a simple solidification model which can explain the local compositional shifts. In this model, we only consider the solidification phenomena after the Y211 particles have piled up in the liquid, and after solute diffusion away from the Y123/liquid interface has been prevented by the large volume fraction of Y211 particles. This model describes, therefore, the extreme case of the local final transient.

We make the following assumptions:

1. the Y123 crystal grows under isothermal conditions
2. the composition at the Y123 liquid/interface is uniform
3. the dissolution of Y211 particles as part of the peritectic reaction occurs at the outer end of the diffusion boundary layer ($z = \delta$)
4. there is no curvature effect of the Y211 particles
5. the Y123 crystal growth interface is flat
6. physical and thermodynamic properties are constant with temperature and composition.

Under these conditions, we consider the non-steady solute diffusion in the diffusion boundary layer and four solute fluxes which affect the boundary conditions at each end of the boundary layer. Figure 24.6 shows a schematic diagram of the diffusion boundary layer for the peritectic reaction. The basic non-steady diffusion equation for each solute in the boundary layer is

$$\frac{\partial C^i}{\partial t} = D_L \frac{\partial^2 C^i}{\partial z^2} + R_1 \frac{\partial C^i}{\partial z}, \qquad (24.1)$$

where C^i is the mole fraction of i solute in the liquid (i = Y, Ba, and Cu), t is the time, D_L is the diffusivity in the liquid within the boundary layer, z is the distance from the Y123 crystal interface and R_1 is the growth rate of the Y123 crystal.

At the Y123/liquid interface ($z = 0$), the boundary conditions are described by the following fluxes:

1. The necessary solute flux J_0^i for Y123 growth is given by

$$J_0^i = (1 - V_{f,123}^{211})(C_{L,123}^i - C_{S,123}^i)R_1, \qquad (24.2)$$

where $C_{S,123}^i$ is the mole fraction of i in the Y123 crystal, $C_{L,123}^i$ is the mole fraction of i in the liquid at the Y123/liquid interface, and $V_{f,123}^{211}$ is the volume fraction of Y211 particles in the Y123 crystal.

2. The solute diffusion flux $J_{1,(z=0)}^i$ to the Y123/liquid interface is given by

$$J_{1,(z=0)}^i = -(1 - V_{f,b}^{211})D_L \frac{\partial C^i}{\partial z}\bigg|_{z=+0}, \qquad (24.3)$$

where $V_{f,b}^{211}$ is the volume fraction of Y211 particles in the boundary layer.

Solidification model

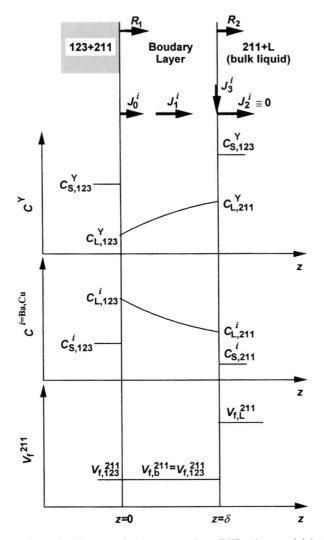

Figure 24.6. Schematic diagram of the non-steady solidification model including the distributions of the solute compositions and the volume fraction of Y211 around the boundary layer.

3. Using these two fluxes, $J_0^i - J_{1,(z=0)}^i$ (the mass balance at the interface) is given by

$$\frac{\partial C_{L,123}^i}{\partial t} \Delta z = J_0^i - J_{1,(z=0)}^I, \tag{24.4}$$

where Δz is the minimal volume at the interface.

In addition, assuming local equilibrium conditions at the interface, the liquid composition at $z = 0$ is on the Y123/liquid liquidus line

$$f_{123}(C^i_{L,123}) = C^Y_{L,123} + \alpha_{123} C^{Ba}_{L,123} + \beta_{123} = 0, \tag{24.5}$$

where f_{123} is a function of the liquidus line obtained from the ternary phase diagram and assumed to be linear.

Similarly, the boundary conditions at the outer end of the boundary layer ($z = \delta$) are described by the following fluxes:

1. The diffusion flux, J^i_2, away from the outer end of the boundary layer is non-zero, but is assumed to be negligible in the limiting case

$$J^i_2 = -(1 - V^{211}_{f,L}) D_{L,e} \left.\frac{\partial C^i}{\partial z}\right|_{z=\delta+0} \cong 0, \tag{24.6}$$

where $V^{211}_{f,L}$ is the volume fraction of Y211 particles in the bulk liquid and $D_{L,e}$ is the effective diffusivity in the bulk liquid.

2. The solute diffusion flux to the Y123/liquid interface is given by

$$J^i_{1,(z=\delta)} = -(1 - V^{211}_{f,b}) D_L \left.\frac{\partial C^i}{\partial z}\right|_{z=\delta-0}. \tag{24.7}$$

3. The solute flux from dissolution of Y211 particles at the outer end of the boundary layer ($z = \delta$) is given by

$$J^i_3 = (V^{211}_{f,L} - V^{211}_{f,b})(C^i_{S,211} - C^i_{L,211}) R_2, \tag{24.8}$$

where $C^i_{S,211}$ is the mole fraction of i in the Y211 crystal, $C^i_{L,211}$ is the mole fraction of i in the liquid at the Y211 interface, and R_2 is the rate of motion of the outer end of boundary layer.

Using these three fluxes, J^i_2, $J^i_{1,(z=0)}$ and J^i_3 the mass balance condition, the compositional change at the outer end of the boundary layer is given by

$$\frac{\partial C^i_{L,211}}{\partial t} \Delta z = J^i_{1,(z=\delta)} - J^i_2 + J^i_3. \tag{24.9}$$

From the local equilibrium condition, the composition at $z = \delta$ should be on the Y211/liquid liquidus line.

$$f_{211}(C^i_{L,211}) = C^Y_{L,211} + \alpha_{211} C^{Ba}_{L,211} + \beta_{211} = 0, \tag{24.10}$$

where f_{211} is the function of the Y211/liquid liquidus line from the ternary phase diagram. In this model, it should be noted that the length of the boundary layer is determined as

$$\delta = \delta_0 - \int_0^t (R_1 - R_2)\, dt, \tag{24.11}$$

where δ_0 is the initial length of the boundary layer.

Modelling results

The equations described in the last section and the material properties summarized in table 24.1, can be used to calculate changes in the growth rate and the compositions in the boundary layer from initial compositions slightly deviated from the line connecting Y211 and Y123. Figure 24.7 shows a typical result obtained from an initial composition slightly deviating from the Y211–Y123 tie-line to the Cu-rich side. This initial compositional shift may be due to several reasons, including non-stoichiometry of the RE123 phase, slight differences in diffusivity of the different solutes and small experimental errors in preparation of the precursor mixed oxide pellets.

As shown in figure 24.7(a), the growth rate of the Y123 crystal interface, R_1, and the rate of motion of the outer end of the boundary layer, R_2, have almost the same value. Both growth rates begin to decrease gradually after 100 s and reduce to almost zero after 300 s. This decrease in growth rate corresponds to the compositional shift at the end of the boundary layer to the composition of the ternary phase equilibrium point, as shown in figure 24.7(c).

The relationship between the growth time and the growth length, is obtained by integrating the growth rate versus time in figure 24.7(a). The result in figure 24.7(b) is similar in form to that obtained experimentally as shown by comparing figures 24.1 and 24.7(b). However, calculations indicate that the growth rate reduces to zero after a period of 300 s, which is 1000 times smaller than the experimental results. The reason for this difference is that the model only considers phenomena after the accumulation of Y211 particles. In addition, the flux, J_2^i, is assumed to be negligible corresponding only to the limiting case of the local final transient in the boundary layer. The value of J_2^i is not zero when the compositional shift begins, and this flux to the bulk liquid tends to restore the compositions in the boundary layer to their initial ones. When the effect of J_2^i is included, the crystal growth rate will show steady-state behaviour for longer, the rate of the compositional shift will be much slower, and it will take much longer to reach the three-phase equilibrium point.

The effects of Y211 particle pushing, and obstruction of solute diffusion by a large Y211 volume fraction in liquid, help to simulate the non-steady-state peritectic solidification of the RE123 crystals.

Table 24.1. Physical properties and parameters used in the solidification model.

Diffusivity in the liquid, D_L	1×10^{-6} cm^2 s^{-1}
Initial composition of Y and Y123/liquid interface, $C_{L,123}^Y$	5.7003×10^{-3}
Initial composition of Y and Y211/liquid interface, $C_{L,211}^Y$	6.0000×10^{-3}
Volume fraction in the liquid, $V_{f,L}^{211}$	0.6
Initial length of the diffusion boundary layer, δ_0	2×10^{-4} cm

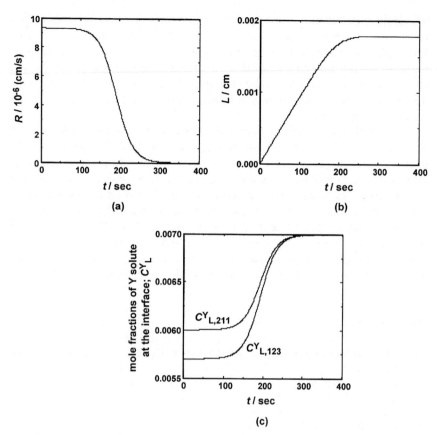

Figure 24.7. Calculations from the non-steady-state growth model as a function of the growth time: (a) growth rates R_1 and R_2, (b) growth length and (c) mole fractions of Y at the Y211/liquid and Y123/liquid interfaces.

Summary

Experimental investigation of isothermal solidification of Y123 crystals shows that the Y123 growth rate decreases gradually and finally reaches zero in the later stages of growth, even at a constant growth temperature. Under all growth conditions Y211 particles accumulate in the liquid near the Y123/liquid interface as the growth rate reduces to zero. These results suggest that the large volume fraction of Y211 particles in the liquid act to obstruct solute diffusion away from the Y123 interface, enhancing the compositional shift to the three-phase equilibrium point in the boundary layer, and causing a final transient locally to reduce the growth rate when the initial composition is not on the Y211–Y123 tie-line.

Based on this idea, we have developed a simple non-steady solidification model in the diffusion boundary layer, which can describe the local final transient phenomenon. Calculations show a non-steady-state growth rate variation with time which has a similar form to experimental measurements of crystal growth rate. The final transient and decrease of growth rate to zero take place readily in the boundary layer from a slight deviation of the initial material composition away from the Y211–Y123 tie-line.

In practical applications of bulk RE123 superconducting oxides, a large single-domain RE123 crystal containing a large volume fraction of RE211 particles as pinning centres is desirable for good magnetic properties. However, the experimental results suggest that large RE211 contents may prevent RE123 crystals from growing very large. A careful selection of initial RE211 content is, therefore, important to grow large RE123 crystals. In addition, control of the size of the RE211 particles and the growth of the RE123 crystal at large undercooling is required to prevent Y211 particles pushing, since accumulation of Y211 particles in the liquid enhances the local final transient and also reduces the RE123 crystal growth rate.

References

[1] Jin S, Tiefel T H, Sherwood R C, van Dover R B, Davis M E, Kammlott G W and Fastnacht R A 1988 *Phys. Rev.* **B37** 7850
[2] Murakami M, Morita M, Doi K and Miyamoto K 1989 *Japan. J. Appl. Phys.* **28** 1189
[3] Fujimoto H, Murakami M, Gotoh S, Oyama T, Shiohara Y, Koshizuka N and Tanaka S 1990 *Advances in Superconductivity II* (Tokyo: Springer) p 285
[4] Izumi T, Nakamura Y and Shiohara Y 1992 *J. Mater. Res.* **7** 1621
[5] Cima M J, Flemings M C, Figuredo A M, Nakade M, Ishii H, Brody H D and Haggerty J S 1992 *J. Appl. Phys.* **72** 179
[6] Mori N, Hata H and Ogi K 1992 *J. Japan. Inst. Met.* **6** 648
[7] Endo A, Chauhan H S, Egi T and Shiohara Y 1996 *J. Mater. Res.* **11** 1996
[8] Kambara M, Yoshizumi M, Umeda T, Miyake K, Murata K, Izumi T and Shiohara Y in preparation

Index

Acoustic optical device (AOD) 299
Activation energy 396
Active feeding 98
Aero-acoustic levitation 392–6
Agglomeration 189, 193
Airslip casting configuration 16–17
Airslip mould technology 22
Al_4C_3 181
Al–Cu, porosity in 126–8
Al–Cu plate castings 128
Al–Fe_4Al_{13} 262–5
Al–Fe alloys 262
Al–$FeAl_6$ eutectic 262–3
Al–$FeAl_m$ eutectic 265
Al–Fe–Si 266
Al–Mg alloy 229
Al–Mg–Zr alloy 271
Al_3Ni 271
Al_3Ni_2 271
Al–Ni alloys
  centrifugal atomization 284
  cooling rates, inert gas atomization 281–4
Al–Ni catalysts 280–1
Al–Ni phase diagram 280–1
$Al_{70}Ni_{13}Si_{17}$ 194
Al–Ni–Si glass 195
$Al_{87}Ni_{10}Zr_3$ 191
Al_2O_3
  clusters floating on low-carbon Al-killed steel melt 321
  critical condition for engulfment of inclusions during solidification of low-carbon Al-killed steel 321

preferred growth of solid–liquid interface toward inclusions 324
pushing of small aggregates of inclusions by advancing solid–liquid interface in low-carbon Al-killed steel 321–3
Al_2O_3/GAP 373–89
  mechanical properties 374–7
  solidification structures 377–8
  stress–displacement curves 377
  three-dimensional configuration of single-crystal GAP 378
Al_2O_3/YAG 373–89
  interlamellar spacing 381
  mechanical properties 374, 375–7
  solidification structures 377–8
  stress–displacement curves 377
  temperature dependence on flexural strength 375
Al_2O_3–Y_2O_3
  cooling curves 384
  eutectic systems 381–5
  phase diagram 382
  scanning electron micrographs 383
Al–Si alloys 228, 251
  calcium addition 332–5
  grain structure 250
  hypereutectic 252–4
  hypoeutectic 181–3, 249–52
  interfacial energy and structure 326–38
  mechanical properties of powders 279
  microstructure 73

415

Al–Si alloys (*continued*)
  refinement of primary aluminium 249–52
  refinement of primary silicon 252–4
  silicon interphase spacing 331
  silicon morphology at solid–liquid interface 332–5, 335
  simultaneous use of duplex casting and inoculants 254
  strontium addition 329–32
Al–Si piston alloys 69–76
  fatigue failure 75
  mechanical properties 75–6
  microstructure 72
Al–Si–Ca alloys 328, 332–5
Al–Si–Ni alloys 266
Al–Si–Sr alloys 328
  influence of cooling rate and tube material in morphology of silicon 336
  morphology of silicon in 329
  silicon interphase spacing 331
Al–Si–Sr–Ca alloys 329
  influence of calcium content and cooling rate on silicon morphology 337
Al–Si–Ti equilibrium diagram 182–3
Al_3Ti 186, 193, 215, 216, 217, 219, 277
Al–Ti alloys 185, 193
Al–Ti–B 214–16, 218, 219, 222–4, 227–8, 232–4, 265
Al–Ti–C 214, 218, 219, 222, 227–8, 234
Al–Ti–C–Fe 229
$Al_3(Ti,Si)$ 193, 196
$Al_5Ti_7Si_{12}$ 182, 193
$Al_3(Ti,Ta)$ aluminide 187
Al–Zr alloys 279
Al_3Zr 192, 217, 271, 279, 280
Al–Zr–Mg alloy 280
Alcan bed filter (ABF) 12
Alpur degasser 9, 10, 11
Aluminium alloys
  4-Hi caster for 40–3
  cold-working 279
  continuous casting 26–47, 258
  controlled solidification 177
  direct chill casting 3–25, 258

entrained droplet melting differential scanning calorimeter 266
experimental development 32
grain refinement 177, 213–14, 242–3, 266
heterogeneous nucleation in 177–98
inoculated 230
intermetallic compounds in solidification 257–70
intermetallic phase selection 268
nucleation mechanisms in undercooled melts 185–7
powder atomization 271–85
wrought 32, 179–81
Aluminium-based glasses 240
Aluminium bronzes 91
Aluminium car parts 67
Aluminium–intermetallic eutectic phases 269
Aluminium shape casting
  casting process selection 133–5
  defects in 121–41
  future 138–40
  numerical simulations 121
Aluminium–transition-metal-lanthanide alloys 239
Analytical models of microporosity formation 123
Analytical systems 19–20
Anisotropy of magnetic energy 290
Array growth
  degree of freedom 156
  of cells and dendrites 157
Array stability limit 151–5, 158
Articulated block casters 27, 28
As-cast equiaxed structures 213
Atomistic models 139
Atomization of aluminium alloy powders 271–85
Automotive alloys 32
Automotive components
  mechanical properties 72
  microstructure 72
Automotive industry 59–68
  cast product design and development 65–6
  casting performance evaluation 66
  casting products 64–5

Index   417

change in materials 62–3
change in structural components 62
change in technologies 60–2
changes in vehicle development 63
cost reduction 59
development lead time 59, 60, 67
endurance test 66
global alliances 59–60
innovative materials and processes 59
market share 60
safety and environmental issues 60
three-dimensional design 65
trends 60–5
vehicle production targets 63–4
weight optimization 59
weight reduction 67

Back-reflection Laue patterns 107
Bamboo structure 240
Banded structure 161, 163–5, 172
Basicity effect on crystallization temperature 51, 52
Batch homogenization 21
BiMn, grain alignment in 291–5
Bi–Mn alloys 291–5
　magnetization curves 293
　scanning electron micrographs 295
　X-ray diffraction patterns 292
Block caster 27
Boride–Al–Ti–Si multi-component systems 193
Borides 179, 187, 188, 190, 191, 195, 197, 267
Bridging 54
Bridgman growth 267–8
Bridgman method 373, 387
Bubble damage 94–6
Bubble trapped by hook 52
Buckling
　defects 36–8, 42
　limit diagram 36–7
　onset of 39
　transition 39
　transition diagram 36–7
Building products 29

Ca–Al alloys 55
Ca–Si alloys 55

Calcium
　addition to Al–Si alloys 332–5
　influence on equilibrium silicon structure 335
Can-end alloys 32
Canstock materials 27
CAP models 129
Car production 60
Carbide additions 179
Carbon steels 91
Carbon tetrabromide–hexachloroethane eutectic 144
Cast gauge
　productivity as function 31
　reduction 30–1, 34, 35
Cast products
　automotive industry 64–5
　evaluation 32
Cast slabs 27
Castex machine 45–6
Casting configurations 13–18
Casting defects, see Defects
Casting length and longitudinal cracks 56
Casting rules 87–105
Casting runner design 66
Casting section thickness and convection damage 100–2
Casting speed 18
　and sump shape 39
　fraction solid as function of 38
Catalysts 281–4
Cd–Sn alloys, periodic structure 173
Cell loss by overgrowth 146
Cell tip splitting 146
Cellular automata (CA)
　continuum model 138
　methods 122
　modelling 228, 228–9, 230
　porosity model 132
　simulations 138, 139
Cellular interface 163
Cellular primary phase 165
Centrifugal atomization
　Al–Ni alloys 284
　equipment 272
　overview 272–3
Centrifugal granulation 271

418 Index

Ceramic foam filters (CFF) 12, 13
Ceramic-matrix composites 373
Ceramics 373
Charge-coupled device (CCD) 300
Chequer-board pattern 242
Cleanliness of molten steel 54, 55
CMS process 82
CMSX-4
   high-pressure turbine blade 113, 114
   microsegregation 116
   single-crystal superalloy 106–7
CO_2 laser irradiation 392, 393
Cold-working, aluminium alloy 279
Collision-limited solid–liquid interface 236–8
Colony boundaries, elimination 380
Columnar structures 208
Columnar to equiaxed transition (CET) 177, 200, 201, 228–30
Columnar to grain-refined transition 212
Competitive growth 261–2
Competitive nucleation 262–4
Computer aided design (CAD) 65, 82, 133
Computer aided engineering (CAE) 60, 67, 79–80
Computer modelling of feeding 99
Computer simulation 386
Confocal laser scanning microscopy 299–300
Conform process 43–5
Congruent compound, phase diagram 291
Constitutional supercooling criterion 307
Containerless processing 391
Continuous casting 286
   aluminium alloys 26–47, 258
   current status 29–30
   experimental development 31–2
   industrial development 32–5
   ratio in Japan 49
   scale of production 27–30
   steels 48–58
Continuous homogenization 22
Continuous uphill advance of meniscus 93–4
Continuum models 138

Continuum numerical solutions of general Stokes flow 124–5
Continuum-stochastic model 132–3
Control technology 55
Controlled solidification of aluminium alloys 177
Convection damage 100–2
Cooling, in DC casting systems 18–19
Cooling rate 184, 223, 226, 234, 235, 260–2
   aluminium alloys 271–85
   and dendrite arm spacing (DAS) 274–5
   effect on insoluble particle distribution 276
   in mould 49
   inert gas atomization 273–5, 281–4
   influence on silicon morphology in Al–Si–Sr alloys 336–7
Copper film 241
Core blows 96–7
Cosworth Process 90, 94, 96
Cracks 92–3
Criterion functions 124
CRU International 29–30
Cryocooled superconducting magnets 286, 287
Crystal growth 217–18
Crystallization temperature
   and heat flux 51
   basicity effect on 51, 52
Crystallographic relationship and high-temperature stability 381
Crystallographic texture 212
Cu–Al alloys 281
Cu–Ni alloys 207
Cu–Ni system 210, 211
$Cu_{30}Ni_{70}$ 208
Cu–O system 210
Cu–Sn system 210
Czochralski process 286

Darcy's law 124
   models 124–8
Davy International 30
Defects 66, 89
   in aluminium shape casting 121–41
   in non-steady-state regions 54–7
   prediction 67

Index 419

Deformation to reduce macroscopic segregation 53–4
Degassing 9
Dendrite arm spacing (DAS) 122, 130, 135, 137
  and cooling rate 274–5
Dendritic grain evolution 130, 131
Dendritic growth, modelling 110–13
Deterministic model for grain nucleation and growth 130–1
Development caster 39–43
Devitrification 234–40
Die-casting
  macroscopic defects in 121
  mechanical and thermal behaviour 79
  mould filling simulation 79–86
  moulds 67
  optimization problems 80
  overview 79–80
  systems 90
Diffusion-controlled growth of spherical crystals 223
Diffusion-limited solid–liquid interface 236–8
Dip tube and float system 13, 16
Direct chill casting 178, 181
  aluminium 200
  aluminium alloy billets 3–25
  aluminium alloys 258
  billet production 4–5
  wrought alloys 213
Directional solidification 200
  see also Unidirectional solidification
Distribution function 289
Droplet solidification 202
Ductile irons 91
Duplex casting 248, 250, 251, 254
  comparison with conventional method 255
Durville-type casting processes 91, 93, 101–2
Dye penetrant 89

Electromagnetic levitation (EML) 391
Electromagnetic stirring 52, 54
Electromagnets 286
Electron back-scattered diffraction (EBSD) 107, 111, 180

Electron beam remelting method 54
Electron beam surface melting 268–9
Electron diffraction pattern analysis 185
Electron microprobe analyser 107
Electron microprobe maps 116
Embrittling temperature range 55, 57
Energy dispersive X-ray analysis 185, 191, 195
Entrained droplet technique 264–6
Epitaxial grain growth 269
Equiaxed grains, solidification rate of 327
Equilibrium partition coefficient 221
Eutectic alloys 160
Eutectic ceramics 373–89
  Al_2O_3/Y_2O_3 381–5
  solidification structures 378–81
  unidirectional solidification 373–89
  Eutectic composites
  compressive flow stress 376
  see also Al_2O_3/GAP; Al_2O_3/YAG
Excess titanium in grain refinement 187–9, 197
Extrusion
  impact on 23–4
  limit diagram 23
  productivity 23

Faceted and non-faceted growth 380
Faceted cells and dendrites 156
Faceted cellular growth 146–7
Fading 189, 193
Fatigue failure, Al–Si piston alloys 75
$FeAl_6$ 263–4, 264
Fe_4Al_{13} 263, 264
Fe–C alloys
  crystal morphology 305
  direct observation 301
  gamma phase formation 311–12
  instability study 303
  peritectic reaction 311–16
  peritectic transformation 311–16
  perturbation at liquid–δ interface 306
  perturbation at liquid–γ interface 305
  phase diagram 306, 307
  high-temperature–low carbon portion 302

Fe–C alloys (*continued*)
  planar cellular transition of δ crystals 304
  planar liquid–δ interface 303, 305, 311–12
  sinsusoidal perturbation developing at liquid–δ interface 304
Fe–C system 298
Fe–C–(Mn) alloys, solid–liquid interface 307
$Fe_{73.5}Cu_1Nb_3Si_{13.5}B_9$ 239, 263, 264
Fe–Nd–B alloys 239
Fe–Ni alloys 207
  peritectic reaction 316
  peritectic tranformation 317
Fe–Si–B system 239
$Fe_{91}Zr_7B_2$ 239
Federal–Mogul 69–73
Feeding systems 97–100
  computer modelling 99
  design 97
  mechanisms 98
  random perturbations from casting to casting 99–100
Fibrous structures 380
Fickian flux 132
Filter provision 90
Filtration methods 12
Finite difference algorithm 133
Finite difference method (FDM) 129
Finite element (FE) algorithm 228
Finite element method (FEM) 59, 60, 129, 138
Finstock 32, 37
Fixed hearth furnaces 5
Flow tube structure 93
Fluctuated structure 161
Fluid flow simulation 66
Focused-ion-beam workstation 242
Foilstock 27, 29, 32
4-Hi caster for aluminium alloys 40–3
4-Hi commercial casting line 43
Fourier equation 129
Fraction solid as function of casting speed 38
Freckling
  prediction 115
  resistance 115

Free growth 217–18
Free-growth modelling 223, 225, 226, 230, 232, 233
Freezing time 100
Front powder 55
Front tension 33

Gadolinium aluminium perovskite (GAP), *see* $Al_2O_3/GdAlO_3$
Gas porosity 132
Gaussian distribution 231
Gaussian size 232
Gibbs free energy 396
Gibbs–Thomson parameter 306
Glass transition temperature 184
Glassy alloys 239
Golden Aluminum 27
Grain alignment in BiMn 291–5
Grain boundaries, pinning by inclusions 321
Grain distributions, comparison of observed and predicted 111–13
Grain growth 138, 200
  deterministic model for 130–1
Grain initiation 213, 230–1
Grain nucleation, *see* Nucleation
Grain refinement 7–9, 22, 179, 248
  aluminium alloys 177, 213–14, 242–3
  basic phenomena and mechanisms 215–19
  by inoculation 213–34
  by rapid quenching and devitrification 234–40
  effective 214
  efficiency 224
  excess titanium in 187–9, 197
  problems 214
  spontaneous 208–13
  thermal modelling of 219–28
Grain refiners 8, 9, 180, 182, 193, 197, 231–4, 233, 266, 277
Grain rotation, kinetics of 290–1
Grain selection in single-crystal superalloy castings 106–20
Grain size 8, 135, 137, 138
  and control in solidification 199–247
  and numbers of grains per unit volume 243

commercial purity aluminium 224, 226
  prediction 134, 230–4
  thin films 240–2
Grain structure 110, 139
  in weld metal 200
Gränges AB 33
Gränges Eurofoil 40
Gravity castings 101
Gravity die-cast aluminium alloys 71
Gravity feeding 98–9
Gravity-filled running systems 95–6
Green sand casting 134
Growth–length relationship 411
Growth model, non-steady-state 412
Growth restriction 178–9, 182
Growth-restriction factor 221, 223, 226, 227
Growth studies 267–9
Growth temperatures 263
Growth–time relationship 411
Growth velocity 262

Hagen–Poiseuille equation 123
Heat flux and crystallization temperature 51
Heat flux density with and without grooves on mould 50
Heat transfer in starter block 108–10
Heat transfer coefficient 38, 109, 126–8, 136
Heat treatment 103
Heterogeneous nucleation 203–6, 235
  in aluminium alloys 177–98
  quantitative modelling 205
Hexagonal devitrification phase 194
HgMnTe 287
HgZnTe 287
High-carbon steel 308
  development of secondary arms 310
  planar to cellular interface transition 309
  tip radius and growth rate of γ crystals 310
High-resolution transmission electron microscopy 190
High-speed thin-strip casting 33, 40, 42–3

High-temperature stability and crystallographic relationship 381
High-velocity effect 155
Holding furnaces 5
Homogeneous nucleation 202–4, 206, 238
Homogenization 257
  of DC cast billet 20–1
Hooks
  bubble trapped by 52
  formation 52
Hot top casting configuration 14–15
Hunt–Jackson criteria 380
Hydrogen concentration 125–8, 133
Hydrogen diffusion 132
Hydrogen-diffusion controlled models 129
Hydrogen removal 9, 11, 22
Hypoeutectic Al–Si alloys 181–3

Ideal gas law 125
Inclusions 52, 54
  counter-measures 54, 56
  pinning of grain boundaries by 321
  solid–liquid interface interaction with 321–4
  variation with casting length 55
Inert gas atomization 271
  cooling rates 273–5
  equipment 274
  overview 273
Infrared imaging furnace 300–1
In-line metal cleanliness 22
In-line metal treatment 6–7
In-line multi-chamber degassing systems 22
Inoculants 214, 254
Inoculation, grain refinement by 213–34
In situ joining 255–6
Insoluble particle distribution, effect of cooling rate on 276
Interface positions, time evolution of 146
Interface shapes for cells and dendrites 149
Interface temperature
  and phase fractions 170–3
  and solute concentrations 170
Interfacial energy in solidification structures 326–8

Intermetallic compounds
  in solidification of aluminium alloys 257–70
  inert gas atomization 281–4
  manufacture 255–6
  precipitated from solution by rapid solidification 279–80
Intermetallic phases
  nucleation and growth 261, 266
  selection in aluminium alloys 268
Investment casting 106
Iron-based glasses 240
Iron-based systems 239
  see also Fe
ISO 9000 87
Ivantsov function 309

Kaiser Micro-Mill process 28–9
Karma model 210, 212
Kurz, Giovanola and Trivedi (KGT) model 111

Ladle exchange 54
Lamellar eutectic 143–5, 156
  analytic solution 143–4
  growth process 145
  spacing selection 145
Lamellar spacing 380–1
Lamellar structure 380
Latent heat release 238–9
Lattice matching 186
Law of mixtures 133
Lead–bismuth alloys 161–4, 169
  banded structure 163–4, 173
  competitive structure 173
  growth morphologies 163, 164
  longitudinal microstructures 163
  phase diagram 163
  unidirectional solidification 162, 164, 173
Levitated drop experiments 211
Lightweight materials 60, 67
Liquid front damage 88–92
Liquid metal front stop 92–4
Liquidus slope 221
Lithographic sheet 37
Location points 103–5
Longitudinal cracks and casting length 56

Lorenz force 286, 288
  macrosegregation control with 286–7
Low-carbon steel
  Al-killed 321–4
  critical conditions for engulfment of Al_2O_3 inclusions during solidification 321
  dihedral angles at δ–γ interphase boundaries formed during $\delta \rightarrow \gamma$ transformation 317–18
  instability at δ–γ interphase boundary during $\delta \rightarrow \gamma$ transformation 320
  morphological instability of δ ferrite–γ austenite interfaces 318–21
  morphological instability of δ–γ interphase boundaries during $\delta \rightarrow \gamma$ transformation 320
  nucleation and growth of δ phase at δ grain boundaries during $\delta \rightarrow \gamma$ transformation 319
  preferred growth of solid–liquid interfaces towards Al_2O_3 inclusions 324
  pushing of small aggregates of Al_2O_3 inclusion by advancing solid–liquid interface 321–3
Low-pressure filling systems 96
Low-temperature gradient effect 155

Macromodelling 129, 135
Macroscopic defects in die casting processes 121
Macroscopic models 138
Macrosegregation control with Lorenz force 286–7
Magnesium alloy pistons 71
Magnesium car parts 67
Magnetic anisotropy energy 290, 292–4
Magnetic fields
  effect on nucleation and growth 291–5
  solidification structure control by 286–97
Magnetic susceptibility and magnetization force 287–8

Magnetization curves 292
  Bi–Mn alloys 293
  magnetic anisotropy energy from 294
Magnetization force 286
  and magnetic susceptibility 287–8
  for structure control 288–90
  melt flow control using 287–8
Master alloys 20
Maximum meniscus velocity 88–90
Maxwell–Hellawell model 219–22, 225
Mechanical properties
  Al–Si piston alloys 75–6
  automotive components 72
Mechanical strength requirements 103
Medium-carbon steel 308
Melt flow control using magnetization force 287–8
Melt growth process 402
Melt quality 88, 96
Melt spinning 234–5, 238
Melting furnaces 5
Mesomodelling 122, 129–33
Mesoscopic models 122, 138
Metal matrix composites 71
Metallic alloy developments 71–3
Metallic glasses 183–5, 265–7
  critical cooling rates 359
  new materials 359
  optimization of composition for large glass-forming ability 360–3
  overview 359–60
  solidification 359–72
Micro/meso-model 122
Microporosity
  analytical models of formation 123
  in as-cast structures 121
  modelling 129
Microprobe dataset 117
Microsegregation
  CMSX-4 116
  numerical treatment 114–15
Microstructure 35–7
  Al–Si alloys 72, 73
  as-solidified 202
  automotive component 72
  control 242
  development on solidification 74–5

modelling 228–30
  optimization 22
Microstructure-selection map 202, 207, 211
Mini-mills 27
Modification mechanism 326, 338
Molten metal
  pre-treatment 6
  turbulence control 23
Molten steel, cleanliness of 54, 55
Monte Carlo techniques 74
Mould cooling 49, 51, 100
Mould-filling patterns 83–5, 137
  comparison between conventional and quasi-fluid model 85
  multicavity problem 84
  simulation 67, 79–86
Mould-flux 49, 51, 55
Moulds
  die-casting 67
  electromagnetic stirring in 52
  heat flux density with and without grooves on 50
  internal observation 66
  inversion 101
  making 66
  mechanically treated 49
  surface treatment 49, 50

Nanophase composites 236–40
Navier–Stokes equations 129
NbAl 341
Nd123 392–400
  cooling curves 393
  growth rates 397–9
  nucleation rates 396–7
  phase diagram 398
  scanning electron micrograph 395
  seeding 399–400
  solidified phases 393–4
  X-ray diffraction patterns 394
Nd422, seeding 399–400
$Nd_4Ba_2Cu_2O_{20}$ 397
Neodymium-YAG laser 241
NiAl 255–6
$NiAl_3$ 281, 282
Ni_2Al_3 281, 282
Ni–Al alloys 271

424 Index

Nickel catalysts 280, 281
Niyama criterion function 122, 124
No-fall requirement 90–2
Non-faceted cells and dendrites 156
 growth 147–52
  maximum spacing 150
  minimum spacing 150–1
   limited by interaction limit 152
  spacing adjustment 150
  spacing selection 149–50
Non-metallic inclusions 22
Nucleation 172–3, 178–81, 201, 213, 228, 264–7
 aluminium alloys in undercooled melts 185–7
 analysis 202–7
 calculation 131
 curves 264
 deterministic model for 130–1
 frequency 203–6
 intermetallic phases 261, 266
 mechanism 183, 185–7
 modelling 196–7
 theory 202
 see also Heterogeneous nucleation
Numerical modelling 30, 38–9
Numerical simulation, aluminium shape castings 121

Orientation imaging microscopy (OIM) 107
Oscillation marks 51
Oxford University 30
Oxide
 defect 94
 entrainment 91
 film defect 92
 films 96
 flow tube 92
 folding-in 89
 peritectics 391–2

Particle size distribution 232, 233
Pattern formation during solidification, see Solidification
$Pd_{40}Cu_{30}Ni_{10}P_{20}$ 360, 363–4
 differential scanning calorimeter curve 367
 homogeneous nucleation frequency 368–9
 incubation time to crystallize a volume fraction and crystallization temperature 367
 time–temperature transformation and continuous cooling transformation curves 368–70
 undercooling behaviour 364–7
$Pd_{40}Cu_xNi_{40-x}P_{20}$ 362
$Pd_{40}Ni_{40-x}Cu_xP_{20}$ 361, 362
$Pd_{40}Ni_{40}P_{20}$ 360
Peclet number 157
Periodic structures 161
 Cd–Sn alloys 173
 growth mechanism 171
 growth morphology 170
 instability of side-by-side growth 167–70
 mechanism of forming 167–70
 peritectic alloys 173
 phase diagram 172
Peritectic alloys 160
 morphological maps 161
 periodic structures 173
Peritectic compound, phase diagram 291
Peritectic growth
 banded structure 170–1
 morphology 160–1
Peritectic microstructures 160–1
Peritectic nucleation mechanism 182
Peritectic phases, morphology and solute concentration profile 168
Peritectic reaction 160, 271, 291
 Fe–C alloys 311–16
 Fe–Ni alloys 316
Peritectic solidification 160–74, 390
 banded structure 172
 competitive structures 173
 superconducting oxides 402–13
 three-phase junction as function of interface temperature 172
 two-phase unidirectional 172
Peritectic systems
 banded structure 171
 competitive structure 171
 oxide 391–2

phase diagram 390
rapid solidification 390–401
single crystals 390
Peritectic transformation 160
  Fe–C alloys 311–16
  Fe–Ni alloys 317
Permanent magnets 286
Permanent mould casting 134, 135
Permeability 127
Phase diagrams
  congruent compound 291
  peritectic compound 291
  solid solution 291
  see also specific systems
Phase field
  calculations 122–3
  models 74–5
Pinning of grain boundaries by inclusions 321
Piston acceleration 69
Piston alloys 69–76
Piston design 69–71
Piston mass 69, 71
Piston materials 71
Planar flow casting 234
Plate thickness 126, 127
Plunger movement 83
Plunger speed 82–3
Pore growth 139
Pore length 135, 137
Pore–microstructure interactions 129
Porosity 54, 135, 137
  in Al–Cu plate castings 126, 127
  in Al–Cu system 128
  models 123–9, 132
  percentage and size 123
  prediction 134
Porous tube filtration 12, 22
Potassium fluoroborate 277
Potassium fluorotitanate 277
Powder metallurgy materials 71
Precursor powder mixtures 402–3
Pressure drop 124
Primary orientation 113–14
Probability function 289
ProCAST software 108
Process control 19–20
Productivity

as function, cast gauge 31
  gains in 31
  limit diagrams 38
Pumped systems 96

Quality assurance/control 21, 24, 66, 67, 88
Quantitative wavelength dispersive spectroscopy 107
Quasi-fluid model 83
Quenching 102–3

Rapid prototyping 65, 66
Rapid quenching and devitrification, grain refinement by 234–40
Rapid solidification 234–40
  intermetallic compounds precipitated from solution by 279–80
  peritectics 390–401
Rayleigh instability 210, 212
RE123 402, 406, 411, 413
RE123/liquid interfaces 406
RE211 402
REBa$_2$Cu$_3$O$_x$ oxide superconductors 402
Recalescence 210, 235, 393
Residual alkaline metals 22
Residual stress 102–3
Rheocasting of TiAl alloys and composites 341–58
Roll bend and sheet profile 42
Roll bend capability 42
Roll casters 27, 30–2
Roll eccentricity compensation 33
Roll gap control 33, 42
Roll-over 101
Rotating-water atomization technique 292
Round products 43–5

Salicylic acid–acetamide system 160
Sand casting 135
Scanning electron micrographs, Bi–Mn alloys 295
Scheil approximation 110
Scheil equation 38
Scheil rules 115, 117
SCORPIO method 107
Secondary cooling 55

Seeding 396, 399–400
Segregation 102
   A type 54
   behaviour 35–6
   limit diagram 35
   macroscopic 53–4
Self-sealing unidirectional solidification 326–9
Semi-commercial caster 33, 34
Semi-solid processing 341
Sequential lateral solidification of silicon thin films 240
SGI Origin2000 138
Sheet buckling 36
Sheet defects 36
Sheet profile and roll bend 42
Shell depth 18
Shell zone 15, 16, 17
Shot Control technique 89
Shrinkage 95, 97–100, 132, 209
Side-to-side variations in gauge 41–2
Silicon
   growth restraint theory 326
   morphology at solid–liquid interface of Al–Si alloys 335
   nucleation restraint theory 326
   poisoning 193–6
   thin films, sequential lateral solidification 240
Simulation techniques 67, 139
Single crystal turbine blades 106
Single-phase polycrystals 234–6
Slow cooling in mould 49, 51
SNIF degasser 9, 10, 11
SOLA-VOF base algorithm 80
Solid feeding 98
Solid–liquid interface 203
  collision limited 236–8
  diffusion limited 236–8
  instability during solidification 302–8
  interaction with inclusions 321–4
Solid–liquid interfacial energy 218
Solid-phase transformations, direct observations 298–325
Solid-solution phase diagram 291
Solidification
   aluminium alloys, intermetallic compounds in 257–70

   control 54
   direct observations 298–325
     at high temperature 299–301
   equiaxed grains 327
   grain size control 199–247
   homogeneity 49
   instability of solid–liquid interface during 302–8
   kinetics 237, 240
   longitudinal irregularities in 54
   metallic glasses 359–72
   microstructure development on 74–5
   pattern formation 99–100, 142–59
     comparison with experiment 153–6
     factors controlling 156–7
   processing 258–61, 290–1
   simulation 66
   steel melts, transition from planar to cellular dendritic 308–11
   time 137, 238
   velocity 263
   see also Peritectic solidification; Rapid solidification
Solidification model 71
   physical properties and parameters used in 411
   Y123 407–10
Solidification structures 143, 157
   Al_2O_3/GAP 377–8
   Al_2O_3/YAG 377–8
   control of 386
     by magnetic fields 286–97
   eutectic ceramics 378–81
   interfacial energy in 326–8
Spherical crystals, diffusion-controlled growth 223
Spherical-cap model 206, 221, 222
Spontaneous grain refinement 208–13
Sprues 95–6
Squeeze casting 134–7
Stainless steels 91
Starter block region 110
Statistical process control (SPC) 99
Steady-state casting 54
Steel melts, transition from planar to cellular dendritic solidification 308–11
Steels, continuous casting 48–58

Steer 33, 41–3
Stefan3D 79–83
Step casting 248–56
Stochastic models 74, 129
Stokes' flow 125
Streamlines as function of tip setback 39
Strip casting 32
Strontium
  addition to aluminium alloys 179, 329–32
  influence on equilibrium silicon structure 335
Structural refinement 380–1
Sump
  recirculation 39
  shape and casting speed 39
Superconducting oxides 391–2, 402–13
Surface cracks 48–52
Surface energy of δ ferrite–γ austenite interface 316–18
Surface temperature 55–6
Surface tension 91
Surface turbulence 93
Synthesis casting 248, 255–6
System 21 control system 33

TCB solid modeller 80–2
Temperature
  control 5
  gradients 178, 201
  measurement 5
Texture
  comparison of observed and predicted 113
  evolution 110
Thermal efficiency 39
Thermal modelling of grain refinement 219–28
Thermal simulations 136
Thermocouples
  Type B 301
Thermodynamic modelling 73–4
Thermomechanical processing 257
Thick-section castings 100
Thick-strip casters 27, 28
Thin films, grain size 240–2

Thin-section castings 100
Thin-strip casters 29
Thin-walled castings 91
Third world/developing countries 29
TiAl alloys and composites 341
  apparatus and operating process 342–4
  rheocasting 341–58
  tensile properties 353–7
$TiAl_3$ 179, 185, 187–90, 196, 197
Ti–Al–ZrC alloys 341, 348–9
TiB_2 179, 182, 186, 190, 195, 197, 214, 215, 217, 219, 222, 233, 234, 265, 271, 276–9
TiBAl grain refiners 277
TiC 7, 9, 214, 218, 219, 227–8, 234
Tilting furnaces 5
Time evolution of interface positions 146
Tin–cadmium alloys 164–7
  banded structure 164–5, 173
  cell structure 165
  fluctuation period 165–6
  growth morphologies 165, 166
  longitudinal sections 165
  microstructure 164, 165, 167, 173
  unidirectional solidification 165, 166, 173
Tip setback, streamlines as function of 39
$TiSi_2$ 195
Tooling points 104–5
Torque sharing capability 33
Total productive maintenance (TPM) 19
TP-1 test 178, 215, 223, 225–7
Transmission electron microscope (TEM) 186, 189, 191, 194, 195
Twin-belt caster 27–9
Twin-roll casting 28, 258–60
Twin theory 326
Type-B thermocouples 301

Undercooling 144–6, 148, 151, 156–7, 192, 196, 203, 205–13, 229, 235, 238, 242, 390–3, 397, 399
Unidirectional solidification 160, 162, 164, 173
Unidirectionally solidified eutectic ceramics 373–89

Unidirectionally solidified eutectic
    ceramics (*continued*)
  computer simulation 386
  control of solidification structure
      386
  future applications 385–7
  manufacturing processes 387
  physical properties 386
  productivity increase 387
  relationship between solidification
      structure and properties 385
  solidification phenomena 386
Uphill filling techniques 90

Vacuum rheocasting 342
Vertical injection squeeze casting
    machine 89
Vogel–Fulcher–Tammann temperature
    dependence of viscosity 237

Wall compositions plotted against
    distance for five dendrites 152
Water cooling 136
Wax replicas 107
Weld metal, grain structure in 200
Work-in-progress times 29
WRAFTS 129
Wrought aluminium alloys 32, 179–81,
    257

X-ray diffraction patterns
  Bi–Mn alloy 292
  yttrium iron garnet (YIG) 395
X-ray radiography 89

Y123
  crystal growth 406–7
  effect of Y211 particles 404–6
  experimental procedure 402–3
  growth rates 397–9, 403–4
  melt synthesis versus sintering 396
  modelling results 411
  nucleation rates 396–7
  phase diagram 398
  relation between crystal growth length
      and growth time 404
  solidification model 407–10
Y123–Y211 tie-line 406
Y123–Y211–liquid 407
Y123/liquid 406
Y211, nucleation rates 396–7
Y211 particles
  distribution 404–6
  effect on Y123 interfaces 407
  volume fraction 405
YBaCuO 288
Yttrium aluminium garnet (YAG), *see*
    $Al_2O_3/Y_3Al_5O_{12}$
Yttrium–barium–copper–oxygen
    superconducting system 160
Yttrium iron garnet (YIG) 392–400
  cooling curves 393
  cross-section in seeded sample 396
  X-ray diffraction patterns 395

Zirconium aluminide 191, 192
Zirconium poisoning 189–93
Zn–Al alloys 91
ZrB_2 190